Baking & Pastry FUNDAMENTALS

College of Culinary Arts
International Baking and Pastry Institute

PEARSON
Prentice Hall

Upper Saddle River, New Jersey
Columbus, Ohio

Johnson & Wales University
Project Manager: *Paul McVety*
Associate Editor: *Debra Bettencourt*
Associate Project Manager: *John Chiaro*
Associate Project Manager: *Wanda Cropper*
Associate Project Manager: *Dr. Robert Nograd*
Associate Project Manager: *Dr. Bradley J. Ware*
Photographer: *Ron Manville*

Pearson Education, Inc.
Editor-in-Chief: *Vernon Anthony*
Acquisitions Editor: *William Lawrensen*
Editorial Assistant: *Lara Dimmick*
Production Coordination: *Carla Kipper, S4Carlisle Publishing Services*
Project Manager: *Kris Roach*
Operations Supervisor: *Deidra Schwartz*
Creative Director: *John Christiana*
Art Director, Interior: *Kathryn Foot*
Art Director, Cover: *Kathryn Foot*
Cover art/image/photo[s]: *Johnson & Wales University*
Director of Marketing: *David Gesell*
Marketing Manager: *Leigh Ann Sims*
Marketing Assistant: *Les Roberts*
Color Imaging Specialist: *Rob Uibelhoer from Pearson Imaging Centers*

This book was published by Pearson Education, Inc., Upper Saddle River, New Jersey 07458.

This book was set in Janson by S4Carlisle Publishing Services and was printed and bound by Courier Kendallville, Inc. The cover was printed by Moore Langen Printing Company, Inc.

Copyright © 2009 by Johnson & Wales University. All rights reserved. Printed in the United States of America. No portion of this book may be reproduced, stored in a retrieval system, or transmitted in any form or by any means, electronic, mechanical, photocopying, recording, or likewise without prior written permission from Johnson & Wales University.

10 9 8 7 6 5 4 3
ISBN-13: 978-0-13-513358-3
ISBN-10:   0-13-513358-0

Contents

Message from the University President xix
Foreword xxi
Dedication xxiii
Acknowledgements xxiii

CHAPTER 1 History of Baking & Pastry 3

　Prehistory and Early Civilizations 4
　The Middle Ages (400–1500) 5
　The Modern Period (1500–1900) 6
　The Contemporary Period (1900–present) 10

CHAPTER 2 Food Safety 15

　Key Terms 15
　Contamination 16
　　Cross-contamination 16
　　Cleaning and Sanitizing 16
　　Foodservice Hazards 16
　　Foodborne Illness 17
　　Potentially Hazardous Foods 17
　　Foodborne Illness Outbreaks 17
　Biological Hazards 17
　　Bacteria 17
　　Viruses 20
　　Parasites 20
　　Fungi 21
　　Growth Requirements 21
　　Time and Temperature Abuse 22
　　Personal Hygiene 23
　Chemical Hazards 24
　　Cleaning Products and MSDS 24
　　Kitchen Cleanliness 25
　Physical Hazards 25
　　Avoiding Physical Hazards 25
　Hazard Analysis Critical Control Point (HACCP) 25
　　The HACCP System 25
　　Establishing a HACCP Plan 26
　Safe Food Handling from Receiving to Service 26
　　Receiving and Storing Food 27
　　Preparation and Cooking 27
　　Holding Food Safely 28

Serving Food Safely 29
Cooling Food Safely 29
Reheating Food Safely 29
Disposal of Food 29

Workplace Safety 30
Personal Protective Clothing 30
Personal Injuries 31
Cleaning Kitchen Equipment 32

Fire Safety 32
Equipment 32
Emergency Procedures 32

First Aid 33
First Aid for Burns 33
First Aid for Wounds 33
First Aid for Choking 34
Cardiopulmonary Resuscitation (CPR) 34
Sickness in the Workplace or Classroom 34

CHAPTER 3 Nutrition 37

Key Terms 37

Nutrients 38
Food Energy 38
Carbohydrates 38
Proteins 41
Fats/Lipids 42
Vitamins 46
Minerals 47
Phytochemicals 49
Water 50
Reducing Sugar in Products 50
Substituting Sugar Syrups for Granulated Sugar 50
Increasing Nutrient Density in Baked Goods and Pastries 50
Celiac Disease 51
Developing Gluten-Free Products 51
Formula Modification 52
Steps in Formula Modification 52
Nutrition Labels 53
Organics 54
Genetically Modified Foods 54
Additives 54
Irradiation 54
Functional Foods 54

CHAPTER 4 Cost Control & Purchasing 57

Key Terms 57

Cost Controls 58
Types of Costs 58
Managing Food Costs 58
Benefits of Standardized Formulas 59

Formula Conversion 61
Considerations When Converting Formulas 61
Formula Costing 63
Determining the Portion Cost 63

iv

Controlling Costs   65
   Purchasing   65
   Receiving Goods   67
   Storing and Issuing Controls   68
   Production Controls   69
   Sales Controls   69

CHAPTER 5   Equipment, Hand Tools, Smallwares & Knives   71

Key Terms   71

Equipment   72
   Large Bakeshop and Pastry Shop Equipment   72

Hand Tools   75
   Selecting Appropriate Tools   75

Smallwares   78
   Measuring Equipment   78
   Selecting Cookware   80

Knives   82

Knife Construction   82
   Types of Knives   83
   Knife Skills   83
   Knife Cuts   84
   Knife Safety and Care   86

STEP PROCESS
Sharpening a Knife   88

STEP PROCESS
Trueing a Knife   89

CHAPTER 6   Fundamentals of Baking & Pastry   91

Key Terms   91

Bakeshop Measurements   92
   Scaling and Volume Measurement   92
   Baker's Percentage   93

STEP PROCESS
Using a Balance Scale   94

Essential Bakeshop Ingredients   95
   Wheat Flour   95
   The Wheat Kernel   95
   Milling Process   96
   Straight Flour   96
   Wheat Flour Treatments and Additives   97
   Hard-Wheat Flour   97
   Soft-Wheat Flour   98
   Flour Blends   99
   Fats   99
   Butter   100
   Margarine   100
   Lard   101
   Sweeteners   101
   Leaveners   103

  Thickeners   105
  Liquids   106
  Flavorings   107
  Salt   108
  Chocolate   109
  Combining Doughs and Batters in the Bakeshop and Pastry Shop   110
 The Mixing Process and Gluten Development   112
  Controlling Gluten   112
  Mixing Method   112
 The Baking Process   113
  Gas Formation and Expansion   113
  Gases Trapped in Air Cells   113
  Gelatinization of Starches   113
  Coagulation of Proteins   113
  Water Evaporation   113
  Melting of Shortenings and Fats   113
  Crust Formation and Browning   113
  Carryover Baking   114

CHAPTER 7   Dairy Products & Eggs   117

Key Terms   117

Milk   118
 Milk Processing Methods   118
 Receiving and Storing Milk, Cream, and Cultured Dairy Products   120
 Foodborne Illness and Dairy Products   121
 Fermented Dairy Products   121
 Cultured Products   121

Cheeses   122
 Common Bakeshop and Pastry Shop Cheeses   122
 Handling Cheese   126
 Ripening Cheese   126
 Receiving and Storing Cheese   126

Butter and Margarine   126
 Butter Grades   127
 Types of Butter   127
 Margarine   128

STEP PROCESS

Clarifying Butter   128

Eggs   129
 The Composition of Eggs   129
 Forms of Eggs   130
 Egg Substitutes   131
 Testing for Freshness   132
 Receiving Eggs   132
 Storing Eggs   132
 Preparing and Cooking Eggs Safely   132

STEP PROCESS

Separating Eggs   133

CHAPTER 8   Fruit   135

Key Terms   135

Fruit   136
   Types of Fresh Fruit   136

STEP PROCESS
Zesting a Lemon   137

STEP PROCESS
Segmenting an Orange   139

STEP PROCESS
Pitting and Cutting Mangoes   149

STEP PROCESS
Peeling, Coring, and Trimming a Pineapple   151
   Purchasing and Storing Fresh Fruit   153
   Purchasing and Storing Preserved Fruit   155

CHAPTER 9   Herbs & Spices/Nuts & Seeds   159

Key Terms   159

Herbs   160
   Using Herbs   160
   Storing Herbs   160

Spices   162
   Using Spices   162
   Storing Spices   162
   Spice Blends and Mixes   162

Nuts and Seeds   164
   Storing Nuts and Seeds   164
   Nuts   164
   Nut Butters   165

CHAPTER 10   Liqueurs, Wines, Spirits, Coffee & Tea   171

Key Terms   171

Liqueurs   172
   Brandy   172

Wines   173

Spirits   174

Coffee   175
   Coffea Arabica   175
   Coffea Robusta   175
   Varietal and Specialty Coffees   175

Tea   176
   Green Tea   176
   Oolong Tea   176
   Black Tea   177

CHAPTER 11 Cooking Techniques 179

Key Terms 179

Heat Transfer 180
   Conduction 180
   Induction 180
   Convection 180
   Radiation 181

Types of Cooking Techniques 181
   Dry Cooking 182
   Moist Cooking 182

Dry Cooking Techniques 182
   Baking 182
   Roasting 183

STEP PROCESS
Roasting Fruit 183
   Sautéing 184

STEP PROCESS
Sautéing Apples 184
   Deep-Frying 185
   Grilling 186

STEP PROCESS
Deep-Frying Apple Fritters 186

STEP PROCESS
Grilling Pineapple 187
   Broiling 188

Moist Cooking Techniques 188
   Poaching 188
   Simmering 189

STEP PROCESS
Poaching Pears 189
   Boiling 190

STEP PROCESS
Making a Wine Reduction 190
   Blanching 191
   Steaming 191

CHAPTER 12 Dessert Presentations 193

Key Terms 193

Plated Desserts 194
   Components of a Plated Dessert 194

STEP PROCESS
Making a Web Pattern with Sauces 195

STEP PROCESS
Using Chocolate Sauce as a Border for Another Sauce 196

STEP PROCESS
Using Two Sauces as a Plate Decoration 197
   Types of Plating 198
   Plating Contrasts 199
   Tips for Plating 200

Planning a Buffet   200

STEP PROCESS

Plate Dusting   202

STEP PROCESS

Making Caramelized Cages   203

STEP PROCESS

Making Caramelized Sugar Decorations   204

STEP PROCESS

Working with Tulip Paste   205

CHAPTER 13 Breads   209

Key Terms   209

An Introduction to Bread   210

The Four Basic Ingredients of Bread   210
- *Flour*   210
- *Yeast*   213
- *Water*   216
- *Salt*   217

Preferments   217
- *Four Yeasted Preferments*   218
- *Levain or Sourdough Starter—Naturally Leavened Preferment*   219

The Steps of Breadmaking   219
- *Scaling*   219
- *Mixing*   219

STEP PROCESS

The Steps of Breadmaking   220
- *Primary Fermentation*   223
- *Dividing*   223
- *Rounding*   224
- *Bench Rest*   224
- *Shaping*   224
- *Final Proof*   224
- *Baking*   225
- *Cooling and Storage*   227

Types of Breads   228
- *Lean-Dough Breads*   228
- *Enriched Breads*   232
- *Decorative Breads*   233

CHAPTER 14 Quick Breads   235

Key Terms   235

Quick Breads   236
- *Chemical Leaveners*   237
- *Physical Leaveners*   237
- *Types of Quick Breads*   238
- *Mixing Methods*   238

STEP PROCESS

The Rubbing, or Biscuit, Method   239

STEP PROCESS

The Blending Method   241

Panning, Baking, and Cooling Quick Breads   242

STEP PROCESS

Panning Quick Breads   242

CHAPTER 15 Laminated Doughs   245

Key Terms   245

An Introduction to Laminated Dough   246

Unyeasted Laminated Dough   246
Yeasted Laminated Dough   248

STEP PROCESS

Making Yeasted Laminated Dough   250

STEP PROCESS

Shaping and Finishing Croissants   253

STEP PROCESS

Shaping and Finishing Danish   255

CHAPTER 16 Pies & Fruit Tarts   257

Key Terms   257

Types of Pie Dough   258

Dough Ingredients   258

Flour   258
Fat   258
Cold Liquid   259
Salt   259
Sugar   259

Making Pie Dough   259

Shaping Pie Dough   259

Rolling and Panning Pie Dough   260

STEP PROCESS

Making Pie Dough   260

Single-Crust Pies   261

STEP PROCESS

Rolling and Panning Pie Dough   261

Two-Crust Pies   262

STEP PROCESS

Making a Single-Crust Pie with a Streusel Topping   262

Preparing Pie Fillings   263

Fruit Fillings   263

STEP PROCESS

Making a Two-Crust Pie with the Cooked-Juice Method   265

Cream Fillings   266
Custard and Soft Pie Fillings   266
Chiffon Fillings   266
Specialty Fillings   267

Fruit Tarts   267

French Short Doughs   268

Fruit Tart with Pastry Cream   268
Linzertorte   268
Swiss Apple Flan   268
Tarte Tatin   268

Baking Pies and Tarts   269

STEP PROCESS
Making Fruit Tarts   269

STEP PROCESS
Making a Linzertorte   270

Cooling and Storing Pies and Tarts   271

STEP PROCESS
Blind Baking   271

CHAPTER 17   Pastry Doughs & Batters   273

Key Terms   273

Pâte à Choux   274
Pâte à Choux Products   274

STEP PROCESS
Making Pâte à Choux   275
Baba and Savarin   276

Stretched Doughs   277
Phyllo Dough   277
Strudel   277
Cannoli   278

STEP PROCESS
Working with Phyllo Pastry   278

STEP PROCESS
Making Apple Strudel   279

Pourable Batters   280
Pancakes   280
Waffles   280
Crêpes   280
Popovers   280

STEP PROCESS
Making Crêpes   281

CHAPTER 18   Custards & Cheesecakes   283

Key Terms   283

Defining the Custard and Cheesecake Family   284

Custards   284
Basic Custard Method of Preparation   284
The Role of Ingredients   285
The Importance of a Water Bath   285
The Custard Family   286

STEP PROCESS
Crème Caramel   287

Cheesecake   290
The Creaming Method   290
Types of Cheeses   290

Types of Cheesecake 291
Flavoring Options 292

STEP PROCESS

Making a New York-Style/Deli-Style Cheesecake 293
Crust Options 294
Finishing Cheesecake 295
Unmolding Cheesecake 295

CHAPTER 19 Creams & Mousses 297

Key Terms 297

Basic Creams 298
Whipped Cream 298
Chantilly Cream 298
Crème Parisienne 299

Cream-Based Desserts 299
Creams Stabilized with Gelatin 299

STEP PROCESS

Making Bavarian Cream 300

STEP PROCESS

Making Charlotte Royale and Charlotte Russe 301
Chocolate Mousse 302
White Chocolate Mousse 302
Fruit Mousse 302

STEP PROCESS

Making Chocolate Mousse 303
Creams That Are Egg Thickened 304

STEP PROCESS

Making Pastry Cream 306

CHAPTER 20 Meringues & Soufflés 309

Key Terms 309

Meringues 310
The Science of Meringues 310
Tips for Success 310
Additional Tips for Success in Making Meringues 311
Stages of Meringues 311
Categories of Meringues 311

STEP PROCESS

Making French/Common Meringue 312

STEP PROCESS

Making Swiss Meringue 313

STEP PROCESS

Making Italian Meringue 314
Baking and Storing Meringues 315

STEP PROCESS

Piping Meringue Nests 316

Soufflés   317
  The Rise and Fall of Soufflés   *317*
  Soufflé Bases   *317*
  STEP PROCESS
  Making Soufflés Using a Roux Base   318
  STEP PROCESS
  Making Soufflés Using an Egg Yolk Base   319
  Tips and Hints for Making Successful Soufflés   *320*
  Prepping Ramekins Prior to Baking Soufflés   *320*
  Serving Soufflés   *320*
  Frozen Soufflé/Soufflé Glacé   *320*

CHAPTER 21 Frozen Desserts   323

Key Terms   323

Ice Cream   324

Types of Ice Cream   324

Frozen Yogurt   324

Ingredients in Ice Cream   325
  Milk and Cream   *325*
  STEP PROCESS
  Making French-Style Ice Cream   325
  Eggs   *326*
  Sweeteners   *326*
  Flavorings   *326*
  Additives in Ice Cream   *327*

Qualities of Ice Cream   327
  Smoothness of Texture   *327*
  Body/Firmness   *327*
  Richness   *327*
  Flavor   *327*

The Production of Ice Cream   328
  Preparing the Base   *328*
  Aging   *328*
  Freezing and Overrun   *328*
  Hardening   *328*

Plating Frozen Desserts   329

Ice Cream Products and Classic Desserts   329
  Coupes and Sundaes   *329*
  Peach Melba   *329*
  Poire Belle Hélène   *330*
  Ice Cream Bombes   *330*
  Baked Alaska   *330*

Still-Frozen Desserts   331
  Parfait Glacé and Soufflé Glacé   *331*
  STEP PROCESS
  Making Baked Alaska   331

STEP PROCESS
Making Soufflé Glacé   332
    Semifreddo   *333*
Ices   333
    Special Problems with Ices   *333*
    Measuring Sugar Density   *334*
    Types of Ices   *334*

CHAPTER 22   Fruit Desserts   337

Key Terms   337
Fruit Desserts   338
    Poached Fruit   *338*
    Roasted and Grilled Fruit   *338*
    Compote   *339*
    Fruit Salads and Salsas   *339*
    Fruit Chutney   *340*
    Fruit Gratins   *340*
    Traditional Fruit Desserts   *340*

STEP PROCESS
Making Individual Peach Cobblers   341
    Flambéed Fruits/Tableside Desserts   *343*

STEP PROCESS
Making Banana Flambé   345

CHAPTER 23   Sauces & Syrups   347

Key Terms   347
Sauces   348
    Types of Dessert Sauces   *348*

STEP PROCESS
Making Crème Anglaise   350

STEP PROCESS
Making Traditional Caramel Sauce   351
Syrups   352

STEP PROCESS
Making Zabaglione   353

CHAPTER 24   Cookies & Petits Fours   355

Key Terms   355
Cookies   356
    Types of Cookies   *356*
    Ingredients   *356*
    Methods of Preparation for Cookies   *357*

STEP PROCESS
The Creaming Method   358
    Varieties   *359*

STEP PROCESS
Making Checkerboard Cookies   361

    Baking   362
    Cooling and Storing   363
Petits Fours   363
    Producing Petits Fours Assortments   363
    Types of Petits Fours   363
    STEP PROCESS
    Making Petits Fours Glacés   365

CHAPTER 25   Buttercreams, Icings & Glazes   367

Key Terms   367

Buttercreams   368
    Buttercream Functions   368
    Types of Buttercream   368
    STEP PROCESS
    Making Swiss Buttercream   370
    Buttercream Hints   371
    STEP PROCESS
    Making Italian Buttercream   371
    Flavoring Buttercreams   372
Icings   373
    Types of Icing   373
Glazes   375
    Types of Glazes   375
    Tips for Glazing Cakes   377

CHAPTER 26   Cakes   379

Key Terms   379

Cakes   380
    Layer Cakes   380
    Pound Cakes   380
    STEP PROCESS
    The Creaming Method   381
    High-Ratio Cakes   381
    STEP PROCESS
    Assembling a Layer Cake   382
    Sponge or Foam Cakes   383
    Genoise Sponge Cake   383
    STEP PROCESS
    The Blending Method   383
    Chiffon Cake   384
    STEP PROCESS
    Making a Genoise Sponge Cake   384
    STEP PROCESS
    Assembling a Torte Using a Ring   385

Angel Food Cake   386

STEP PROCESS

Preparing Angel Food Cake   386

Baking Cakes   387
Scaling Batter   387
Preparation of Cake Pans   387
Baking Cakes   387
Cooling and Storing Cakes   388

High-Altitude Baking   388

STEP PROCESS

Making Buttercream Decorations and Borders   389

STEP PROCESS

Assembling and Decorating a Wedding Cake   390

STEP PROCESS

Piping Chocolate Decorations and Words   391

CHAPTER 27 Chocolate   393

Key Terms   393

An Introduction to Chocolate   394
History   394
Cacao Growth   395
The Cacao Bean   395

The Manufacturing of Chocolate   395
Preparation of Cocoa Beans   396
Mixing   396
The Addition of Soy Lecithin   396
Refining   396
Conching   396
Tempering   397

STEP PROCESS

Tempering Chocolate   400

Why Chocolate Thickens (Seizes Up) When Melted   401

Enemies of Chocolate   401
Moisture   401
Excessive Heat   401

Chocolate Storage   402

Buying Chocolate   402

Melting or Reheating Chocolate   402
Water Baths   402
Ovens   402
Warming Cabinets   403
Microwaves   403

Chocolate Equipment List   403

Couverture   404

Pralines and Other Fillings   404
Pralines (Bonbon au Chocolat, Chocolate Candies)   404
Types of Fillings   404

STEP PROCESS
Making Ganache   405

STEP PROCESS
Making Chocolate Truffles   407

STEP PROCESS
Preparing and Working with Butter Ganache   408

STEP PROCESS
Preparing Gianduja Filling for Pralines   409
- *Methods for Incorporating Fillings*   411
- *Items That Should Not Be Used for Fillings in Pralines*   412
- *Working Conditions for the Dipping of Pralines*   412

Molded Chocolates   413
- *Procedures for Making Molded Pralines*   413

STEP PROCESS
Making Pralines   413

STEP PROCESS
Molding Chocolate Pralines   415
- *Storage and Shelf Life of Pralines*   416

Chocolate Showpieces and Decorations   416
- *Tempered Chocolate*   416
- *Molded Chocolate*   417

STEP PROCESS
Using a Transfer Sheet   417
- *Piping Chocolate*   418

STEP PROCESS
Molding Chocolate   418
- *Modeling Chocolate (Chocolate Paste)*   419
- *Confectioner's Coating (Compound Chocolate)*   420
- *Chocolate Carving*   420
- *Cocoa Painting*   420
- *Postproduction*   421
- *Coloring Chocolate*   421
- *Storage*   421

STEP PROCESS
Making a Chocolate Box   422

CHAPTER 28   Pastillage, Sugar Artistry & Marzipan   425

Key Terms   425

An Introduction to Showpieces/Centerpieces   426

Pastillage   426
- *Procedure for Rolling, Cutting, Sanding, and Gluing Pastillage*   427

STEP PROCESS
Making a Pastillage Showpiece   428
- *The Coloring of Pastillage*   429
- *Other Uses of Pastillage*   429
- *Storage of Pastillage Showpieces*   429

Icings   429
- *Royal Icing*   429
- *Flow Icing (Run-out Icing)*   430

Fondant   430

Rolled Fondant   430
Procedure for Covering a Cake with Rolled Fondant   430

Sugar Artistry   431
Sugar Equipment   431
Ingredients for Poured Sugar   432
Mise en Place for Cooking Sugar   432
Boiling Point and Sugar Concentration   434
Coloring the Sugar   434
Appropriate Surfaces for Poured Sugar   435
Molds for Poured Sugar   435
Gluing Poured Sugar Pieces   436
Pulled Sugar   436

STEP PROCESS
Making a Sugar Showpiece   437
Blown Sugar   438

STEP PROCESS
Pulling Sugar   441

STEP PROCESS
Making Pulled Sugar Roses   442
Other Mediums and Techniques Used in Sugar Artistry   443

STEP PROCESS
Making Pulled Sugar Leaves for Flowers   443

STEP PROCESS
Making Pulled Sugar Ribbons and Bows   444
Finishing Touches   447
Storage   448

Marzipan   448
Storage and Handling of Marzipan   449

STEP PROCESS
Making a Marzipan Alligator   450

STEP PROCESS
Making Marzipan Roses   451

Appendix   452
Bibliography   455
Glossary   458
Index   469
Credits   483

Message *from* *the* University President

Dear Student,

The College of Culinary Arts at Johnson & Wales University is committed to your success as a foodservice professional. The revised curriculum of the College of Culinary Arts is a reflection of this commitment. This carefully designed, flexible curriculum provides you with an exciting challenge to learn and excel in your chosen career.

*Johnson & Wales University delivers a multidisciplinary educational experience for students who are serious about success. JWU was the **first** to offer a four-year Culinary Arts degree in 1993, the **first** to offer a four-year Baking & Pastry Arts degree in 1997, and the **first** to offer a bachelor's degree in Culinary Nutrition in 1999. A JWU education integrates rigorous academics and professional skills, community leadership opportunities, and our unique career education model. Students graduate from our university with the knowledge and skills necessary to achieve success in a global economy. Opportunities for practical and cooperative education complement laboratory and related classroom studies and provide valuable on-the-job experiences.*

Your formal education in baking and pastry is only the beginning of your lifelong study. However, this education is the foundation on which your future will be built, a foundation established by Baking & Pastry Fundamentals. From this text you will learn that foodservice is a multifaceted field that is constantly growing and changing to respond to the needs and desires of the customer. In studying the fundamentals, you will learn the facts and theories of preparation and presentation. These fundamentals will prepare you to explore, discover, and create your own food frontiers.

The faculty and staff of the College of Culinary Arts have spent countless hours and dedicated years of their knowledge and experience to the development of this curriculum. I encourage you to take advantage of this combined wisdom. Within the cover of Baking & Pastry Fundamentals you will discover a truly enjoyable learning experience.

John J. Bowen '77
University President

Foreword

Karl Guggenmos
University Dean of Culinary Education

Over 40 years ago I selected a career in culinary arts and, to this day, I have never regretted my choice. The numerous opportunities afforded me in this field have been both professionally and personally rewarding. As you embark on your culinary education journey, I ask you to remember that "the choices you make determine the kind of life you lead."

The Johnson & Wales culinary program offers an "education without compromise" that will empower you to achieve in a competitive global market. This broad, hands-on, experientially-based, and globally-minded education will allow you to successfully compete in the 21st century workplace. I encourage you to take full advantage of the resources available so that you might build upon this solid foundation to achieve professional readiness.

Our textbook, Baking & Pastry Fundamentals, is one of the tools that will provide you with the knowledge necessary in this great profession. It is a valuable collective resource created by talented professionals, faculty, and administrators. Whether you pursue employment as a pastry chef at a fine dining establishment or at a prestigious hotel, or elect to own a business, you will find that the contents of this text will be invaluable.

Numerous knowledgeable individuals have collaborated to produce this very current and easy to use resource. They have invested their time in creating a book that you will come to appreciate and value. A special thanks to Paul McVety and his team who spearheaded this project, and to all the culinary professionals and writers involved in making this text a success.

Dean Karl Guggenmos M.B.A., Certified Master Chef, A.A.C.

Dedication

This book is dedicated to the students and faculty members of the International Baking & Pastry Institute at Johnson & Wales University College of Culinary Arts. We wish you much success and happiness in your professional careers.

Acknowledgements

Professionals who truly love educating students have written Baking & Pastry Fundamentals. I wish to thank all the faculty, administration, students, and friends from the University for their support and participation in this tremendous undertaking.

A special thanks to Deb Bettencourt, John Chiaro, Wanda Cropper, Dr. Robert Nograd, Dr. Bradley J. Ware and Dr. Claudette Lévesque Ware for their tireless efforts and dedication in producing this textbook.

Paul McVety
Dean of Culinary Academics
College of Culinary Arts
Project Manager

EDUCATIONAL TASK FORCE

Paul McVety Project Manager
Debra Bettencourt Associate Editor
John Chiaro Associate Project Manager
Wanda Cropper Associate Project Manager
Dr. Robert Nograd Associate Project Manager
Dr. Bradley J. Ware Associate Project Manager
Dr. Claudette Lévesque Ware Senior Editor

WRITING TEXT

Charles Armstrong
Cynthia Coston
Jean-Luc Derron
Amy E. Felder
Dean Lavornia
Richard Miscovich
Stephanie Miscovich
Dr. Bradley J. Ware

READING AND REVIEWING TEXT

Sadruddin Abdullah
Jeffery Alexander
Michael Angnardo
Johannes Busch
Elena Clement
Jerry Comar
Valerie Ellsworth
Robert Epskamp
Armin Gronert
Gilles Hezard
Andrew Hoxie
Jacqueline Keich
Ronald Lavallee
Stephanie Miscovich
Kim Montello
Dr. Mary Etta Moorachian
Harry Peemoeller
Ronald Pehoski
Suzanne P. Vieira
Susan Wallace

REVIEWING AND TESTING FORMULAS

John Maas
Richard Miscovich
Megan Lambert
Martin Lovelace
Mitchell Stamm

PHOTOGRAPHS

Charlie Armstrong
Cynthia Coston
Jean-Luc Derron

Christina L. Harvey
Mark Harvey
Edward Korry
Jean-Louis M. Lagalle
Richard Miscovich
David Ricci
Gary Welling
Robert Zielinski

PHOTOGRAPHS / STUDENT ASSISTANTS

Kate Good
Linda Melfi
Eric Rivett

PHOTOGRAPHER

Ron Manville

FORMATTING THE TEXT AND FORMULAS

Debra Bettencourt

CURRICULUM AND FORMULA CONTRIBUTORS

Michael Angnardo
Wanda Cropper
Ciril Hitz
Gary Welling

COLLEAGUES, FRIENDS, AND ASSOCIATES

North American Millers' Association (NAMA)
Reggie Dow, Director of Food Service Procurement, Johnson & Wales University, College of Culinary Arts
Sid Wainer & Sons Specialty Produce
The Bread Bakers Guild of America
The Culinary Archives & Museum at Johnson & Wales University

EQUIPMENT VENDORS

Ach Food Companies, Inc.
American Baking Systems, Inc.
August Thomsen Corporation / Ateco
Barry Callebaut
Baxter Manufacturing
Blodgett Oven Company
Cambro Manufacturing
Cardinal Scale Manufacturing Co./Detecto

Carlisle FoodService Products

Carpigiani

Demarle USA

Dexter-Russell, Inc.

Enodis USA

Ericka Record LLC

Friedrick Dick Corporation

Garland Group

Gemini Bakery Equipment

Hatco Corporation

Hobart Corporation / ITW

JB Prince Company

John Boos & Co.

Matfer Bourgeat Incorporation

Norton Abrasives/Saint Gobain

Ohaus Corporation

Robot Coupe USA

Sid Wainer & Son Speciality Produce-Speciality Foods

Taylor Precision Instruments

The Vollrath Company, L.L.C.

Vulcan-Hart Company

CHAPTER 1

History of Baking & Pastry

W hile history, science, and politics have all played an important role in the development of Baking and Pastry Arts, the most influential contributor to this evolution has been technology. This chapter offers a chronological overview of the events and innovations that have allowed for the product quality, variety, and sophistication that is seen today.

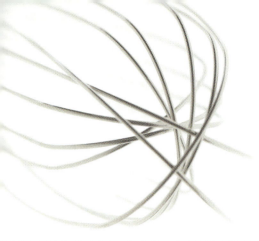

Prehistory and Early Civilizations

12,000 B.C. Some of the earliest evidence of man as "baker" is seen at this time. Tribesmen first grind the seeds of wild grasses to produce a type of flour. Grains are also roasted and then ground and moistened with water to create a more digestible gruel that is either thick or thin depending upon the amount of water added.

1680 B.C. Leavened bread is made in Egypt during this time. Only the very rich can afford bread made with wheat flour. Less expensive alternatives such as barley and sorghum are used for the general populace.

1500–400 B.C. All the major food plants are now cultivated somewhere in the world, except for sugar beets.

The Olmec civilization along the Mexican Gulf coast draws hieroglyphs in tombs that make reference to cacao, a beverage enjoyed almost exclusively by the upper class.

776 B.C. The first recorded mention of cheesecake appears when it is served to athletes participating in the Olympic Games.

350 B.C. Wheat is introduced to Greece from Egypt. At least 80 different kinds of breads and cakes are made by both professional bakers and home bakers in Ancient Greece. The bakers' ovens are built at the outskirts of the village to alleviate the fear of fire spreading to homes. Villagers supply the baker with the raw materials for the bread and laws are in place that regulate the amount of bread a customer can receive per weight of raw material supplied. Breads differ by region and by the occasion for which they are intended. Wedding cakes in Ancient Greece are made by the bride herself as part of the marriage ceremony. The cake is then burnt as an offering to the gods.

325 B.C. An admiral in the army of Alexander the Great makes one of the first mentions of sugar cane. He describes it as a plant that "produces honey, although there are no bees."

170 B.C. Commercial bakers are considered to be the first professional chefs by the Romans. They make bread only for the elite and wealthy while other classes of society continue to bake their own bread as well as grind their own flour.

History of Bread Making

The development of bread baking is vital to the history of baking and pastry making. Food historians have found that there are four basic stages in the history of bread making. In the first stage the grain is eaten in its natural state, with little or no preparation involved. In the second stage the grain is ground into either a fine or a rough flour to which a bit of water is added to form a mash that is eaten raw. A dough is made in the third stage by combining ground flour and a small amount of water. The result is a thick mass that is then shaped and placed directly on top of hot stones or in the embers of a fire. In the last stage a leavener is incorporated. The leavener is usually a piece of dough taken from an earlier batch, as in a sourdough. The dough is allowed to rise and is then cooked in an enclosed dome.

101 B.C. The Romans begin to use water power to mill flour.

80 B.C. Alexandria, Egypt, is now the largest trading port for spices in the Eastern Mediterranean, and is referred to as "Pepper Gate."

30 B.C. Rome has approximately 329 bakers. The bakers are part of an exclusive guild that requires a baker's son to follow in his father's profession. Roman wedding cakes, at this time, are most often made of spelt flour and constructed by professional bakers. The cakes are now eaten by the guests because the tradition of burning the cake in an offering to the gods is no longer practiced.

14 A.D. The world's first cookbook is published in Rome, *Of Culinary Matters* by Marcus Gavius Apicius.

32 A.D. Roman bakers are producing dozens of varieties of breads. The breads are both leavened and unleavened and are made from various flours and meals. Leavened wheat bread is prepared for only the extremely wealthy.

234 A.D. After centuries of distributing grains to the poor, Roman Emperor Severus Alexander now dispenses ready-made loaves of bread. In the next 40 years, the daily bread ration reaches 1½ pounds per person.

250–900 A.D. This era marks the height of the Mayan empire. Mayan gods are pictured in hieroglyphics holding cacao pods. Chocolate, at this time, is a drink. It is made by roasting cocoa beans, grinding them between two stones, and then mixing the results with cold water, cornmeal, chili peppers, and other indigenous herbs. This beverage is thought to come from the gods and plays an integral part in many Mayan rituals.

The Middle Ages (400–1500)

406 Butter is introduced in Europe and slowly replaces the use of olive oil.

618 China begins to use a technology earlier seen in India to produce sand sugar by reducing the juice from sugar cane. Within the next 30 years, China grows sugar cane and manufactures sugar extensively.

701 Sailors from both Arabia and Persia visit the Spice Islands. Arab merchants bring Oriental spices to Mediterranean markets. Two hundred years later (901), England receives its first shipments of spices. Throughout the Middle Ages spices are used for their medicinal benefits rather than for flavoring. Due to the lack of refrigeration and the inferior quality of meats, spices are utilized primarily to mask the taste of meat that is spoiled or rancid. Spices are used with a heavy hand and are mostly seen in the homes of the very wealthy.

Confections in the Middle Ages

Confections in the Middle Ages consist of a blend of sugar and spices. They are considered therapeutic in aiding digestion and in calming stomach upsets caused by an unbalanced diet of foods that are either rancid or on the verge of spoiling. Peter the Great of Aragon passes a law that requires the constant presence of an apothecary in his court. The apothecary is responsible for producing therapeutic confections or comfits. Spices such as cloves, ginger, and juniper berries, as well as nuts are dipped into melted sugar and then cooked until they are lightly caramelized. This process marks the birth of the modern dragée, a sugarcoated nut. Guests often attend dinner parties bringing their own personal box of comfits. Overnight guests sometimes find small gifts of comfits in their bedroom. This tradition is still practiced today when wedding guests are presented with gifts of Jordan Almonds (sugarcoated nuts).

Separating the wheat from the chaff

1000 European crusaders "discover" sugar while traveling in the East. They return home and soon spread the word of the wonders of this "new spice." Salt at this time is unrefined and gray in color, thus sugar is often referred to as "white salt." Sugar too is a luxury enjoyed by only the extremely wealthy and is often used to mask the strong flavor of medicinal herbs as well as the flavor of poison.

1202 England enforces laws that regulate the price of bread and limit the profit that bakers can make. These statutes are extremely stringent and remain in place for the next 600 years. Edicts prompt the birth of trademarks on the bottom of loaves to identify the creator of those loaves that fall short of the prescribed legal weight.

1298 Marco Polo returns to the West from his travels in China. His descriptions of the vast variety of spices grown in the East stimulate Western interest in establishing direct trade routes to the Orient.

1319 Sugar is imported to England for the first time. By 1324 it reaches Denmark and in 1390 Sweden.

1440 This year marks the beginning of the corporation of pastry cooks in France, who are primarily making savory pies of meat and fish. All cake making is now in the hands of pastry cooks. Many 15th and 16th century banquets feature the pastry cooks' art that focuses on extravagant sugar work, gum paste, and marzipan sculptures.

1456 Sugar is now used in food products and in medicines. Sweetmeats are common and consist of any product that is heavily sweetened including vegetables and proteins.

1492 Spice traders and explorers such as Columbus, Cortés, Magellan, and da Gama introduce an extensive array of products such as sweet potatoes, peppers, plantain, allspice, corn, tomatoes, vanilla, and chocolate to Europe.

The Modern Period (1500–1900)

1501 Portugal now controls the spice trade. Its monopoly does not last but does mark the beginning of years of feuding between the Portuguese, and the Dutch, and the English. In 1594 the Dutch East India Company is founded when Portugal refuses to trade spices with England and the Netherlands. The Company's first ship sails the following year and continues shipping to the Orient until its charter expires in 1799.

Cortés arrives on the shores of the Mayan empire where he is greeted by Montezuma. Cortés asks to see Montezuma's treasures and is led to a room full of cacao beans. The cold, spicy, and gritty drink produced from the beans does not initially appeal to the Spaniard. Additionally, achiote seeds that are sometimes added to the mixture turn the liquid bright red, convincing the visitor that the Mayans appear to be drinking blood. Taste and flavor aside, Cortés soon realizes that the cacao beans are worth their weight in gold, and cocoa trees come to be known as "trees bearing money."

1533 Catherine de Medici of Florence marries Henri II of France, and raises the level of sophistication of food served in France. The double boiler, pastries such as macaroons, and the introduction of sorbets are all accredited to Catherine de Medici. The French come to love sorbets, and in Paris alone, there are 250 master ice makers within the next 40 years.

1544 Chocolate makes its debut in the Old World when Mayans are brought to the court of Prince Phillip of Spain to prepare the chocolate drink for him. The first official shipment of cacao beans from Mexico to Spain takes place in 1585.

The arrival of chocolate to Europe also marks its transformation. Chocolate is now consumed hot rather than cold, and sweetened with cane sugar. Although the beans are still roasted and ground on stones, spicy and unfamiliar New World spices such as chili peppers are replaced with the familiar Old World spices of anise seed and black pepper.

Chocolate gains popularity as a medicine and is believed to reduce fevers and to relieve discomfort in hot weather.

1550s Sugar is widely used and its demand throughout Europe leads to the development of sugar plantations. These plantations are also built in the New World and slave labor is used to maintain a constant labor force. Portuguese settlers build sugar plantations on islands off the coast of Africa and employ slave labor from the neighboring African countries.

1569 Slave traffic begins between the African Coast and the Brazilian Bulge, 1,807 miles away. The slave trade introduces new food stuffs such as maize, cassava, peanuts, beans, and sweet potatoes to Africa.

1660 Louis XIV of France marries Maria Theresa of Spain. It is said that after the King, chocolate is the Queen's true passion. The consumption of morning chocolate, particularly in bed, becomes an integral part of aristocratic life. Chocolate is enjoyed by nobility and royalty throughout Europe. By 1693 the sale of cocoa and chocolate is available to everyone.

1718 France passes a law stating that only pastry cooks can use butter, eggs, and sugar to make cakes for sale. French pastry cooks later introduce ingredients and products such as glucose, corn flour, icing sugar, and biscuits. Pastry making becomes a large industry.

1747 Sugar beets are recognized as a source of sugar. Europe develops this source quickly as British blockades restrict the importation of cane sugar to the continent.

1753 The chocolate tree is assigned its botanical name, *Theobroma cacao*. *Theobroma* in Latin means, "food of the gods."

1793 The first industrial chocolate manufacturing firm is established in France. At this time, chocolate is still a gritty beverage because technology does not exist to separate the cocoa powder from the cocoa butter. The beverage is served with a fatty scum on the top.

1796 *American Cookery* written by Amelia Simmons is the first cookbook written and published in America. Its recipes include a wide variety of native American ingredients such as pumpkin, corn, and squash.

1824 Cadbury chocolate is founded in Birmingham England. John Cadbury, a Quaker, becomes interested in chocolate because he believes that it is a nourishing and healthy drink. He sees chocolate as a viable alternative to Dutch Gin and hopes that the poor will begin drinking chocolate instead of gin. Cadbury builds entire communities to provide his workers with a better life. He is one of the first businessmen to offer paid holidays and a five-day workweek. His care and concern continue even after his workers retire. This holistic trend of caring for employees is echoed by the American Milton Hershey in the late 1800s.

1828 Conrad J. van Houten, a Dutch chemist, develops a hydraulic press that separates cocoa butter from cocoa powder. This process allows the large scale manufacturing of affordable chocolate for the general public. Cocoa powder is treated with an alkaline that makes it easier to dissolve in water (a boon for chocolate drinkers). This process, known as "dutching," darkens the color of the cocoa powder and lightens its flavor.

1832 Franz Sacher introduces the Sachertorte; a sweet cake of two chocolate layers, filled with apricot jam and covered with chocolate glaze.

1847 Nancy Johnson develops the first hand crank ice cream maker and receives a patent for it. She later sells her patent rights for $200.00. The first wholesale ice cream business opens in Baltimore, MD, four years later.

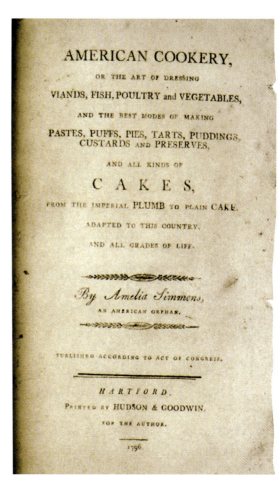

1850 Thomas Masters of London invents the first oven thermometer. It is only years later, however, that this technology is readily available to home cooks. Almost 90% of all bread at this time is made at home, even though there are approximately 2,000 professional bakers in America.

1851 Oliver Chase of Boston receives a patent for a machine that makes powdered sugar.

1852 Self-rising flour, the first of its kind, is produced by Croton Flour Mill.

1854 Gail Borden produces condensed milk in Connecticut. Fresh milk at this time carries great health risks due to the bacteria from cows and a lack of reliable refrigeration. Borden seeks to create a healthier alternative and introduces condensed milk. Although, initially not well received, in 1861, when the Union army uses the milk as part of its soldiers' rations in the Civil War, there is a broader public acceptance of the milk that contributes to the company's future success.

1855 The gas stove is introduced by Smith and Phillips of England. The cost of cooking fuel is prohibitive and the stove is not widely used for another 20 years.

1860 Louis Pasteur uses low heat (250°F/121°C) to sterilize milk. He later creates the "pasteurization" process to prevent milk-borne diseases. Pasteurization is slow to catch on because it is said to slightly alter the taste of milk.

- French chemist, Hippolyte Mège-Mouriés, invents margarine, an alternative to butter with a longer shelf life, which becomes increasingly popular throughout the world. Initially margarine contains a large amount of animal fat and a small amount of vegetable fat. Today most margarine consists solely of vegetable fat.

1861 Domingo Ghirardelli establishes the Ghirardelli Chocolate Company in San Francisco. He buys his chocolate equipment with money that he has made as a supplier of tents and sundries to the miners in the California gold rush.

1866 Cadbury uses van Houten's technology and mixes a blend of cocoa powder and sugar with melted cocoa butter. The mixture is then formed into molds and is one of the first eating chocolates.

1867 Arm and Hammer Baking Soda is founded in Brooklyn, NY.

- Boston candy maker, Daniel Fobes invents mocha using equal parts of powdered coffee and cocoa with cocoa butter. The resulting mix is used in both candies and beverages.
- Baked Alaska is served for the first time at Delmonico's restaurant in New York. The dessert, consisting of a block of ice cream surrounded by meringue, is designed to celebrate the purchase of Alaska.
- The Swiss chemist, Henri Nestlé, develops powdered milk that is vital in making milk chocolate, while attempting to create a form of shelf-stable milk to mix with his brand of children's breakfast cereal.

1868 Charles Fleischmann begins the production of compressed yeast in Cincinnati, OH. Fleischmann delivers the yeast to Cincinnati housewives via a horse and wagon. Two years later, Fleischmann begins to package the yeast in its trademark tinfoil wrappings to keep the yeast fresher for longer and to allow its shipment outside of the Cincinnati area.

- Cadbury markets his first box of chocolate candies. The box is extensively decorated and the chocolates are extremely popular.

1870 Timothy Earle of Smithfield, RI, is granted a patent for a rotary egg beater. The use of the mechanical egg beater is quickly embraced by housewives who spend hours whipping eggs for baked goods. Angel food cake, leavened with egg whites, becomes extremely popular.

1871 C. A. Pillsbury & Co. is established in Minneapolis, MN. The company introduces Pillsbury's Best XXXX Flour in the following year.

1874 Margarine is introduced to America. Made in New York, it is marketed as "artificial butter." In the years that follow, many states prohibit the sale of margarine, or allow it to be sold only in a white, colorless form. Many homemakers add yellow food coloring to the margarine at home. The last state to repeal restrictions on the sale of margarine is Wisconsin in 1967.

1875 The collaboration of two Swiss companies Nestlé (the inventor of sweetened condensed milk) and the Peters Chocolate Company leads to the development of milk chocolate. The mixture of sweetened condensed milk and chocolate becomes an immediate success and Nestlé eventually grows to be the largest food company in the world.

1879 Rodolphe Lindt of Switzerland develops "conching," in which chocolate liquor, sugar and milk (if used) are ground between large granite rollers to give chocolate a smooth, rich mouthfeel. Conching soon becomes a standard step in the chocolate making process.

1880 "Philadelphia" brand cream cheese is introduced in New York. It is named after the city of Philadelphia because of its reputation for superior dairy products.

1886 After his New York caramel business goes bankrupt, Milton Hershey tries his hand again at candy making in Lancaster, PA. He produces tissue wrapped caramels and enjoys modest success for a few years. In 1891 he nearly declares bankruptcy again, but is assisted by family members who loan him additional capital.

- American farmers begin to use hand separators to separate cream from milk. This process brings a greater consistency to dairy products and increases dairy productivity for farmers who do not live near a creamery as they are now able to perform this process at home.
- Budapest baker, József Dobós, creates the Dobos Torte. The torte consists of five thin cake layers filled with chocolate buttercream and topped with a caramel covered cake glaze.

1889 Aunt Jemima Pancake Flour is the first commercially produced self-rising flour mix for pancakes.

- This year also marks the creation of William M. Wright's Calumet Double-acting Baking Powder.

1890 American law now states that all milk must be pasteurized.

- New York Biscuit Company is created by the merger of 26 bakeries.

1893 Milton Hershey becomes interested in chocolate after viewing chocolate making machinery at the Chicago World's Fair. He combines his confectionary skills and chocolate to produce chocolate caramel candies. Eight years later he sells his caramel factory for one million dollars and moves to rural Pennsylvania to build Hershey Village. The village is modeled after the village developed by Cadbury in England at the beginning of the century.

- Chef Auguste Escoffier creates Peach Melba while working at London's Savoy Hotel. The dessert consists of vanilla ice cream topped with a poached peach half and covered with raspberry sauce and almond slivers. The dish is named for Madame Nellie Melba, a popular opera singer.

1895 Swans Down Cake Flour is introduced. It is produced by Addison Weeks Igleheart in Evansville, IN. The finely ground flour is the first packaged cake flour to be nationally advertised. Ads for this product appear throughout America in the *Ladies' Home Journal*.

- Fannie Farmer's *Boston Cooking School Cookbook* is published. The book introduces a precise system of measuring and weighing out ingredients. Soon after the book's publication standardized cups and spoons are manufactured. The concept of formulaic recipes and detailed methods of preparation is born. Written recipes replace recipes simply passed orally from mother to daughter.

The Contemporary Period (1900–present)

1900 Milton Hershey sells the caramel production portion of his company to focus his attention solely on chocolate making. In 1903 he starts to build his chocolate factory and a town for his workers. His chocolate making begins in 1905 and Hershey kisses are produced in 1907. Hershey develops a chocolate with a high melting point by substituting cocoa butter with solid vegetable shortenings. He succeeds in producing a 6-ounce bar that is issued in K rations during the First World War. Hershey's name becomes synonymous with chocolate, and at the turn of the century, chocolate consumption is a part of American culture.

1904 The ice cream cone is invented through the collaboration of a pastry maker and an ice cream concessionaire at the 1904 St. Louis Exposition. When the ice cream stand runs out of dishes in which to serve ice cream, a neighboring pastry stand owner begins to roll wafer cookies into cone shapes. The pastry maker sells the cones to the ice cream vendor and they are received with great excitement by the crowds.

1911 Crisco is introduced by Proctor & Gamble. It does not gain popularity until World War II when the availability of butter and lard is limited.

1914–1918 World War I

1916 The first mechanical refrigerator is offered for home use. It is prohibitively expensive and as costly as an automobile at the time.

1926 Joseph Draps starts Godiva Chocolates in Belgium. Draps introduces America to high-end chocolate candies. The candies are unique at the time as they are made in intricate molds rather than simply dipped.

1928 The commercial bread slicer is developed by Frederick Rohwedder of Battle Creek, MI. The introduction of packaged sliced bread leads to an increase in sales of butter, cheese, jams, and other spreads.

1933 Innkeeper Ruth Wakefield inadvertently creates the Toll House cookie by adding pieces of semisweet bar chocolate into her cookie batter. The chocolate remains in chunks resulting in the Toll House cookie. Nestlé purchases her recipe and includes it on its packages of chocolate chips. Toll House is the name of Wakefield's inn in Whitman, MA.

1936 *The Joy of Cooking* by Irma Rombauer is published. Rombauer, a housewife with no professional cooking experience, produces a book with a wealth of information for the novice cook. Her recipes include depth and detail. The book is continuously revised and is still published today.

Transporting sugarcane in Cuba

1937 Margaret Rudkin begins to make bread that becomes the foundation of the Pepperidge Farm Bread Company. Believing that homemade wheat bread will cure her son's asthma, Rudkin begins to make whole grain breads at home using recipes from Fannie Farmer's *Boston Cooking School Cookbook*. Soon Rudkin is selling bread to small markets in the surrounding area and word of mouth leads to increased sales. She establishes Pepperidge Farm Bread (named after her family's farm in Connecticut) and is soon making over 4,000 loaves of bread a week. The creation of the popular Pepperidge Farm cookies soon follow.

1938 *Larousse Gastronomique* by Prosper Montagne is published in France. The book is still recognized today as a definitive culinary reference text. Its scope is impressive, covering culinary history, techniques and terminology, descriptions of foods and menu items, and recipes.

1939–1945 World War II

1941 M&M candies are introduced. Their hard candy shell prohibits the chocolate from melting at high temperatures. Believing that "eating candy makes for more effective soldiering," the U.S. Army buys large quantities of candy throughout the course of the war. The candies are an immediate hit with servicemen. Peanut M&Ms are later introduced in 1954.

1943 The government releases the Required Daily Allowances (or RDAs) for the first time. Throughout the years these allowances change to accommodate the general health and fitness of Americans.

1948 General Mills introduces a chiffon cake that substitutes vegetable oil for solid shortening. The Minnesota company advertises the cake as "the cake of the century."

1950 The first Pillsbury Bake Off, in which contestants are challenged to develop a recipe using Pillsbury flour, takes place. The list of sanctioned ingredients increases over the years as Pillsbury expands its product line.

1951 Duncan Hines introduces its cake mix in Omaha, NE. The mix can be adapted to three flavors (white, yellow, and chocolate) depending upon the additional ingredients incorporated. By 1957, almost half of all cakes made at home are made using a cake mix.

1953 Charles Lubin, founder of Sara Lee in Chicago, IL, develops a line of frozen baked goods. Sara Lee's frozen cheesecakes are extremely successful. Lubin is also responsible for the development of foil pans. His products can be baked, frozen, and distributed in the same pan to reduce the cost for both manufacturer and consumer. Within the following year Lubin distributes his frozen baked goods in 48 states.

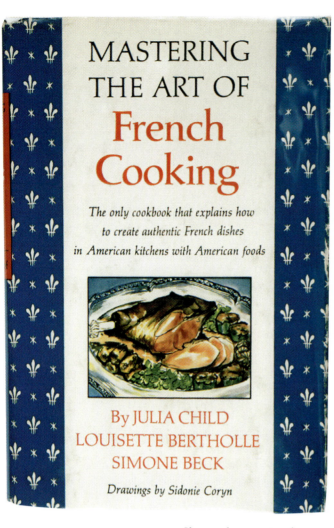

1957 An article in the *Journal of the American Medical Association* makes a clear connection between diets high in saturated fat and heart disease. Health concerns related to diet continue to remain a topic of concern for Americans.

1958 Williams-Sonoma opens its first store in California. The store specializes in traditional European and French cookware previously unavailable to American cooks. The availability of these products allows for the creation of tarts, soufflés, charlottes, and madeleines in the home.

1959 America is introduced to the world of a premium ice cream called Häagen Dazs. Although the name sounds European, Häagen Dazs is the creation of Reuben Mattus from the Bronx, NY. His ice creams have a high fat content and a low amount of overrun. Initially only three flavors are available: vanilla, chocolate, and coffee.

1961 *Mastering the Art of French Cooking* by Julia Child and Simone Beck is published in the United States. The book demystifies the art of French cooking, laying out recipes in exquisite detail. It is a tremendous success with housewives who try to duplicate French specialties. The book is followed two years later by Julia Child's cooking show on public television in Boston.

1971 Chez Panisse opens in Berkeley, CA. Owner Alice Waters uses only the freshest ingredients procured from local farmers and encourages American chefs to use seasonal and regional foods in their restaurants. The first pastry chef at Chez Panisse is Lindsey Shere, who remains there until her retirement in 1998.

1973 The Cuisinart food processor is introduced at the National Housewares Exposition in Chicago, IL. Although its high price tag deters many potential customers, chefs such as Julia Child, James Beard, and Craig Claiborne praise this Sontheimer creation to help boost sales.

1978 Ben & Jerry's ice cream begins in a remodeled gas station in Burlington, VT. The ice cream rivals Häagen Dazs and is soon available throughout the country.

1985 Cookbook author and gourmet, James Beard, dies at 81. He is regarded as the father of American cookery. He emphasizes simple cooking techniques such as grilling.

- Lindsey Shere of Chez Panisse publishes *Chez Panisse Desserts*. The book focuses on the use of seasonal fruits in desserts.

1989 The inauguration of the Pastry World Cup competition is held in Lyons, France. Twelve teams compete in the first competition. Teams consist of three members: a pastry chef, a chocolatier, and an ice carver. The competition is open to the public, and for the first time pastry chefs become showmen. Over the years the competition's prestige increases.

1990 A mandatory plan for nutrition labeling is announced by the Bush administration. All products must now include their nutritional breakdown as part of their label.

1995 The French Pastry School is established in Chicago, IL. The school is founded by pastry chefs Jacquy Pfeiffer and Sebastien Canonne. It offers classes for those wishing to become pastry chefs and continuing education classes for professionals.

1997 François Payard opens up his pâtisserie, Payard's, in New York City. Originally from France, Payard worked as a pastry chef at both Le Bernardin and Boulud's before realizing his dream of having his own pastry shop.

1999 This year marks the publication of *The Making of a Pastry Chef* by Andrew MacLauchlan. The author conducts extensive interviews with pastry chefs throughout America, and discusses the source of their inspiration, both on a personal level and in terms of the ingredients they use.

- The American Bread Bakers Guild astounds the world by winning the World Cup of Baking. This competition is held every three years.

2000 Jacques Torres opens a chocolate factory specializing in uniquely flavored handmade chocolates. Before opening his chocolate shop, Torres served as the executive pastry chef at Le Cirque in New York City. Torres is also the Dean of Pastry Studies at New York's French Culinary Institute.

2001 Publication of *The Last Course* by Claudia Fleming appears. Fleming writes her book about the development of flavorful desserts while a pastry chef at Gramercy Tavern in New York City. In the same year she is named pastry chef of the year by the James Beard Society.

2002 This year marks the publication of *Sweet Seasons* by Richard Leach. Leach is the executive pastry chef at the Park Avenue Café in New York City. His book focuses on the introduction of plated desserts.

2004 The National Bread and Pastry Competition in Atlantic City is started. The competition is unique because teams consist of both a pastry chef and a baker. Sadruddin Abdullah and Ciril Hitz of Johnson & Wales University win the first place award.

2005 The Bread Bakers Guild Team USA 2005 wins the World Cup of Baking. The team is made up of three members who compete in the four categories: viennoiserie, artistic design, baguette and specialty breads, and savory selection.

CHAPTER 2

Food Safety

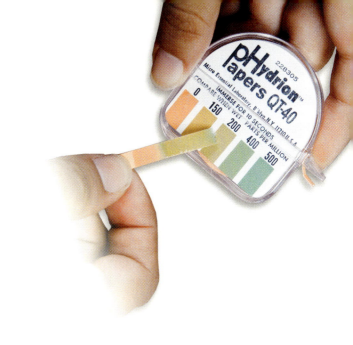

S erving safe food is a top priority for bakers and pastry chefs in the foodservice industry. Many cases of food contamination and food-borne illness that could easily be prevented occur because of inadequate food safety precautions.

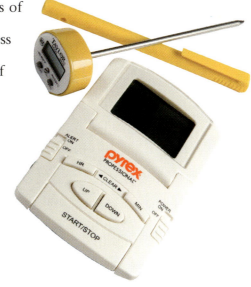

KEY TERMS

contamination
cross-contamination
cleaning
sanitizing
potentially hazardous food
biological hazards
bacteria
putrefactive
pathogen
infection

intoxication
toxin-mediated infection
virus
parasite
fungus
mold
yeast
aerobic
anaerobic
facultative
water activity (a_W)

temperature danger zone
lag phase
log phase
stationary growth phase
death phase
chemical hazard
physical hazard
Hazard Analysis Critical Control Point (HACCP)

critical control points (CCPs)
flow of food
First In, First Out (FIFO)
first aid
abrasion
laceration
avulsion
puncture wounds

15

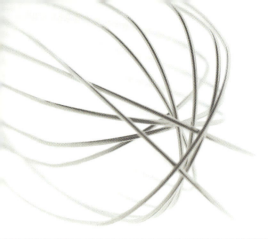

Contamination

Foodborne illnesses are diseases caused by microorganisms, such as bacteria and viruses, in food. Foodservice professionals who follow industry standards for safety can reduce the threat of these serious diseases.

Contamination occurs when harmful substances such as microorganisms adulterate food, making it unfit for consumption. Biological, chemical, and physical hazards can contaminate food.

Cross-contamination

Foods may also become adulterated by cross-contamination. *Cross-contamination* involves the transfer of harmful bacteria from one food to another through physical contact. Although cross-contamination is most commonly caused by people, rodents and insects may also transfer contaminants to food.

Cleaning and Sanitizing

Cross-contamination can be prevented to some degree by maintaining a clean and sanitary environment. *Cleaning* removes all visible soil residue and grime but does not kill microorganisms. *Sanitizing* reduces the number of pathogenic microorganisms to a safe level with the use of chemicals and/or moist heat. To ensure food safety, both procedures must be conducted.

Foodservice Hazards

Foodservice hazards can be biological, chemical, or physical in nature and can cause food contamination. These hazards are usually caused by improper foodhandling and inadequate safety practices. Contaminated food must be discarded and therefore increases food cost. Dissatisfied customers, fines, summonses, lawsuits, injuries, medical bills, and increased absenteeism may also occur and contribute to additional expenses.

Biological

Biological hazards include bacteria, viruses, parasites, and fungi. Some foodborne bacterial illnesses include *Salmonella* spp., *Listeria monocytogenes*, *Shigella* spp., *Campylobacter jejuni*, and *Clostridium botulinum*.

Chemical

Chemical contamination results from environmental conditions and improper food safety procedures. Chemical contaminants include cleaning supplies, polishes, pesticides, soap residue, fertilizers, food additives, and toxic metals.

Physical

Physical hazards that pose a risk include hair (human and rodent), dirt, metal shavings, stones, insects and insect parts, glass, and shards of bone. These may cause injury or discomfort to the consumer.

Foodborne Illness

Microorganisms can grow in and on food that is improperly handled or cooked. Cross-contamination, poor personal hygiene, and illness may produce these microorganisms as well. Chemical spills or residue and physical hazards that can come in contact with food may also contaminate food and produce foodborne illnesses.

Certain sectors of the population are at greater risk for contracting foodborne illnesses when contaminated food is consumed. Children, the elderly, pregnant women, and those who are chronically ill or who have weakened immune systems are at highest risk. It is imperative that foodservice establishments prescribe and follow strict procedures for handling food safely.

Potentially Hazardous Foods

Some foods are considered *potentially hazardous foods* (time/temperature control for safety) because they present a greater risk for harboring or supporting the growth of microorganisms that cause foodborne illnesses. These foods are typically high in protein and moisture and have a low pH—the amount of acid or alkaline in a substance. Dairy products, eggs, meats, poultry, fish, and shellfish are all considered to be potentially hazardous food as well as cooked rice and pasta, beans, potatoes, sliced melons, and garlic and oil mixtures. Particular attention and precautions must be taken when receiving, storing, issuing, preparing, cooking, holding, serving, cooling, and reheating any of these items.

Foodborne Illness Outbreaks

A foodborne illness outbreak occurs when two or more people experience the same illness after eating the same food. Symptoms of foodborne illness may vary from person to person but typically include nausea, vomiting, abdominal pain, and diarrhea. Dizziness, fever, headache, chills, dry mouth, and prostration may also occur. These symptoms can appear within minutes to several days after consuming the infected food. An outbreak of foodborne illness must be reported to the Department of Health, which then investigates the outbreak in an effort to determine the cause.

Biological Hazards

Disease-causing microorganisms, such as bacteria, viruses, parasites, and fungi, are *biological hazards* that result in foodborne illnesses. *See Figure 2-1.*

Bacteria

Bacteria are tiny, single-celled microorganisms that are in the air and water, on the ground, on food, in our bodies, and on our skin. They are a major cause of foodborne illness. Three categories of bacteria are:

- ***Beneficial*** Most bacteria are harmless, living in the intestines and aiding in digestion. They may provide nutrients as well. Beneficial bacteria are used in food products such as cheese, yogurt, and sauerkraut.
- ***Putrefactive*** Undesirable, or *putrefactive* bacteria cause food to spoil and produce off-flavors, odors, slimy surfaces, and discoloration.

Figure 2-1

Foodborne Illnesses

SALMONELLOSIS (NON-TYPHOID)—BACTERIA (SALMONELLA SPP.)

Symptoms

Nausea, vomiting, abdominal cramps, diarrhea, headache

Associated Foods

Raw poultry and poultry products such as eggs, milk, shrimp, fish, frog legs, yeast, coconut, sauces, salad dressings

Preventive Measures

Keep foods refrigerated; avoid cross-contamination; cook poultry to a minimum of 165°F (74°C) for at least 15 seconds and other foods to minimum internal temperatures; avoid food and surface contamination by following good personal hygiene practices; avoid pooling eggs.

CAMPYLOBACTERIOSIS—BACTERIA (CAMPYLOBACTER JEJUNI)

Symptoms

Diarrhea, vomiting, fever, abdominal pain, headache, muscle pain, bloody stools

Associated Foods

Meats from cattle, poultry, and sheep, and milk from those animals; unpasteurized milk and dairy products

Preventive Measures

Pasteurize milk; use only treated water; cook all foods to safe internal temperatures; avoid cross-contamination.

SHIGELLOSIS (BACILLARY DYSENTERY)—BACTERIA (SHIGELLA SPP.)

Symptoms

Abdominal pain, nausea, diarrhea, vomiting, fever, chills, dehydration

Associated Foods

Protein salads, lettuce, raw vegetables, poultry, shrimp, milk and milk products

Preventive Measures

Use only sanitary food and water sources; avoid cross-contamination; practice good personal hygiene; cool foods rapidly; control flies.

LISTERIOSIS—BACTERIA (LISTERIA MONOCYTOGENES)

Symptoms

Fever, chills, headache, nausea, vomiting, diarrhea, backache, meningitis, encephalitis, affects people with impaired immune systems

Associated Foods

Ice cream, frozen yogurt, unpasteurized milk and cheese, raw vegetables, poultry, meat, seafood, prepared and chilled ready-to-eat foods (deli meats, hot dogs, soft cheese, paté)

Preventive Measures

Avoid cross-contamination; properly clean and sanitize work areas; ensure that all milk and dairy products are pasteurized; cook all foods to safe internal temperatures.

BOTULISM—BACTERIA (CLOSTRIDIUM BOTULINUM)

Symptoms

Neurotoxic symptoms, constipation, diarrhea, vomiting, fatigue, vertigo, dry mouth, paralysis, death

Associated Foods

Underprocessed foods, canned low-acid foods such as vegetables, sautéed onions in butter sauce, baked potatoes, untreated garlic and oil products, sausages

Preventive Measures

Do not use home-canned products; cool leftovers rapidly; use proper time and temperature control for large, bulky, and reduced oxygen packaged items; refrigerate garlic-and-oil mixtures; sauté onions to order.

HEMORRHAGIC COLITIS (HEMOLYTIC UREMIC SYNDROME)—BACTERIA (SHIGA TOXIN-PRODUCING ESCHERICHIA COLI INCLUDING 0157:H7 AND 0157:NM)

Symptoms

Severe abdominal cramps, bloody diarrhea, vomiting, mild fever, kidney failure

Associated Foods

Raw ground beef, undercooked meat, roast beef, dry salami, unpasteurized milk and apple cider or juice, commercial mayonnaise, lettuce, melons, fish from contaminated waters

Preventive Measures

Avoid cross-contamination; thoroughly cook ground beef to a minimum of 155°F (68°C) for 15 seconds; practice good personal hygiene.

STAPHYLOCOCCAL GASTROENTERITIS—BACTERIA (STAPHYLOCOCCUS AUREUS)

Symptoms

Nausea, retching, cramps, diarrhea, headache

Associated Foods

Meats, poultry, protein foods, sandwiches, dairy products, eggs, salad dressings, reheated foods

Preventive Measures

Practice good personal hygiene, especially proper hand-washing procedures; do not allow employees with skin infections to handle or prepare foods; rapidly cool prepared foods; refrigerate foods properly.

CLOSTRIDIUM PERFRINGENS GASTROENTERITIS—BACTERIA (CLOSTRIDIUM PERFRINGENS)

Symptoms

Abdominal pain, diarrhea, nausea, dehydration, vomiting

Associated Foods

Meat, poultry, gravy, stew, meat pies, beans, leftovers

Preventive Measures

Use proper time and temperature control to cool and reheat cooked foods.

BACILLUS CEREUS GASTROENTERITIS— BACTERIA (BACILLUS CEREUS)

Symptoms

Vomiting and nausea or watery diarrhea and abdominal cramps

Associated Foods

Rice products, starchy foods, puddings, casseroles, pastries, meats, milk, dairy products

Preventive Measures

Practice careful time and temperature control particularly when holding, cooling, and reheating food. Cook to required minimum internal temperatures.

HEPATITIS A—VIRUS (HEPATOVIRUS OR HEPATITIS A VIRUS)

Symptoms

Begins with loss of appetite, vomiting, fever, headache. Develops into jaundice and darkened urine.

Associated Foods

Water, ice, salads, cold cuts, sandwiches, shellfish, fruit, fruit juices, milk and milk products, vegetables

Preventive Measures

Use only sanitary water sources; ensure that all shellfish comes from approved sources; practice good personal hygiene; avoid cross-contamination; clean and sanitize all equipment and food-contact areas.

NOROVIRUS GASTROENTERITIS—VIRUS (NOROVIRUS) FORMERLY CALLED NORWALK VIRUS

Symptoms

Nausea, headache, fever, vomiting, diarrhea, abdominal pain, cramps

Associated Foods

Water, raw vegetables, fresh fruit, salads, shellfish, uncooked ready-to-eat foods

Preventive Measures

Use only sanitary, chlorinated water sources; ensure that all shellfish comes from approved sources; practice good personal hygiene; avoid cross-contamination; cook all foods to safe internal temperatures.

ROTAVIRUS GASTROENTERITIS—VIRUS (ROTAVIRUS)

Symptoms

Abdominal pain, diarrhea, vomiting, mild fever

Associated Foods

Hors d'oeuvres, water, ice, salads (green leaf), fruit, and ready-to-eat foods

Preventive Measures

Cook all foods to safe internal temperatures; practice good personal hygiene; use only sanitary, chlorinated water.

CYCLOSPORIASIS—PARASITE (CYCLOSPORA CAYETANENSIS)

Symptoms

Watery diarrhea, loss of appetite and weight, cramps, gas, bloating, nausea, vomiting, fatigue, muscle aches

Associated Foods

Fish, raw milk and produce, water

Preventive Measures

Purchase from reputable sources. Use only sanitary water; practice good personal hygiene; wash produce thoroughly.

ANISAKIASIS—PARASITE (ANISAKIS SIMPLEX)

Symptoms

Tickling in throat, coughing; vomiting, nausea, diarrhea, and abdominal cramps

Associated Foods

Raw, undercooked, or improperly frozen saltwater fish

Preventive Measures

Obtain seafood from approved sources; carefully inspect fish; freeze in accordance with the Food Code.

GIARDIASIS—PARASITE (GIARDIA DUODENALIS) ALSO CALLED GIARDIA LAMBLIA

Symptoms

Intestinal gas, diarrhea, abdominal cramps, nausea, weight loss, fatigue, watery stools

Associated Foods

Contaminated water, ice, salads, ready-to-eat foods

Preventive Measures

Use sanitary water supplies; ensure foodhandlers practice good personal hygiene; wash raw produce thoroughly.

TOXOPLASMOSIS—PARASITE (TOXOPLASMA GONDII)

Symptoms

Swollen lymph glands, fever, muscle aches, blindness, brain damage, mental retardation, death

Associated Foods

Contaminated water, raw meats and vegetables

Preventive Measures

Proper personal hygiene especially washing hands that have been in contact with soil, animal feces, or raw meats; avoid raw meats; cook meat to minimum internal temperatures.

- **Pathogens** *Pathogens* are the disease-causing bacteria responsible for many serious foodborne illnesses. Because they cannot be detected by sight or smell, they are especially dangerous.

These bacteria include *Salmonella* spp., *Staphylococcus aureus*, *Listeria monocytogenes*, and *Clostridium botulinum*. They grow by multiplying, and can double in number as often as every 20 minutes. In the right environment, one bacterium can split and multiply to numbers in the billions within a 12-hour period. Food, oxygen, moisture, pH level, temperature, and time all impact the bacteria's ability to grow.

Bacteria cannot move by themselves. They must be transported from one location to another by hands, coughing, sneezing, foods, equipment, utensils, air, water, insects, or rodents. Pathogens can cause foodborne illness through infection, intoxication, and toxin-mediated infection.

Infection

Infection occurs when pathogens grow in the intestines of an individual who has consumed contaminated food. Listeriosis is such an infection.

Intoxication

Intoxication results when the pathogen produces toxins that cannot be seen, smelled, or tasted. It is not the toxin-producing bacteria that are harmful, but the toxins they produce. Handling food properly is the only way to prevent intoxication. Botulism is an example of intoxication.

Toxin-Mediated Infection

Toxin-mediated infection occurs when a person eats food contaminated with pathogens. These pathogens establish colonies in the consumer and produce toxins. This type of infection is particularly dangerous for children, the elderly, and those with weakened immune systems and can result in inflamatory conditions such as gastroenteritis, hemorrhagic colitis, and clostridium perfringens.

Viruses

Viruses are the smallest known form of life and are responsible for the majority of foodborne illnesses. Unlike bacteria, viruses require a host, such as an animal, plant, or human, in order to grow. They can be transmitted from person to person through poor hygiene, from people to food-contact surfaces, and from people to food. Although viruses can be present on any foods, salads, sandwiches, milk, baked goods, raw fish, shellfish, and other unheated food products are especially susceptible. Hepatitis A, Norovirus, and Rotavirus are examples of viruses.

Parasites

Parasites, such as protozoa, roundworms, and flatworms, are tiny organisms that must live in or on a host to survive. Often found in water, poultry, fish, and meats, parasites enter the animal through contaminated feed or water and then settle in the intestines and grow. Common foodborne parasitic illnesses include *Giardia duodenalis*, caused by contaminated water and undercooked infected game, and anisakiasis, caused by roundworms in raw or undercooked fish. Parasitic illnesses can be prevented in several ways:

- Utilize proper cooking and freezing techniques.
- Use only sanitary water supplies.

- Avoid cross-contamination.
- Guarantee that all employees utilize proper hand-washing procedures.

Fungi

Fungi are microorganisms found in plants, animals, soil, water, and the air. They can be tiny, single-celled plants or large, multi-celled organisms such as giant mushrooms. Fungi also are present in some foods. Molds and yeast are two forms of fungi.

Molds

Molds can grow in any environment and are commonly found growing on bread or cheese. *See Figure 2-2.* The algaelike spores produced by molds are visible to the naked eye. Molds mostly affect the appearance and flavor of foods through discoloration, odor, and off-flavors. Some molds create mycotoxins. Heating mold cells and spores can destroy the mold but not the toxins. Discard foods that contain mold unless the mold is a natural part of the product, such as mold on blue cheese.

Yeast

Yeast is most often associated with bread and the baking process, and is considered to be beneficial. However, yeast can cause spoilage when present in other foods. To survive, yeasts require water and a starch or sugar. Because carbon dioxide and alcohol are by-products of yeast, some foods spoiled by yeast smell of alcohol. Others appear slimy or have a pink discoloration. Discard any foods spoiled by yeast.

Growth Requirements

All microorganisms require a specific environment in which to survive and thrive. Viruses need a living cell in which to reproduce, and are not affected by oxygen, moisture, or pH levels. They can be destroyed by heat. Parasites depend on the host to survive and can be destroyed either by heating the food to high temperatures or by freezing it for several days. Fungi can grow in any environment and are destroyed by heat.

Bacteria, however, have the most complex growth requirements, which include the presence of food, oxygen, moisture, pH level, temperature, and time.

Figure 2-2 Mold is often seen growing on bread.

Food

Food provides energy for bacteria to grow. Hazardous foods, in particular, provide the ideal environment for bacterial growth.

Oxygen

Bacteria can be *aerobic* (requiring oxygen to survive); *anaerobic* (growing best without oxygen); or *facultative* (surviving with or without oxygen). Most pathogenic bacteria are facultative in varying degrees.

Moisture

Bacteria cannot survive without moisture. This moisture is referred to as *water activity (a_W)*. The amount of water activity in food is measured on a scale of 0 to 1.0. (Distilled water has a water activity of 1.0.) At .85 a_W or greater, bacteria can generally grow. Dry foods and those high in salt or sugar do not easily support bacterial growth.

pH level

A balanced, or neutral, pH environment supports bacterial growth. The pH scale ranges from 0 to 14.0—0 to 7.0 (acidic); 7.0 to 14.0 (alkaline); 7.0 (neutral). Bacteria grow well in foods with a pH level between 4.6 and 7.5. Growth is unlikely at a pH level of 4.6 or lower. Highly acidic foods such as lemon juice and vinegar do not actively promote the growth of bacteria.

Temperature

Temperature is the easiest factor to control of all the bacterial growth requirements. Bacteria thrive in the *temperature danger zone,* or TDZ (41°F–135°F/5°C–57°C), and they are destroyed when exposed to high heat for a specified amount of time. Freezing does not destroy bacteria; it only slows its growth. *See Figure 2-3.*

Time

Bacteria require time to adjust to a new environment. This duration is called the *lag phase,* and typically lasts anywhere from one to four hours. The lag phase allows foodhandlers to quickly and safely work with food. A period of accelerated bacterial growth known as the *log phase* is next. This phase continues until overcrowding is reached and competition exists among bacteria for food and space. At this point in time, the bacteria population comes to a *stationary growth phase,* in which bacteria grow and die at the same rate. When more bacteria die than grow, a population decline called the *death phase* is reached.

Time and Temperature Abuse

Most foodborne illnesses result from time and temperature abuse by foodhandlers. Leaving potentially hazardous foods in the temperature danger zone (41°F–135°F/5°C–57°C) allows for the growth of bacteria that cause foodborne illnesses. It is vital that staff minimize the time that food is kept in the temperature danger zone. Potentially hazardous foods held more than 4 hours in the temperature danger zone must be discarded. Typically, the four hour time zone begins when food is received. However, food can be time and temperature abused before arriving at its destination. Food can be contaminated at the point of production. Every time the food is within the temperature danger zone, the four-hour time zone is decreased and does not start again.

Figure 2-3

Product	Temperature
Injected meats, including brined ham and flavor-injected roasts	155°F (68°C) for 15 seconds
Poultry, stuffed meats and pasta, casseroles, stuffings, and other dishes combining raw and cooked foods, eggs to be held	165°F (74°C) for 15 seconds
Beef, pork, veal, and lamb roasts	145°F (63°C) for 4 minutes
Beef, pork, veal, and lamb steaks/chops	145°F (63°C) for 15 seconds
Hamburger, ground pork, sausages, flaked fish	155°F (68°C) for 15 seconds
Fish	145°F (63°C) for 15 seconds
Fresh eggs for immediate service	145°F (63°C) for 15 seconds
Eggs, poultry, meat, or fish cooked in a microwave	165°F (74°C) and then let food stand for 2 minutes

Internal Cooking Temperatures

Thermometers

Both bimetallic stem and digital thermometers are used in the foodservice industry to measure the internal temperature of food. *See Figure 2-4.* When using a thermometer, place it in the thickest part of the food, and take at least two readings in different locations. Thermometers should be thoroughly cleaned, sanitized, and air-dried after each use to avoid the risk of cross-contamination. Thermometers should be calibrated and accurate to within +/−2°F or +/−1°C, after an extreme temperature change.

Personal Hygiene

Microorganisms on equipment, hand tools, smallwares, and cooking surfaces that come into contact with the hands of kitchen staff can be transported to food, contaminating the food. To avoid contamination, practice good personal hygiene, including bathing, proper hand washing, wearing single-use gloves, using utensils when handling ready-to-eat food, and maintaining good personal health.

Bathing

Foodservice employees should shower or bathe before arriving at work. Fingernails should be clean and short.

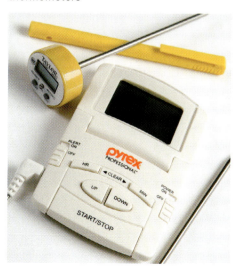

Figure 2-4 Bimetallic stem and digital thermometers

Washing Hands

The transmission of microorganisms occurs mostly by hands, so it is vital to wash hands frequently—after eating, drinking, smoking, chewing gum, and using the restroom; touching the face, hair, or skin; and coughing, sneezing, or wiping the nose. Hands should also be washed before and after handling raw food, between performing tasks, before putting on single-use gloves, prior to using preparation equipment, and after removing trash or clearing tables.

Clothes

Clean clothes should be worn daily to work. If uniforms are provided, food-handlers should change their clothes, if possible, in the establishment after arriving at work. This helps lower the risk of cross-contamination.

Jewelry

Jewelry can harbor harmful microorganisms and poses a physical hazard as well. Food-handlers should remove all jewelry, except for a plain ring such as a wedding band, prior to starting a shift.

Hair and Facial Hair

Hair should be cleaned daily prior to arriving at work. Dirty hair can be a breeding ground for pathogens that may cause cross-contamination. Hair should also be held back in a hair restraint to prevent touching the hair and then touching food. Hair restraints also keep hair from falling into food. Men with beards should wear beard restraints for the same reasons.

Chemical Hazards

Chemical hazards—chemical substances such as cleaning supplies, polishes, pesticides, food additives, and toxic metals from worn cookware—can cause food contamination when they come into contact with food. When using such chemicals, follow manufacturer directions. Chemical hazards are a common cause of foodborne illness. Any chemical that is known to have chronic or acute effects on health, or is flammable or unstable, is considered a hazardous chemical.

Cleaning Products and MSDS

A variety of cleaning products and sanitizers are commonly used in food service. Material Safety Data Sheets (MSDS) that list a material's chemical and common names as well as its possible hazards and ways to avoid these hazards must be kept on file for each cleaning product. Products should be used and stored properly and always kept in their original containers. Chemical containers should be properly labeled and never reused for any purpose. Discard all packaging when the product is finished. Cleaning products and sanitizers should never be used near food.

Detergents

Detergents are used to clean walls, floors, prep surfaces, equipment, and utensils and to cut through grease.

Hygiene Detergents

These detergents are made specifically to clean, deodorize, and disinfect floors, walls, and tabletops.

Degreasers

Degreasers are solvents used on range hoods, oven doors, and backsplashes to remove grease.

Abrasive Cleaners

Abrasives are used to clean hard-to-remove soil on floors and burned-on food from pots and pans.

Acid Cleaners

Acid cleaners remove mineral deposits in dishwashers and steam tables and should always be used with care.

Kitchen Cleanliness

Keeping the kitchen clean and sanitary decreases the risk of contamination. Equipment should be cleaned as work is performed, and followed by sanitizing. Pots, pans, dishes, and food-contact surfaces should be cleaned before preparing another food product or if hand tools become contaminated.

Some commercial kitchens use designated color-coded cutting boards, knives, cloths, and containers for each particular type of food product to further prevent the risk of contamination. For example, green may be the designated color for vegetables and fruits. All utensils, including knives, tongs, and cutting boards, as well as cleaning equipment that come into contact with produce are green. Red is typically the color designated for raw meat, brown is for cooked meat, yellow for poultry, blue for raw fish and seafood, and white for dairy. *See Figure 2-5.*

Pesticides are frequently used in food storage and preparation areas to control pests, and must be used properly to prevent chemical contamination. Only a trained Pest Control Operator (PCO) should apply pesticides. Before pesticides are applied, all foods should be properly stored and wrapped. Store pesticides away from food and in a locked or secure area, and clearly label the containers to avoid any confusion.

Figure 2-5 Color-coded cutting boards, knives, and other kitchen equipment can be used to prevent the risk of contamination.

Physical Hazards

Physical hazards are foreign particles such as glass chips, metal shavings, tooth picks, soil residue, stones, hair, jewelry, fingernail polish chips, and bone shards that can contaminate food. Most physical hazards are the direct result of poor food handling and can cause injury or discomfort to the consumer.

Avoiding Physical Hazards

The easiest way to avoid physical hazards is by practicing proper food handling procedures. These procedures include:

- carefully inspecting foods upon receiving them
- avoiding physical contamination during the flow of food

Hazard Analysis Critical Control Point (HACCP)

The *Hazard Analysis Critical Control Point (HACCP)* system is used to monitor the flow of food through a foodservice operation. HACCP allows operators to identify potential hazards and to correct them before they become a problem.

The HACCP System

The HACCP system was developed by the Pillsbury Company for the National Aeronautics and Space Administration (NASA) in the early 1960s to ensure safe food in outer space. The foodservice industry has adopted parts of this system of self-inspection in addition to the regular inspections performed by local health departments.

HACCP ensures food safety by combining food-handling procedures, monitoring techniques, and record keeping. The system focuses on the flow of food through the foodservice facility at critical control points. It helps employees to identify foods and procedures that are likely to cause foodborne illness, to develop facility procedures that will reduce the risk of foodborne illness, to monitor procedures to keep food safe, and to ensure that it is served at the proper temperature.

Establishing a HACCP Plan

In establishing a HACCP plan, a foodservice operation may rely on research, food regulations, and past experience. A good HACCP plan should have standards that are measurable and observable, such as temperature and time. Set standards for each critical control point should be put in place and adhered to by all employees and documented in a log.

The Seven Steps of HACCP

A typical HACCP system has seven steps: analyzing hazards, determining critical control points, establishing critical limits, monitoring the system, taking corrective action, verifying procedures, and record keeping.

1. *Analyze hazards*   The first step of HACCP is to identify and analyze potential food safety hazards. The flow of food is checked to identify where potential biological, chemical, and physical hazards may occur. All potentially hazardous foods are identified as well as steps in the production process where food may become contaminated and where microorganisms may survive and multiply.
2. *Determine critical control points*   Determine **Critical control points (CCPs)** by identifying the points in the flow of food where hazards can be prohibited, eradicated, or reduced to safe levels. Common CCPs are cooking, hot or cold holding, and cooling.
3. *Establish critical limits*   For each critical control point identified, establish standards or thresholds that each food item must meet to be considered safe. Critical limits must be measurable and should be based on science, regulation, and/or expert opinion.
4. *Establish procedures to monitor the system*   Foodservice operators and employees are responsible for monitoring the systems in place and creating checks and balances to ensure that potential problems are eliminated. Procedures should establish the who, what, when, and how items are monitored.
5. *Take corrective action*   Taking corrective action involves enacting predetermined steps if a critical limit is not met.
6. *Establish verification procedures*   An important aspect of HACCP is that of determining whether the set procedures are working. Critical control points, such as cooking, holding and cooling food, and critical limits can be checked by the chef or shift manager to ensure that they are adequate. Follow-up must be conducted on all corrective action taken to ensure the reduction of foodborne illness. Furthermore, check that established procedures are being followed by all employees. An audit from an outside agency is sometimes warranted to verify that the HACCP system is working.
7. *Establish record keeping*   Record keeping is important in maintaining an effective self-inspection system. Record-keeping systems, which are simple and easy to maintain, include flowcharts, policy and procedure manuals, written records, and temperature readings.

Safe Food Handling from Receiving to Service

The *flow of food* is the process by which food items move through a foodservice operation, from receiving to service.

Receiving and Storing Food

The flow of food begins when food items are received. Safety and sanitation procedures also begin here. All food items must be carefully inspected for damage and to ensure that they have been maintained at the proper temperatures during transit. Potential problems that might be encountered in receiving include thawed and refrozen foods, insect infestation, damaged foods or containers, repacked or mishandled items, and foods shipped at incorrect temperatures. Products that do not meet food safety standards should be rejected.

Storing food, whether dry, refrigerated, or frozen, is another control point at which improper handling can result in contamination. All foods should be stored properly and in the appropriate location, to prevent contamination, spoilage, and the growth of harmful bacteria. Storage areas should always be kept clean and dry, and the temperature should be monitored.

Dry Storage

Foods that have a low moisture content such as flour, salt, sugar, and canned foods can be placed in dry storage. The temperature in a dry storage area is 50°F–70°F (10°C–21°C) with 50%–60% humidity.

Cold Storage

Keep food products that need refrigeration at or below 41°F (5°C). The *First In, First Out (FIFO)* inventory program should be used for food storage, moving older items to the front and storing fresher items behind them. Cooked or ready-to-eat foods should be stored above raw ingredients to prevent cross-contamination. Drip pans should be placed under raw ingredients to catch any spills. Foods being thawed should be placed below prepared foods, leaving sufficient space for air to circulate around the food. Cooked, ready-to-eat potentially hazardous foods should be date marked and discarded within 7 days after preparation (as long as food is kept 41°F(5°C) or below.) Also routinely check unit and food temperatures.

Frozen Storage

Frozen foods should be stored at 0°F (−18°C) or below. Frozen dairy products can be stored between 6°F and 10°F (−14°C and −12°C).

- Label and date all containers, and ensure that wrappings are airtight to prevent freezer burn.
- Check unit and food temperatures.
- Defrost the freezer on a regular basis.
- Use cold curtains to maintain temperature control.
- Use FIFO.
- Open the freezer only when absolutely necessary.

Preparation and Cooking

After properly receiving and storing foods, the flow of food moves to preparation and cooking. Wash all fresh fruits and vegetables before preparation, and never prepare them on the same cutting boards used to prepare uncooked meats. Each type of food product prepared is susceptible to a different kind of contamination. Designated color-coded cutting boards, knives, and cloths can cut down the risk of cross-contamination.

When preparing and cooking foods, use clean, sanitized cutting boards, knives, and utensils. Remove refrigerated food products only when needed. Wash, rinse, and sanitize the work areas each time a different food product is prepared, and prepare or cook foods immediately.

Ready-to-eat foods should not be handled with barehands. Tongs or spatulas, deli tissue or single use gloves should be used when handling ready-to-eat foods. Ready-to-eat foods are defined as follows:

- raw, washed, or cut fruits and vegetables
- food that is ready for consumption (no further washing or cooking is required)
- unpacked, potentially hazardous food that has been cooked according to the required time and temperature specifications

Wear gloves to create a barrier between hands and food when nothing else (tongs, deli tissue, and so on) is being used. *See Figure 2-6.* Hands should be washed before putting on gloves, when removing gloves, and between glove changes. Hands need to be washed thoroughly with soap and water following proper handwashing procedures. Foodservice gloves should be changed when they become damaged or torn, after handling any raw food, or when a job interruption occurs. The type of single-use gloves worn in food service depends on the task to be accomplished. Glove choices include:

- nitrile powder-free gloves
- plastic disposable gloves
- powder-free gloves
- uniseal gloves
- vinyl gloves

Although latex gloves are also available, they should be used sparingly because of common allergic reactions to latex. As of July 2001, the state of Rhode Island and Johnson & Wales University have prohibited the use of natural rubber latex gloves in any foodservice establishment.

Holding Food Safely

Foods may be cooked and served immediately or prepared in advance and held for service. Because foods are extremely susceptible to microorganism growth during holding,

Figure 2-6 Single-use gloves are used when handling ready-to-eat food.

it is important to follow proper procedures. Keep foods covered to reduce contamination, and take the internal temperature regularly, a minimum of every two hours. Cooked foods should be held at 135°F (57°C) or higher. If the temperature drops below 135°F (57°C), reheat the food to 165°F (74°C) for 15 seconds within a two-hour time period, and hold again at 135°F (57°C) or higher. Hot foods that are being hot held should also be stirred regularly to distribute the heat throughout the food.

Cold food should be held at 41°F (5°C) or lower. Except for fruits and vegetables, do not store cold food directly on ice. Place it in a container, such as a hotel pan. Use self-draining drip pans for displays.

Serving Food Safely

Microorganisms are easily spread. Never touch ready-to-eat food or the surfaces of glasses, dinnerware, or flatware with bare hands when serving food. Hold dishes by the bottom or an edge, cups by their handles, glassware by the lower third, and flatware by the handles.

Plates of food should never overlap when dishes are served to guests. Instead, use a tray to carry the plates. When serving beverages, use scoops (not bare hands) to pick up ice.

Cooling Food Safely

The FDA recommends a two-stage method for cooling food safely. Cooked foods require cooling from 135°F–70°F (57°C–21°C) within two hours and temperatures of 135°F–41°F (57°C–5°C) or lower within a total of 6 hours .

- *The Rapid Kool™*   The Rapid Kool™ is a container of water that can be frozen and placed directly in stock to accelerate cooling.
- *Ice-water bath*   An ice-water bath allows smaller portions of stock to be placed in ice water in a food preparation sink or large pot.
- *Blast chiller*   Blast chillers are units used to cool food, by moving it through the temperature danger zone in less than two hours. After food is chilled in a blast chiller, it can be safely refrigerated.
- *Ice or cold paddles*   Ice or cold paddles are directly added to the product to reduce the temperature. The paddles, which are hollow, are filled with ice and are then used to stir the product.
- *Metal cooling pins*   Metal cooling pins are inserted into a food item to cool it by transferring heat from the food to cold air.

Reheating Food Safely

When reheating foods for hot holding, the internal temperature must reach 165°F (74°C) for 15 seconds within a two-hour period after having been removed from the refrigerator.

Disposal of Food

Cleaning and sanitizing are key to the proper disposal of food. Dishes, flatware, smallwares, hand tools, and equipment must be cleaned and sanitized. To do this, remove leftover food by scraping it into a garbage receptacle. Rinse items over the disposal before washing. Dishes, flatware, smallwares, and hand tools can be washed either manually or mechanically. Chemical sanitizers are used in sinks and dishwashers to prevent the growth of bacteria.

Manual Dishwashing

A commercial sink with three compartments is used for manual dishwashing. To wash items by hand, scrape, pre-rinse, and then wash items in at least 110°F (43°C) water and detergent. Follow up by rinsing with clear water. Change the water as needed to keep it clear and hot. Sanitize according to the manufacturer's instructions. Each chemical sanitizer has different levels of concentration and temperature requirements, i.e., chlorine above 75°F (24°C) or below 115°F (46°C).

Mechanical Dishwashing

Commercial dishwashers include single-compartment, multicompartment, carousel, recirculating, and conveyor machines. When using a dishwasher, scrape and rinse soiled dishes, and presoak flatware. Clean utensils by following manual dishwashing procedures. Pre-rinse dishes to remove all visible food and soil, and then load the dishwasher. Always run the dishwasher through a full cycle, and check its cleanliness throughout the workday. Follow the manufacturer's operation instructions to ensure clean and properly sanitized wares.

Drying and Storing Items

After cleaning and sanitizing either manually or mechanically, allow the items to air dry. Do not touch the surfaces that will come in contact with food, and wash your hands before storing the items.

Workplace Safety

Kitchen staff are particularly at risk for injuries that result from workplace accidents because of fatigue, poor kitchen design, and insufficient training. Yet, despite the possibility of workplace accidents, they can be controlled. The Occupational Safety and Health Administration (OSHA) plays a large role in keeping the workplace safe by enforcing workplace standards outlined in the Occupational Safety and Health Act. Employers must post OSHA safety and health information in their facilities, and employees are required to follow OSHA regulations.

The Environmental Protection Agency (EPA) also insists that foodservice operations track how they handle and dispose of hazardous materials such as cleaning products, sanitizers, and pesticides.

Personal Protective Clothing

Personal protective clothing, such as uniforms, aprons, and gloves, keep foodservice workers safe from injury. Personal protective clothing should be kept clean to avoid contamination.

Proper Uniforms

Uniforms should be neat and clean before each shift begins. Change into uniforms at work, when possible, to reduce the risk of cross-contamination. Uniforms should be kept in clean lockers.

Proper Shoes

Protective shoe requirements vary throughout foodservice operations, but all should be sturdy, slip-resistant, and closed-toe.

Protective Gloves and Mitts

According to the Bureau of Labor Statistics, most hand injuries recorded in food service are cuts, punctures, and burn injuries, which could have been prevented by wearing the proper protective gloves or mitts. Protective clothing for hands also guards hands against the effects of chemical compounds. There are a variety of safety gloves and mitts from which to choose:

- knife gloves—used to protect hands from cuts and amputation when cutting a variety of food products

- cut-resistant—heavy gloves, used to protect hands from cuts and amputation
- terrycloth heat-resistant—worn to protect hands whenever the potential for burns or scalding is present

Personal Injuries

Employees are responsible for preventing slips and falls, cuts, burns and scalds, and other personal injuries that can occur in the kitchen. The guidelines that follow can help minimize such workplace injuries.

Slips and Falls

Although most slips and falls can be avoided, they are common work-related injuries. Prevent such accidents by performing tasks carefully and keeping walking areas uncluttered and free of spills, especially around exits, aisles, and stairs.

When working in the kitchen, walk, do not run. Wipe up spills immediately, and use slip-resistant floor mats. Many falls occur on wet floors slick with water and cleaning products. Use care when walking through the kitchen. Also use "Caution" or "Wet floor" signs after mopping or cleaning to warn others to be careful. Never use a chair or box for climbing, use safe ladders or stools. Always close drawers and doors, and use a cart when moving heavy objects.

Cuts

The risk of cuts in a commercial kitchen is high but can be minimized by following safety guidelines. Unplug appliances before cleaning them, wear protective gloves and cuff guards when cleaning slicers, and use proper knife skills.

Knives are the most common cause of cuts in a commercial kitchen. Always use knives for their intended purpose. Never use them to open plastic overwrap or boxes. Knives should be carried at the side of the body with the blade tip pointed toward the floor and the sharp edge facing back. Always reach for the knife handle, and never wave hands when holding a knife. Use a firm grip on the handle when holding a knife, and cut away from the body, not toward the body. If a knife is dropped, do not try to catch it.

In addition, never leave a knife handle hanging over the edge of a work surface. Keep knife handles and hands dry when using knives, and keep knives sharp. Dull knives require applying more pressure, possibly causing slippage.

Wash sharp tools separately from other utensils; never leave knives in the sink. Throw away broken knives or knives with loose blades, and store knives in a knife kit or a rack.

Burns and Scalds

Burns and scalds present yet another risk when working in commercial kitchens. To prevent these accidents, remove lids by tilting them away from the body to let steam escape. Use dry potholders or oven mitts, as wet cloth forms steam when it touches hot pots and pans. The handles of pots and pans should be turned away from the front of the range, thus preventing spills that could cause burns. Use caution when filtering or changing shortening in fryers, and wear gloves and aprons because splatters can cause serious injury.

Back Injuries and Strains

Back injuries from improper lifting and bending are one of the most common workplace injuries, accounting for more than 20%–25% of all workers' compensation claims. When lifting heavy items, use rollers under the objects, or ask for help. When lifting heavy objects, follow these guidelines:

- Bend at the knees.
- Keep your back straight.

- Keep your feet close to the object.
- Center your body over the load.
- Lift straight up without jerking.
- Do not twist your body to pick up or move the object.
- Set the load down slowly, keeping your back straight.

Cleaning Kitchen Equipment

When cleaning any piece of kitchen equipment, turn switches to the "off" position, unplug the machine from its power source, and follow the manufacturer's instruction manual and the food establishment's cleaning directions.

Maintenance and Repairs

Maintaining and repairing equipment require compliance with OSHA's lockout/tagout procedure. All necessary switches on electrical equipment must be locked out and tagged when malfunctioning to prevent the equipment from being used.

Fire Safety

Fires can cause substantial property and equipment damage, injuries, and death. Flames and high heat sources in the commercial kitchen associated with the foodservice industry increase the probability of fires.

Equipment

Foodservice operations rely on a variety of fire extinguishers and hood and sprinkler systems as standard fire safety equipment.

Fire Extinguishers

Fire extinguishers are the most common type of fire protection equipment used in foodservice operations. The type, number, and location of fire extinguishers needed in a commercial kitchen may vary, but each workstation must have a working fire extinguisher within reach. Fire extinguishers use several types of chemicals to fight different kinds of fires. Use the appropriate class of extinguisher to fight a fire properly. Class A extinguishers (for wood, paper, cloth, and plastic), class B (grease, oil, and chemicals), class C (electrical cords, switches, and wiring), class D (combustible switches, wiring, metals, and iron), and class K (for fires in cooking appliances involving vegetable and animal oils or fats).

Hood and Sprinkler Systems

Hood and sprinkler systems are standard equipment in commercial kitchens. A properly ventilated hood system should remove excess smoke, heat, and vapors caused by fire, while the sprinkler systems aid to extinguish flames. Hoods and ducts should be cleaned regularly, and products and supplies should be properly stored.

Emergency Procedures

Every foodservice establishment has fire emergency procedures with which employees should be familiar. Fire exit signs must be posted in plain view above exits, and a meeting place outside should be designated so that a head count can be conducted. Employees should be able to direct customers out of the building and keep them calm during emergencies. If a fire is discovered, regardless of its size, call the fire department immediately. Then quickly and calmly assist customers and co-workers in exiting the building.

First Aid

The definition of *first aid* is assisting an injured person until professional medical help can be provided. "First aid" involves

- checking the scene and staying calm.
- keeping the victim comfortable and calm.
- calling the local emergency number for professional medical help.
- caring for the victim by administering first aid according to the first-aid manual.
- keeping people who are not needed away from the victim.
- completing an accident report that includes the victim's name, the date and time of the accident, the type of injury or illness, the treatment given, and the amount of time it took for assistance to arrive.

The American Red Cross (ARC) offers courses that teach hands-on practical information about first aid in the workplace. Foodservice workers are encouraged to take an ARC first-aid course. Contact your local chapter of the ARC for additional information.

First Aid for Burns

All burns require immediate treatment. *See Figure 2-7.* For more serious burns call the local emergency number for medical assistance. Until help arrives, follow these general guidelines:

- Remove the person from the source of the heat.
- Cool the burned skin to stop the burning by applying cold water over the affected area. Use water from a faucet or apply soaked towels. Do not use ice or ice water.
- Never apply ointments, sprays, antiseptics, or remedies unless instructed to do so by a medical professional.
- Bandage the burn as directed in the first-aid manual.
- Minimize the risk of shock. Keep the victim from getting chilled or overheated. Have the victim rest.

First Aid for Wounds

There are four types of open wounds or cuts to which foodservice employees are susceptible: abrasions, lacerations, avulsions, and punctures. Each requires immediate medical attention.

    Abrasions   An *abrasion* is a scrape and is considered a minor cut.
    Lacerations   A *laceration* is a cut or tear in the skin that can be quite deep. A knife wound is a type of laceration.

Figure 2-7

Types and Characteristics of Burns

First-degree burns
A mild burn where the skin becomes red and sensitive but does not blister.

Second-degree burns
A burn that blisters causing painful damage to the skin.

Third-degree burns
A severe burn where the skin and underlying tissue are destroyed. This burn requires immediate medical attention.

Avulsions An *avulsion* occurs when a portion of the skin is partially or completely torn off.

Punctures *Puncture wounds* occur when the skin is pierced with a pointed object, such as an ice pick, making a deep hole in the skin.

For minor cuts, the first-aid provider should follow these guidelines:

1. Put on disposable gloves to protect against infection.
2. Clean the cut with soap, and rinse it under water.
3. Place a bandage over the cut. Use sterile gauze if possible.
4. Apply direct pressure over the sterile gauze or bandage to stop the bleeding.
5. If bleeding does not stop, elevate the limb above the heart to reduce the amount of blood going to the cut area.
6. Follow instructions in the first-aid manual.

To administer emergency relief for lacerations, avulsions, and punctures, the first-aid provider should:

1. Put on disposable gloves to protect against infection.
2. Control the bleeding by applying pressure, using sterile gauze or a clean cloth towel. Do not waste time washing the wound first. Elevate the area while applying pressure.
3. Cover the wound with clean bandages. Continue to apply pressure.
4. Wash your hands thoroughly after treating the wound.

First Aid for Choking

In case of a choking incident, the Heimlich maneuver may have to be performed. To do this:

1. Stand behind the victim. Wrap your arms around his or her waist.
2. Locate the victim's navel.
3. Make a fist with one hand. Place the thumb side of your fist against the middle of the abdomen. Position your hand just above the navel and below the bottom of the breastbone.
4. Place your other hand on top of your fist.
5. Press your hands into the victim's abdomen. Use quick inward and upward thrusts. Each thrust should be a separate and distinct action.
6. Repeat this motion as many times as it takes to dislodge the object or food from the victim's throat. Note that a conscious victim can become unconscious during this maneuver if the object is not dislodged.

The Heimlich maneuver should be used only on someone who is conscious and choking. If the person can cough or speak, or is unconscious, do not perform this maneuver; doing so can cause physical injury. The Heimlich maneuver should never be performed on a pregnant woman as doing so could harm the baby.

Cardiopulmonary Resuscitation (CPR)

Unresponsive victims—those who are unconscious because of choking, cardiac arrest, stroke, or heart attack—will need cardiopulmonary resuscitation (CPR) until emergency help arrives. CPR helps keep oxygen flowing to the brain and heart, improving the chances of survival. It is advisable that foodservice workers learn CPR. Contact your local chapter of the American Heart Association or the American Red Cross for additional information.

Sickness in the Workplace or Classroom

Employees in food service must maintain their health and be free of illnesses when reporting for work or attending classes. Individuals who are ill can transmit harmful microorganisms to other foodservice employees or to the food they are preparing or serving.

Prior to working with food, a student will sign a Food Handlers Agreement. If an employee or student exhibits any of the symptoms listed, he or she should inform the supervisor or instructor and health care professional of the condition, and not report to work or attend class.

An employee or student who becomes ill while at work or in class, should inform the supervisor or instructor and then leave immediately to report to a doctor or to university health services. Any surfaces with which that employee came into contact should be cleaned and sanitized to prevent cross-contamination.

All medications taken by employees should be stored with the employee's or student's personal belongings to ensure that they do not get mixed in with food. Cuts, sores, wounds, or infections must be covered with a clean, dry bandage. If the bandage is on the hand, gloves or a finger cot should be worn while at work. Employees with these types of injuries may need to be reassigned to another duty.

CHAPTER 3

Nutrition

The demand for healthier, more nutritious, functional, and tasty specialty foods is on the rise. Bakers and pastry chefs have a responsibility to respond to consumer requests by creating products that meet these market needs and standards. The use of healthier foods containing less fats and trans fats, and increased whole grains is important. The creation of gluten- and lactose-free products is also a challenge for today's baking and pastry chefs.

Key Terms

- nutrient
- calorie
- carbohydrate
- glucose
- simple carbohydrate
- monosaccharides
- disaccharides
- complex carbohydrate
- polysaccharides
- soluble fiber
- insoluble fiber
- protein
- amino acid
- complete protein
- incomplete protein
- complementary protein
- lipid
- saturated fat
- monounsaturated fat
- polyunsaturated fat
- hydrogenation
- trans-fatty acid
- cholesterol
- high-density lipoprotein (HDL)
- low-density lipoprotein (LDL)
- vitamins
- minerals
- phytochemicals
- celiac disease
- xanthan gum
- daily values
- genetically modified organism (GMO)
- additive
- irradiation
- functional food

37

Nutrients

The body requires chemical compounds called *nutrients* to survive. Nutrients consist of macronutrients and micronutrients. These two groups are then further broken down into three classes. Macronutrients consist of carbohydrates, proteins, and fats (lipids), while micronutrients are made up of vitamins, minerals, and water.

Food Energy

Food energy is measured in units of heat called *calories*. One calorie is equivalent to the amount of heat required to raise the temperature of 1 kg of water by 1°C. Macronutrients supply energy and calories to the body.

- One gram of carbohydrate supplies 4 calories/gram.
- One gram of protein supplies 4 calories/gram.
- One gram of fat supplies 9 calories/gram.

Interestingly enough, micronutrients, vitamins, minerals, and water do not supply calories or energy.

Fruits and vegetables are excellent sources of a wide variety of nutrients. All fruits and vegetables provide carbohydrates and many provide vitamin A (arytenoids), vitamin C, foliate, fiber, and potassium. *See Figure 3-1* for sources of nutrients.

Carbohydrates

Carbohydrates are compounds composed of *glucose* units that are found only in plants and dairy products. All carbohydrates contain the three chemical elements: carbon, hydrogen, and oxygen. Carbohydrates are ideal sources of energy, and experts recommend that 50%–60% of calorie intake come from carbohydrates. The body breaks down carbohydrates into glucose. Glucose is the only sugar that plants and animals use for energy. Plants store extra glucose as starch, and animals store additional glucose as glycogen.

Simple Carbohydrates

Simple carbohydrates are sugars found in milk, honey, fruit, maple syrup, refined table sugar, and unrefined sugar. Simple carbohydrates can help retain moisture, and some such as the sugar in custard actually help delay protein coagulation. Other simple carbohydrates add texture and sensory appeal, aid in caramelization, and act as leavening agents. In addition, there are those that help preserve foods and stabilize egg foams. Simple sugars are often referred to as concentrated sweets and can be formed from monosaccharides or disaccharides.

Figure 3-1

SOURCE OF VITAMIN A (CAROTENOIDS)

- *Orange vegetables such as carrots, sweet potatoes, and pumpkins*
- *Dark-green leafy vegetables that include spinach, collards, turnip greens*
- *Orange fruits such as mangoes, cantaloupes, apricots, and red grapefruits*
- *Tomatoes*

SOURCES OF VITAMIN C

- *Citrus fruits that include lemons, limes, grapefruits, oranges, kiwis, strawberries, and cantaloupes*
- *Broccoli, peppers, tomatoes, cabbage, and potatoes*
- *Leafy greens such as romaine lettuce, turnip greens, and spinach*

SOURCES OF FOLATE

- *Cooked dry beans and peas, and peanuts*
- *Oranges, orange juice*
- *Dark-green leafy vegetables such as spinach and mustard greens, romaine lettuce*
- *Green peas*

SOURCES OF POTASSIUM

- *Baked white or sweet potatoes, cooked greens (such as spinach), winter (orange) squash*
- *Bananas, plantains, dried fruits such as apricots and prunes, oranges*
- *Cooked dry beans and lentils*

Sources of Nutrients

MONOSACCHARIDES *Monosaccharides* are composed of a single sugar carbohydrate unit. Monosaccharide sugars include glucose, fructose, and galactose. Glucose and fructose are present in some fruits and vegetables. Glucose is the main ingredient of corn syrup, a common ingredient used in baked goods. High-fructose corn syrup can be artificially manufactured by a process that changes glucose to fructose. High-fructose corn syrup is often used in soft drinks. Galactose is not found alone in natural foods but is part of lactose (milk sugar). See Figure 3-2.

DISACCHARIDES *Disaccharides* are two or more monosaccharides linked together. Sucrose, lactose, and maltose are three important disaccharides. Sucrose, which is created from one molecule of glucose and one of fructose, is commonly known as table sugar. Lactose, or milk sugar, is made up of glucose and galactose. It is found naturally in milk and in whey, a by-product of cheese. Maltose is made up of two molecules of glucose and is a component of corn syrup.

Complex Carbohydrates

Complex carbohydrates are composed of starch and fiber units that help maintain the digestive system's function. Digestive enzymes in the body slowly break down these carbohydrate strands into smaller glucose molecules. Complex carbohydrates also contain other necessary nutrients such as vitamins and minerals. Complex carbohydrates such as whole-wheat flour and oat bran can add structure and texture to foods. Grains that have the starch, bran, or germ intact are commonly labeled "whole grains."

Figure 3-2 When more than two monosaccharides join, a disaccharide is formed. *Mono-* means "one" and *di-* means "two."

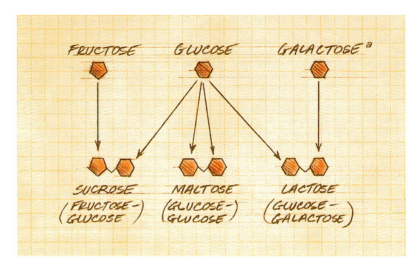

STARCH Starch is a *polysaccharide*. It is a compound made up of many long strands of sugar units linked together. Plants store starch in their seeds and roots. Whole grains, pasta, flour, and vegetables such as potatoes, corn, and legumes all contain starch. During the digestive process, the body breaks down the starch into smaller glucose units. *See Figure 3-3.*

FIBER Fiber is found in complex carbohydrates. Some fiber molecules are not humanly digestible because the body does not have the enzymes to break them down. There are two types of fiber: *soluble fiber,* which dissolves in water to form a gel, and *insoluble fiber*, which absorbs water. Soluble fiber is found in foods such as oat bran, beans, peas, and certain fruits. Soluble fiber has been shown to help reduce blood cholesterol levels and reduce the risk of heart disease by binding to cholesterol and removing it from the body. Without soluble fiber, cholesterol is reabsorbed into the bloodstream. Foods high in soluble fiber slow down the digestive process and produce

Figure 3-3 A polysaccharide is composed of many long branches of sugar units that are linked together. *Poly-* means "many."

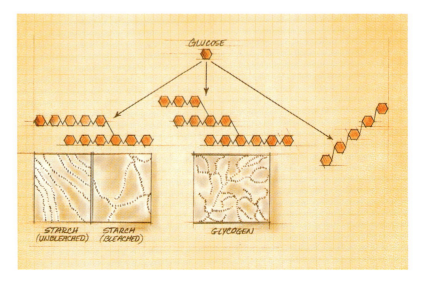

satiation to reduce the desire for calorie-dense fats and sweets. Foods with adequate fiber may be evaluated by glycemic measure. High-fiber foods are digested and absorbed at a slower rate for a lower glycemic response. Weight conscious individuals and those with diabetes can profit considerably from a fiber-dense diet. *See Figure 3-4* listing whole grains. Soluble fiber is found in foods such as oat bran, beans, peas, and some fruits.

Insoluble fiber is found in foods such as brown rice and whole-wheat flour. Consumption of insoluble fiber helps to reduce the risk of colon cancer by binding or diluting cancer-causing materials in the colon, and helping to expel them from the digestive tract.

Proteins

Proteins contain nitrogen as well as the chemical elements carbon, oxygen, and hydrogen. The human body uses protein to build, maintain, and repair body tissues. Proteins

Figure 3-4

List of Whole Grains

Barley
Barley is available as a whole grain with only the outer hull removed. Scotch barley that has been husked and slightly ground, and pearl barley, a polished form with the hull and bran removed are examples. Barley has a chewy texture and a slightly sweet flavor.

Cornmeal
Cornmeal can be white, yellow, or blue, depending on the type of corn from which the meal is ground. It can be used to make baked goods.

Kasha
Kasha is hulled, roasted, and crushed buckwheat (also called buckwheat groats). Kasha has a strong nutty flavor.

Millet
Millet is ancient grain used in bread. It is commonly used in countries such as China, Africa, India, and Nigeria.

Oats
Oats are the berries of oat grass, available as oatmeal, also called rolled oats; as a whole grain, called oat berries or oat groats; and as a toasted and cracked grain form known as steel-cut oats. All forms of oats can be used in baked goods.

Quinoa
Quinoa is a small, bead-shaped grain with white or pink kernels. High in protein and commonly eaten in Colombia, Argentina, and Chile, quinoa is used in breads and can also be ground into flour for baked goods.

Rye
Rye is a grain used primarily in the United States as rye flour. It is heavy and dark and is used in breads.

Spelt
Spelt is an ancient variety of wheat whose small brown grains can be cooked like rice or ground into flour for use in baking. It has a mild, somewhat nutty flavor.

Wheat berries
Wheat berries are whole, unprocessed kernels of wheat. Wheat berries can be added to bread.

help the body to fight disease and are essential for healthy muscles, skin, bones, eyes, and hair. It is suggested that protein make up 12%–15% of daily calorie intake. Proteins for the human body are supplied by plants (in nuts and seeds), and animal products (such as meat and eggs). The body also makes its own protein.

Amino Acids

Amino acids are the building blocks of proteins. Approximately 20 amino acids make up all the proteins in both the human body and in food. Although all amino acids contain the chemical elements carbon, nitrogen, hydrogen, and oxygen, they do not all have the same structure. The body is able to produce about half of the necessary amino acids but must obtain the others from food sources.

Complete and Incomplete Proteins

Proteins that supply all of the essential amino acids are called *complete proteins.* Animal proteins such as meat, fish, poultry, cheese, eggs, and milk are complete proteins. Most plant proteins found in grains, legumes, nuts, and vegetables are known as *incomplete proteins* because they lack one or more of the essential amino acids. The body is able to combine amino acids derived from a variety of foods to create *complementary proteins.* Legumes, for example, provide certain types of amino acids that grains do not have. The consumption of legumes and grains on the same day provides essential amino acids that a body requires.

Protein's Function in Food

Proteins help to tenderize foods and retain moisture. Proteins such as beaten egg whites function as leavening agents, while gluten serves to add structure to flour. Protein also aids in the browning process and in producing gels and foams.

Fats/Lipids

Fats and oils belong to a class of chemical compounds called *lipids*. Both plant and animal foods contain fat. Fat plays an important role in the body by storing energy, cushioning vital organs, and insulating the body from extreme temperature. Fat also acts as a carrier of the fat-soluble vitamins A, D, E, and K and provides the essential fatty acids the body needs. Fats provide a rich mouthfeel to foods, make baked products tender, and serve to conduct heat during cooking.

Fats are classified by the chemical structure of their carbon, oxygen, and hydrogen bonds. Fatty acids are long carbon chains that are either saturated, monounsaturated, or polyunsaturated. The greater the number of hydrogen atoms linked to carbons, the more saturated the fatty acid is. The degree of saturation affects the temperature needed to melt a fat, and the "hardness" of that fat. Lard is saturated, while corn oil is unsaturated.

Saturated Fats

Saturated fats have all their carbon atoms linked, and each link is bonded to a hydrogen atom. *See Figure 3-5.* Saturated fats are solid at room temperature and are found in animal products including lard, butter, whole and reduced-fat milk and cheese, and meats. Tropical oils, such as coconut, palm, and palm kernel oil, are also saturated fats. A diet high in these fats has been linked to an increased risk for cardiovascular disease, cancer, and obesity.

Figure 3-5 Carbon atoms link to form fatty acids.

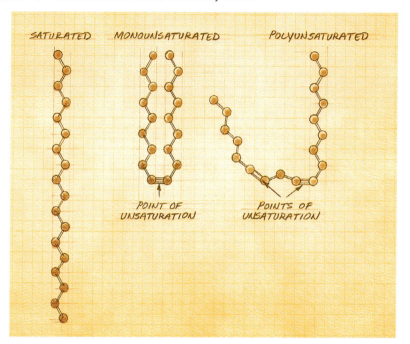

Monounsaturated Fats

Monounsatured fats lack a hydrogen link in their chain. These fats are liquid at room temperature. Olive, canola, and peanut oils are examples of monounsaturated fats. Recent research indicates that olive oil may actually lower the risk of heart disease by increasing the levels of beneficial blood lipids and lowering the levels of harmful blood lipids. The popular Mediterranean Diet includes olive oil rather than saturated fats such as butter, cream, and lard.

Polyunsaturated Fats

Polyunsaturated fats have more than one hydrogen bond missing in their chain. These fats are liquid at room temperature. Safflower, sunflower, corn, soybean, and cottonseed oils are examples of polyunsaturated fats. These fats are generally found in seeds, nuts, and in some fish.

HYDROGENATION *Hydrogenation* is a process that alters the fatty acid chains in polyunsaturated vegetable oil to extend the freshness of oil, raise its smoking point, and change it from a liquid to a solid state. Many food product manufacturers use hydrogenated fats to keep foods fresher longer. Consumers often prefer softer, spreadable hydrogenated margarines that have a longer shelf life than butter. Hydrogenated fats provide none of the health benefits supplied by polyunsaturated oils. Hydrogenated and partially hydrogenated oils are more saturated than the oils from which they are made.

When polyunsaturated oils are hydrogenated and hardened, some of the fatty acids are altered into unusual shapes to become *trans-fatty acids.* Researchers have found a strong correlation between a diet rich in trans-fatty acids and cardiovascular disease. Fast foods, potato chips, hard margarines, doughnuts, cookies, peanut butter, and many other commercially processed foods contain trans-fatty acids. As of January 1, 2006, food manufacturers must list trans fat content separately on nutrition labels. Food manufacturers are demonstrating some effort in decreasing the number of products containing trans fat.

Cholesterol

Cholesterol is a waxy substance found in both animals and humans. Although it is a necessary substance in the body, in excess it may create problems that can lead to cardiovascular disease. Dietary cholesterol is only found in animal products such as meat, organ meats, egg yolks, and dairy products.

High-density lipoproteins (HDL) carry lipids to and from the body's cells. HDL, sometimes called "good cholesterol," carries excess cholesterol in the blood to the liver, and from the body. High concentrations of HDL are associated with a reduced risk of heart attack.

Low-density lipoproteins (LDL) transport excess cholesterol to tissues, where it slowly builds up as a thick, hard deposit called plaque. If too much LDL cholesterol circulates in the blood, plaque can slowly build up in the walls of the arteries that circulate through the heart and brain, to eventually impede the flow of blood. Research indicates that elevated levels of LDL increase the risk of heart attack, a trait that gives LDL its nickname: "bad cholesterol." Foods that are high in saturated fat or trans-fatty acids can cause two problems:

- an increase in LDL levels
- a decrease in HDL levels.

It is best to eat foods that are high in monounsaturated fats to help lower LDL cholesterol levels and increase HDL cholesterol levels.

Essential Fatty Acids

The cell protecting qualities associated with diets that include polyunsaturated fats known as omega-6 fatty acids and omega-3 fatty acids have been shown to lower the rate of heart disease. Vegetable oils, nuts, seeds, and whole grains provide omega-6 fatty acids. These acids contain linoleic acid, which plays a key role in cell membrane structure and body function.

Omega-3 fatty acids are found in fish and oils. These fatty acids are important to brain development and vision. Research suggests that omega-3 fatty acids also help to lower the risk of heart attacks. The linolenic acid of the omega-3 fatty acid family and two types of omega-3 fatty acids, EPA and DHA, are found in high concentrations in cold-water salmon, mackerel, tuna, sardines, and lake trout. Experts say that a balanced diet that includes fish and select vegetable oils two or three times a week provides an adequate supply of omega-6 and omega-3 fatty acids. *See Figure 3-6* for sources of omega-6 and omega-3.

Reducing Total Fats

A reduction of the total fat and/or oil in a formula can affect the taste of the finished product. Because taste is just as important as texture in baked goods and desserts, take time to experiment when reducing these formulas. Begin by working with formulas that have low yields and fewer ingredients to make it easier to calculate and determine the effect of removing some of the fat or oil.

Reducing Saturated Fats

In baking or pastry, substitute vegetable oils such as corn, soy, and canola for saturated animal fats. A reduction in the amount of saturated fat in baked goods or pastry can still allow for the retention of texture and tenderizing properties.

Figure 3-6

Sources of Omega-6 and Omega-3

Sources of Omega-6 Fatty Acids

Linoleic Acid:
leafy vegetables
seeds
nuts
grains
corn oil
safflower oil
soybean oil
cottonseed oil
sunflower oil

Sources of Omega-3 Fatty Acids

Linolenic Acid:
canola oil
soybean oil
walnut oil
wheat germ
margarine/shortening from soybean and canola oil
butternuts
walnuts
soybean kernels
soybeans

EPA and DHA
human breast milk
shellfish
mackerel
salmon
bluefish
mullet
sablefish
menhaden
anchovy
herring
lake trout
sardines
tuna

Replacing Fat

Bakers and pastry chefs can create healthier products by replacing fats with alternatives. Before substituting fats in baked goods and desserts fully understand the function of the fat in the item, which will vary from one item to another.

Use of Natural Fat Replacers

In instances where fat is used to create a creamy mouthfeel, ingredients such as starchy fruits or vegetables (pumpkin, banana, bean purées, and tofu) can be used as substitutes. Fat is used to tenderize, and can be replaced by nonfat and lowfat cheese products (cottage, cream, ricotta, and Neufchâtel). Nonfat cream cheese combined with a fraction of the original fat can be used in pie or tart dough. Although the texture of the dough will change slightly, it will still be a tender finished product with a lot less fat.

Making icings and buttercreams lighter and heavier is more of a challenge. It is difficult to replace butter in buttercream and still retain quality, creaminess, and mouthfeel. An alternative to a very fatty buttercream is a fat-free boiled, or meringue-type icing. Meringues are, by their very nature, fat free. They have a light and delicate texture and a sweetness that lends itself to a lighter, healthful product.

Nonfat or lowfat cream cheese can replace some of the butter in a traditional buttercream, while still retaining texture and mouthfeel. Remember to keep the flavor and texture of the item as close, as possible, to the original.

When working with chocolate, reduce the amount of cocoa butter in a mousse by using cocoa powder in combination with a small amount of vegetable oil. Rich looking, tasty chocolate desserts and baked goods can still be achieved without using solid forms of chocolate.

Vitamins

Vitamins are organic compounds that work in conjunction with enzymes and hormones to regulate bodily functions and to maintain health. Vitamins in themselves do not supply energy. Vitamins may be either fat-soluble or water-soluble.

Fat-Soluble Vitamins

Vitamin-rich foods are often those that are the most vibrant in color. Ripe red strawberries and fresh green broccoli are rich in vitamins and provide color to make a dish visually appealing. Vitamins A, D, E, and K are fat-soluble, and occur in the fats and oils of foods. They are absorbed by the body and stored in the liver and need not be consumed daily. *See Figure 3-7.*

Water-Soluble Vitamins

Vitamin C and all the B vitamins are water-soluble; they dissolve in water. The body easily absorbs water-soluble vitamins and efficiently excretes any surplus intake in the urine. Water-soluble vitamins are not stored for long in the tissues, and therefore rarely reach toxic levels. They must be consumed daily. Special precautions should be taken in preparing foods to avoid the destruction of water-soluble vitamins. *See Figure 3-8.*

Figure 3-7

Fat-Soluble Vitamins

VITAMIN A

Function
Strengthens the immune system

Sources
Fortified dairy foods, eggs; yellow-orange fruits and vegetables; dairy products; liver; and egg yolks

VITAMIN D

Function
Bone mineralization

Sources
Fortified milk; fatty fish such as salmon; liver; egg yolks. Exposure to sunlight causes the body to produce vitamin D

VITAMIN E

Function
Protects and helps create muscles and red blood cells

Sources
Polyunsaturated plant oils; green leafy vegetables; nuts; seeds; and whole grains

VITAMIN K

Function
Synthesis of blood clotting

Sources
Egg yolks; green leafy vegetables such as spinach; soybeans, vegetable oils

Figure 3-8

THIAMIN (VITAMIN B$_1$)

Function

Helps the body to use carbohydrates for energy; nervous system function; promotes normal appetite

Sources

Pork and other meats; whole and enriched grains; legumes and nuts

RIBOFLAVIN (VITAMIN B$_2$)

Function

Keeps skin and eyes healthy; part of coenzyme involved in energy metabolism

Sources

Dairy products; meat, liver, and fish; whole and enriched grains; eggs, leafy green vegetables

NIACIN (VITAMIN B$_3$)

Function

Part of the coenzyme involved in energy metabolism

Sources

All protein-containing food such as meat, poultry, fish, liver, and shellfish; dry beans; nuts; whole and enriched grains

VITAMIN B$_6$

Function

Helps make red blood cells; assists the body's use of carbohydrates and proteins; helps convert triptopan to niacin

Sources

Meat, poultry, fish, liver, and shellfish; dry beans; potatoes, whole grains; some fruits and vegetables

VITAMIN B$_{12}$

Function

Aids the body in building red blood cells; helps maintain the nervous system; part of coenzyme involved in new cell development

Sources

Animal products such as eggs; meat, poultry, and fish; dairy products, and shellfish

FOLATE (FOLIC ACID)

Function

Helps in the synthesis of new DNA; helps the body in synthesizing new cells and tissue

Sources

Green leafy vegetables; legumes; seeds; whole and enriched grains; fruits; peanuts

VITAMIN C (ASCORBIC ACID)

Function

Antioxidant; helps increase infection resistance; keeps blood vessel walls; bone growth

Sources

Citrus fruits such as oranges and grapefruit; kiwi, strawberries, broccoli, tomatoes, cantaloupe, green peppers, potatoes, cabbage-type vegetables, lettuce, mangoes

BIOTIN

Function

Helps several enzymes involved in energy metabolism

Sources

Several varieties of foods, including dark green leafy vegetables; liver; egg yolks; whole grains

PANTOTHENIC ACID

Function

Helps the body in using carbohydrates, fats, and proteins for energy

Sources

Several varieties of foods, including dry beans; meat, poultry, and fish

Minerals

Minerals are naturally occurring substances that the body requires in very small amounts. Although minerals, such as vitamins, supply no energy, they regulate certain body processes and are essential to the structure of bones and teeth. Minerals are divided into two categories: major minerals and trace minerals. Major minerals are those present in the body in amounts greater than 5 grams. Calcium and potassium are major minerals. Trace minerals occur in less than 5 grams, and include iron, zinc, and copper. See Figure 3-9 and Figure 3-10.

Figure 3-9

Major Minerals

CALCIUM

Function

Major mineral for renewing bones and teeth; important for muscle contraction and heart function; aids in blood clotting

Sources

Milk and milk byproducts, soy products, enriched juices and cereals; green leafy vegetables such as kale; turnips; small fish with bones

MAGNESIUM

Function

Assists the nervous system and helps muscles work; builds and renews bones

Sources

Peanuts, legumes, green leafy vegetables; fish and shellfish; dried fruit

PHOSPHORUS

Function

Strengthens bones and teeth; helps the body maintain the proper balance of cell fluids; builds and renews tissues; assists the body to extract energy from nutrients

Sources

Milk and milk byproducts; nuts; dry beans; whole grains; poultry, meat, and fish; egg yolks, cola beverages

POTASSIUM

Function

Maintains a healthy blood pressure and heartbeat, and fluid balance in the body

Sources

Meats; many fruits such as bananas, oranges, and cantaloupes; meats, poultry and fish; dry beans; vegetables; whole grains

SODIUM

Function

Aids in regulating blood pressure; maintains fluid balance in the body also in acid-alkaline balance

Sources

Salt and foods that contain salt; soy sauce; processed foods; deli meats; naturally occurring table salt, pickles

Figure 3-10

Trace Minerals

CHLORIDE

Function

Works as a part in acid-base balance and works with formation of gastric juice

Sources

Salt and foods that contain salt; soy sauce; meats; milk; processed foods

IRON

Function

Assists cells in using oxygen; helps the blood carry oxygen, component of hemoglobin myoglobin

Sources

Protein-containing foods such as meat, fish, and shellish; dry beans; egg yolks; dried fruit; whole and enriched grains; green leafy vegetables

IODINE

Function

Aids in the regulation of the metabolic rate

Sources

Iodized salt; marine fish and shellfish; bread; dairy products

ZINC

Function

Assists formation of protein and genetic material; helps heal wounds and form blood; helps metabolize carbohydrates, fats, and proteins; taste perception and smell

Sources

Protein-contining foods such as whole grains; poultry, fish, and shellfish; legumes; dairy products; eggs

COPPER

Function

Involved in hemoglobin formation; keeps nervous system, bones, and blood vessels healthy

Sources

Organ meats, seafood, drinking water, fish, and shellfish; whole grains; nuts and seeds; dry beans; coffee

FLUORIDE

Function

Maintenance of teeth and bone structure; prevents tooth decay

Sources

Drinking water, tea, marine fish, and seafood

SELENIUM

Function

Helps the heart to function normally; helps with enzymes that protect against oxidation

Food Sources

Seafood; eggs; whole grains, vegetables

Phytochemicals

Phytochemicals are nonnutritive chemicals made by plants. A diet that is high in fruits, vegetables, legumes, and grains contains phytochemicals and may help guard against cardiovascular disease, hypertension, diabetes, and cancer. Scientists have identified more than 900 different phytochemicals in plants. The chart below contains information about some of the most widely studied phytochemicals and the foods in which they are present.

Sources of Phytochemicals

CRUCIFEROUS VEGETABLES
Cruciferous vegetables such as cauliflower, broccoli, cabbage, and brussels sprouts are rich in the phytochemicals known as isothiocyanates and indoles. These phytochemicals can activate the production of enzymes that stop carcinogens from damaging DNA.

ALLIUM VEGETABLES
Allium vegetables include onions, garlic, leeks, and shallots that contain the phytochemical allyl sulfide. Allyl sulfide may slow the production of enzymes that cause carcinogens and increase the production of enzymes that destroy carcinogens.

BERRIES
Berries such as cranberries, blackberries, and strawberries contain the phytochemical ellagic acid. Ellagic acid is known to deactivate carcinogens and prevent other chemicals from causing mutations in bacteria.

HOT PEPPERS AND SPICES
Hot peppers and their derivative spices contain the phytochemical capsaicin. Capsaicin is identified in reducing blood clots. It is also known to prevent carcinogens from binding to DNA.

TOMATOES AND WATERMELON
The phytochemical lycopene is found in tomatoes and watermelon. Lycopene functions as an antioxidant, reducing the risk of cancer and heart disease.

Water

Water makes up about 60% of an adult's body weight and is essential for sustaining life. An individual can survive for only a few days without water. Water cleans toxins from the body, cushions and lubricates joints, regulates body temperature, and transports nutrients throughout the body. In recent years, the inclusion of bottled specialty waters on menus has become a dining trend.

Reducing Sugar in Products

Sugars and sweeteners have many purposes in baking and pastry. They contribute flavor and sweetness, add color through the carmelization process during baking, serve as food for yeast, and aid in the retention of moisture. They also break up gluten and tenderize products, serve as a base for icings and buttercreams, and help to stabilize egg foams (meringues). The most common sweetener in the bakeshop or pastry shop is standard granulated sugar, also called table sugar, a derivative of refined sugarcane or sugar beets. A variety of sweeteners are also available. When producing baked goods or pastries that require small quantities of refined sugar, try to incorporate alternative sweeteners.

Substituting Sugar Syrups for Granulated Sugar

To substitute a sugar syrup (honey, maple syrup, or corn syrup) for granulated sugar: use syrup in place of granulated sugar pound for pound and reduce water or other such liquid in the formula by 2 ounces per pound of sweetener. Although this will not reduce the total amount of simple carbohydrates in the formula, it will affect the flavor.

Naturally Sweet Fruit

Bakers and pastry chefs can easily reduce the amount of refined sugar in formulas by employing sweet fruits for a portion of granulated sugar. A quantity of the granulated sugar can be replaced with ripe banana purée for sweetness and added nutritional value. Add dried or fresh fruit pieces to a formula to decrease the sugar needed. Before using dried fruits reconstitute them in hot water to soften and make them more palatable. When fresh or dried fruit is not accessible, canned fruits may be used. The most commonly used canned fruits in the bakeshop or pastry shop are: apples, peaches, pears, pineapple, and cherries. These fruits are commonly preserved in water, fruit juice, or heavy, medium, or light syrup. They may also be packed in a solid pack can that contains little or no water.

High-Intensity Sweeteners

High-intensity sweeteners, also called artificial sweeteners, are typically two hundred times sweeter than sugar. They are used to sweeten baked goods and pastries and are unsuitable in products that rely on sugar for functions other than sweetness. The three most common artificial sweeteners are saccharin (Sweet'n Low); aspartame (Equal); and sucralose (Splenda).

Increasing Nutrient Density in Baked Goods and Pastries

Nutrient density is a measure of the nutrients that a food provides. Foods low in calories and high in nutrients are considered nutrient dense. Nutrient-dense foods should be consumed as often as possible. A partial list of nutrient-dense foods to include in the production of healthier baked goods and pastries follows:

- ***Nuts and seeds***—almonds, cashews, flax seeds, peanuts, pumpkin seeds, sesame seeds, sunflower seeds, walnuts, and olive oils
- ***Fruits***—apples, apricots, bananas, blueberries, cranberries, figs, lemons/limes, oranges, Bartlett pears, pineapple, prunes, raisins, raspberries, and strawberries
- ***Whole grains***—barley, buckwheat, yellow–corn, oats, brown rice flour, rye, spelt, and whole wheat
- ***Lowfat dairy products***—lowfat cheeses, 2% cow's milk, goat milk, lowfat yogurt
- ***Natural sweeteners***—blackstrap molasses, cane juice, honey, maple syrup

Celiac Disease

Celiac disease (also called celiac sprue or gluten intolerance) is a disease in the intestinal tract brought about by the consumption of gluten (more specifically the gliadin in the gluten). This disease is caused by an intolerance rather than an allergy, and can be very serious and have long-lasting health implications. Individuals with celiac disease, who consume gluten, even very small amounts can damage the small intestine, where nutrients are absorbed by the body. Without proper absorption of nutrients, persons with celiac disease become malnourished and may develop a range of symptoms related to intestinal distress or poor nutrition. Those with celiac disease cannot tolerate any amount of gluten, and must adhere to a gluten-free diet for their entire lives. Products containing barley, rye, oat, and wheat (BROW) are to be avoided.

Numerous recent studies, however, have produced mixed results concerning the effect of oats on celiac patients. While some indicate that the majority of celiac patients can tolerate small quantities of pure oats (50 g per day), others state that a minority of individuals are too sensitive and demonstrate intestinal damage. Celiac patients are urged to discuss their individual situation with a knowledgeable health professional before including oats in the diet. Frequent follow-up blood tests and endoscopies over time are essential in assessing the effect of oats. Bakers and pastry chefs should be aware that this issue remains controversial and that many celiac patients may be reluctant to try products prepared with pure oats.

Celiac disease is genetic, passed down from one generation to the next. It is the most common genetic disease in Europe—affecting one out of every 250 Italians. Because of immigration, it is likely that many Americans also have celiac disease. While celiac disease remains largely undiagnosed in this country, diagnosis is available through blood test or a biopsy of tissue from the small intestine.

As awareness increases about celiac disease, more gluten-free products are being developed. The preparation of gluten-free baked goods can be challenging, but not impossible. Gluten-free products that contain a combination of rice, potato, and/or tapioca and potato starches can replace wheat flour.

Developing Gluten-Free Products

Gluten-Free Flour Mix

To produce gluten-free items that have a texture similar to that of wheat flour, bakers and pastry chefs must develop a gluten-free flour mixture that mimics the properties of wheat flour. Gluten-free flour mix is made up of white rice flour, brown rice flour, soy flour, chickpea flour, tapioca starch, and potato starch. The flours are combined in varying ratios depending on product requirements. A small percentage of flour weight is made up of starches (usually 25% of the flour weight) that give the flour elasticity to hold it together in the finished product. ***Xanthan gum*** is a gum that is produced when the microorganism *(Xanthomonas campestris)* undergoes fermentation. Bakers and pastry chefs use a small amount (approximately 1 teaspoon to 4 cups of gluten-free flour/starch mixture) of xanthan gum in gluten-free mixes to help simulate the gluten found in wheat flour. Some experimentation is necessary to find a mix that has the right balance of flour to starch and gum. A sample gluten-free flour mix that works well in basic muffins, crumb cakes, and cookies consists of the following:

- 2 cups brown rice flour
- 2 cups white rice flour
- ½ cup potato starch
- ½ cup tapioca starch
- 1 teaspoon xanthan gum

This gluten-free mix can replace wheat flour requirements in a 1 to 1 ratio.

Formula Modification

Individuals consume foods because they taste and look good. Formula modification of bakery or pastry products must consider taste, texture, and appearance.

Steps in Formula Modification

There are four steps to follow when modifying a formula:

1. ***Identify the ingredients that need adjustment.*** The adjustment can be an increase or a decrease in quantity or a substitution of one ingredient for another. Concentrate on increasing complex carbohydrates and dietary fiber, and reducing total fat, saturated fat, cholesterol, and sodium.
2. ***Determine the function or purpose of the ingredients in the formula.*** Each ingredient in a formula contributes to the final product. An ingredient may add flavor, bind substances together, provide color, or tenderize.
3. ***Modify ingredients as appropriate.*** Alter the amount of a particular ingredient or substitute it with a new ingredient. An ingredient replacement must give results similar to those of the original. Also consider modifying the method used for preparation or baking.
4. ***Analyze the formula for sensory desirability.*** Ideally, the modification should create a product that is visually appealing and retains the taste, texture, aroma, and mouthfeel of the original item.

Fruit Muffins

Ingredients	U.S. Standard	Metric
Sugar, granulated	5 lbs.	2268 g
Salt	1.5 oz.	43 g
Dry milk solids (DMS)	1 lb.	454 g
Eggs, whole	2 lbs.	907 g
Oil, vegetable	48 oz.	1.361 (1361 ml)
Water	80 oz.	2.271 (2268 ml)
Extract, vanilla	2 oz.	57 ml
Lemon compound	2 oz.	57 g
Flour, high-gluten bread	9 lbs., 8 oz.	4309 g
Baking powder	6.5 oz.	184 g
Fruit, fresh or frozen	4 lbs, 8 oz.	2041 g

Possible Modifications for Fruit Muffins

Lowfat variation   Cut the amount of oil in half and replace that half with applesauce.

No cholesterol variation   Omit whole eggs and use only egg whites.

Dairy (lactose)-free variation   Replace DMS and water with rice, soy, oat, or almond milk and adjust water accordingly.

Low sugar variation   Reduce up to one half the granulated sugar and/or consider an artificial sweetener.

Gluten-free variation   Omit bread flour and replace with a gluten-free flour mix.

Low sodium variation   Leave out salt and add the zest of one lemon.

Nutrition Labels

The Nutrition Labeling and Education Act of 1990 required that most foods include nutrition labels indicating the serving size, number of calories per serving, number of calories from fat per serving, and the nutrient content. Product nutrients are measured in grams and as a percentage of daily values. **_Daily values_** refer to the daily nutrients needed on the basis of a 2,000-calorie diet. As caloric needs vary from person to person, the daily values merely serve as a guide. Labels also show the recommended daily values of several important nutrients and list the number of calories per gram of fat, protein, and carbohydrate.

Hints on How to Read a Nutrition Facts Panel

- Serving sizes are defined by the USDA to reflect the amounts typically consumed.
- In keeping with guidelines stating that fat should provide no more than 30% of daily caloric intake, the Nutrition Facts Panel shows the calories in one serving that come from fat, and the amount of saturated and trans fat.

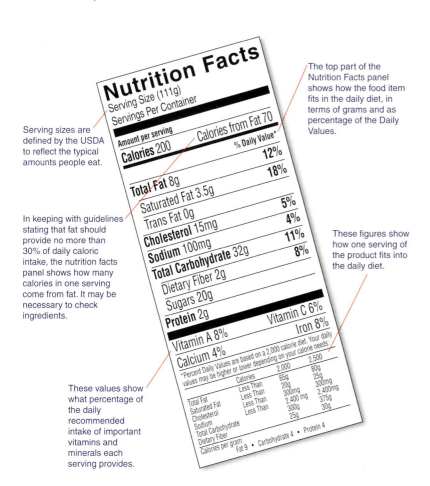

Serving sizes are defined by the USDA to reflect the typical amounts people eat.

In keeping with guidelines stating that fat should provide no more than 30% of daily caloric intake, the nutrition facts panel shows how many calories in one serving come from fat. It may be necessary to check ingredients.

These values show what percentage of the daily recommended intake of important vitamins and minerals each serving provides.

The top part of the Nutrition Facts panel shows how the food item fits in the daily diet, in terms of grams and as percentage of the Daily Values.

These figures show how one serving of the product fits into the daily diet.

- The panel goes on to show how the food item fits in the daily diet, in terms of grams and as a percentage of the Daily Values.
- Cholesterol, sodium, potassium, carbohydrate, and protein content percentages follow in relation to their recommended daily value.
- The percentage of the daily recommended intake of important vitamins and minerals that each serving provides are then listed.

Organics

To avoid the possible health effects of chemicals or other processes used to grow and produce foods, some individuals choose foods that have been organically grown. Farmers produce organic food in an effort to protect the quality of the environment. They do this by using renewable resources to conserve both soil and water. Organic meat, poultry, eggs, and dairy products come from animals that are not given antibiotics or growth hormones. Organic food is also grown without the use of most conventional pesticides; fertilizers made with synthetic ingredients or sewage sludge, bioengineering, or ionizing radiation.

Genetically Modified Foods

Genetically modified organisms (GMO) are foods that have undergone gene modification in a laboratory to enhance specific traits, such as improved nutritional content, and resistance to spoilage or insect damage. Debate still exists about the safety and long-term effects of genetically modified foods. Currently, manufacturers are not bound by law to label food that has been genetically modified. Examples of GMOs are rice with increased iron and vitamins and bananas that produce human vaccines to fight infectious diseases.

Additives

Additives are substances placed in foods to improve characteristics such as flavor, texture, and appearance, or to extend shelf life. The FDA regulates the additives that can be added to foods by evaluating their benefits and risks. A list of approved food additives can be found on the FDA's GRAS list (Generally Recognized As Safe).

Irradiation

Irradiation is the application of ionizing radiation to foods to kill insects, bacteria, and fungi, or to slow the ripening or sprouting process. Irradiation limits or eliminates the use of pesticides and protects consumers from foodborne illnesses. Some fear irradiation, believing that irradiated foods are not safe to eat, or that the radioactive substances used to perform the procedure pose a danger to workers and the environment.

Functional Foods

Functional foods are those that provide health benefits beyond their basic nutrients. Fruits and vegetables at their most basic level are natural examples of functional foods because they contain phytochemicals. Orange juice fortified with calcium and cereal fortified with soy protein are functional foods. Some functional foods are sold as dietary supplements.

chapter 3 Nutrition

Chapter 4
Cost Control & Purchasing

The success of a foodservice establishment, bakeshop, or a pastry shop depends upon the production of a consistently good product that meets customer expectations to guarantee sales. Sales is the amount of money brought into the foodservice operation by customers paying for goods and services. Focus on the amount of money paid out by a business to produce food and to serve its customers is the other important consideration in evaluating profitability.

Key Terms

food cost	standard yield	edible yield	product specification
beverage cost	standard portion	invoice	open market
cost of sales	formula conversion	invoice cost per unit	single-source buying
labor cost	conversion factor	formula cost per unit	purchase order
overhead cost	trim	Q factor	receiving
operating expenses	shrinkage	formula cost	parstock
formula	edible yield percentage	portion cost	periodic ordering method
baker's percentage	yield test	plate cost	perpetual inventory
standardized formula	AP weight	preliminary selling price	requisition
edible portion	trim loss	economies of scale	daily production report

Cost Controls

The success of a foodservice operation, bakeshop, or pastry shop is the result of proper cost controls and sound purchasing procedures. Both of these areas are vital to overall profitability.

Types of Costs

Food cost is the total dollar amount spent to purchase the food and beverage products needed to prepare menu items intended for sale. Beverage items are included in food cost when the beverage is an ingredient in the formula or recipe, such as wine used for poaching. The amount of money spent to purchase products may vary from one type of foodservice establishment, bakeshop, or pastry shop to the next, depending on the demands of the market segment being served. Food cost typically ranges from 20%–35% of the money earned through food sales.

Beverage cost is the total dollar amount spent by a foodservice operation, bakeshop, or pastry shop to purchase all the ingredients needed to produce a beverage item for sale. Beverage cost may include alcoholic beverages, nonalcoholic beverages, and food items needed to produce beverage formulas or recipes. Beverage costs typically represent 15%–20% of money earned through beverage sales. When food cost and beverage costs are combined, they are called *cost of sales.* Cost of sales is the total amount spent to purchase all food and beverage products needed to produce total sales.

Labor cost is the cost of paying employees wages, salaries, and benefits. Labor cost normally represents 25%–35% of the total sales. *Overhead costs* include all other expenses. Examples of overhead costs include equipment maintenance, utilities, linens, mortgage, paper goods, and glassware. Overhead costs frequently represent about 15%–25% of the total sales. The combination of labor and overhead costs are known as *operating expenses.*

A foodservice bakeshop or pastry shop operation must bring in enough sales to pay these major costs. If sales are greater than costs, an establishment has a profit. If sales are less than costs, the operation sees a loss. In bakeshops or pastry shops there are three types of costs: food and beverage, labor, and overhead.

Managing Food Costs

In baking, a standardized recipe is referred to as a *formula*. In some formulas, precise measures of ingredients are seen as percentages, often called a baker's percentage. The *baker's percentage* includes the percentage of each ingredient in relation to the weight of the flour in the final baked product. Baker's percentages make it easy to increase or decrease the quantity of individual ingredients.

Figure 4-1 The standardized formula is an important cost-control tool.

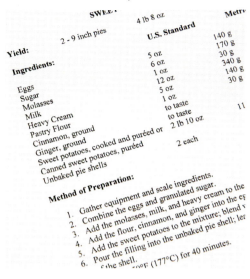

The most important cost-control tool used in a commercial bakeshop or pastry shop is a standardized formula. A *standardized formula* is a written procedure customized to meet the needs of a bakeshop or pastry shop operation. The standardized formula describes the quality and quantity of ingredients, as well as the method of preparation, technique, and temperatures required to make an item. *See Figure 4-1.* The use of a standardized formula allows for the creation of a consistent product. Before standardized formulas are used, they should be tested repeatedly for consistency and quantity. Directions should be easy to follow and the ingredients must be listed accurately.

The quantities of ingredients in a standardized formula are stated in an edible portion form. *Edible portion* refers to the quantities of ingredients in their ready-to-use state. If a standardized formula calls for 42 ounces of sweet potatoes, 42 ounces of peeled, washed, and usable sweet potatoes are needed. By using the standard quantity and quality of ingredients, the standardized formula yields a consistent amount of product, called *standard yield.* The product is then divided into *standard portions,* the amount of food that is served for each order.

Portion control is an essential component of controlling food costs, and it is also important for customer satisfaction. Serving too large a portion reduces profits, while serving one that is too small may drive away customers. A set method of preparation helps staff to prepare a consistent quantity and quality of food. By using standards in the food production cycle, an establishment can serve a consistent product that provides customer satisfaction and profitability. ***(See Chapter 6: Fundamentals of Baking and Pastry for more information on formulas and baker's percentages.)*** See Figure 4-2.

Benefits of Standardized Formulas

Standardized formulas guarantee consistency, cost control, and quality control.

Consistency

A key reason to use standardized formulas is that customers equate consistency with quality. When customers order a dessert item, they expect that the item they order today

Figure 4-2 Mocha torte with evident lines for marking portions and garnish designed to reinforce proper portioning.

will be of the same quality and quantity as the one they enjoyed last week. Many successful bakeshop or pastry shop operations adhere to a simple, well-defined menu that is consistently executed.

Cost Control

Cost control is another benefit of standardized formulas. By specifying the ingredients and portion size of a formula, bakers and pastry chefs are able to forecast the costs and profits associated with an item. Adding more or using less of an ingredient than the standardized formula requires can increase costs, distort inventory, and displease customers.

Quality Control

Quality control is achieved by using standardized formulas that carefully specify the quality and quantity of ingredients. Customers need to get what they expect.

Truth-In-Menu

Accurate information about the ingredients used, the preparation and cooking methods, and the portion size are important to customer satisfaction and operational success. The National Restaurant Association's (NRA) *Accuracy in Menu* paper identifies guidelines that bakeshop or pastry shop operators can use in menu planning and preparation.

Truth-in-menu includes proper identification of an item's quantity, quality, weight, and ingredients. Adherence to ingredient specifications in a standardized formula ensures that what is supposed to be in the formula is, in fact, in the formula so that customers can make informed decisions. People with food allergies need to know what is in their food to avoid life-threatening situations. Failing to properly inform someone who asks about a specific ingredient can also have serious financial implications.

Formula Conversion

Changing a formula to produce a new amount or yield is called *formula conversion*. To change the yield of a standardized formula, a conversion factor has to be established. The *conversion factor* is derived by dividing the desired yield (the quantity of portions needed) by the existing standard yield.

Desired yield ÷ Standard yield = Conversion factor

Increasing the desired yield of a formula results in a conversion factor greater than 1. Using the standardized formula found in Figure 4-1, a standardized formula for sweet potato pie yields 16 portions. If the desired yield is 48 portions, the conversion factor is calculated as follows:

Desired yield (48) ÷ Standard yield (16) = Conversion factor (3)

The conversion factor is 3. Each ingredient quantity is multiplied by the conversion factor to obtain the new quantity needed to produce the desired yield. For example, if the formula calls for 42 ounces of sweet potatoes, the formula is converted as follows:

42 oz. sweet potatoes × 3 = 126 oz. sweet potatoes

Decreasing a formula results in a conversion factor of less than 1. To reduce the desired yield of the formula in Figure 4-1 from 16 to 8, use the following formula:

Desired yield (8) ÷ Standard yield (16) = Conversion factor (0.5)

The conversion factor is 0.5. Each ingredient quantity is multiplied by the conversion factor to obtain the new quantity of ingredients needed to produce the desired yield. If a formula calls for 42 ounces of sweet potatoes, the formula would be converted as follows:

42 oz. sweet potatoes × 0.5 = 21 oz. sweet potatoes

Formula conversion is a skill used daily in the commercial bakeshop or pastry shop. Accuracy and consistency in making conversions is important for maintaining quality.

Considerations When Converting Formulas

Mathematical formula conversions do not take into account adjustments that must be made for certain ingredients, equipment, cooking temperatures, trim, and shrinkage that may occur. When such adjustments are needed, they should be noted in the formula.

Adjustments to Ingredients

Ingredients such as spices are often stated in standardized formulas as to taste, pinch, dash, as needed, and so on. Expertise must be used to carefully adjust the quantities of these ingredients to maintain a consistent product.

Adjustments to Equipment

Formulas usually indicate the size of equipment needed to prepare the food product. If a formula is increased or decreased, the size of the equipment may become a factor. The wrong equipment can affect the outcome of a formula. Always use the proper equipment to accommodate the formula conversion. *(See Chapter 5: Equipment, Hand Tools, Smallwares, and Knives for more information on equipment.)*

Baking Temperatures

Baking temperatures can be affected by a change in equipment. The use of a convection oven rather than a conventional oven requires a change in cooking time because convection ovens cook more quickly.

Trim and Shrinkage

Trim is the percentage of food lost during the preparation of a product. ***Shrinkage*** is the percentage of food lost during cooking. The amount of trim and shrinkage affects not only the edible yield of the product but also the portion cost. Knowing the trim and shrinkage loss that can be expected allows for the use of the correct amount of ingredients per portion. For example, bacon generally shrinks approximately 69% when it is cooked. If 5 ounces of cooked bacon are needed in a quiche Lorraine, 1 pound of uncooked bacon must be used to achieve the forecasted portion.

Edible portion is the amount of food product that is servable after preparing the ingredient for use. Waste, such as peelings, stems, and trim, and shrinkage in cooking are factors that affect product yield. The percentage of a food product that is left after trim and/or shrinkage is called the ***edible yield percentage.*** Yield tests or cooking loss tests are often performed to determine how much product will be lost through preparation or cooking.

Yield Test

Yield tests are performed on food products that are prepared for use or service in their raw form. For example, a mango is cut, pitted, and then cubed for use in a mango trifle, or a pineapple is prepared for inclusion in a pineapple upside-down cake.

Follow these steps to conduct a yield test:

1. Weigh the product before trimming to determine the As-Purchased weight. ***AP weight*** is the weight of the product when purchased.
2. Trim and prepare the product for use. Weigh the by-product material that was trimmed from the purchased product to arrive at the ***trim loss.***
3. Weigh the usable product. This result is the ***edible yield;*** the weight of the product in its edible portion form.
4. Divide the edible yield weight by the AP weight to arrive at the edible yield percentage.

A yield test for 64 ounces of sweet potatoes includes first weighing the potatoes in the as-purchased form (64 oz). After washing and peeling them, the usable product is weighed again (51.2 oz.). The trim loss is the difference between the AP weight and the edible yield weight (12.8 oz.). The edible yield weight (51.2 oz.) is then divided by the AP weight (64 oz) to arrive at an edible yield percentage of 80% (0.8). The edible yield percentages will vary from one bakeshop or pastry shop to another depending on the standard established for trimming products.

AP weight (64 oz.) − Edible yield weight (51.2 oz.) = Trim loss (12.8 oz.)
Edible yield weight (51.2 oz.) ÷ AP weight (64 oz.) = Edible yield percentage (80% or 0.8)

Cooking Loss Test

A cooking loss test is performed on products that must be cooked before the edible yield weight is determined. To determine how cooking affects the yield percentage, follow these steps:

1. Weigh the product before trimming to determine the AP weight.
2. Trim and prepare for cooking.
3. Determine precooked weight.
4. Cook, following methods of preparation on the standardized formula/recipe.
5. Trim after cooking.
6. Weigh the edible portion.
7. Divide the edible portion weight by the AP weight. A cooking loss test is done primarily for meats such as prime rib and is rarely used in the bakeshop or pastry shop.

Formula Costing

When costing a standardized formula, the following items must be well understood by staff and should be listed on the formula cost card. *See Figure 4-3.*

- ***Formula name***—The name of the formula is the same as the one listed on the menu.
- ***Standard yield***—The standard yield is the number of servings that one preparation of the formula produces.
- ***Standard portion***—Standard portion is the standard amount of a food item that is served to each customer, usually stated in ounces or count.
- ***Ingredients***—The items needed to produce a product. This section of the formula cost card specifies the ingredients and the proper amount of ingredients needed to produce the standard yield.
- ***Edible yield percentage***—The part of the purchased product that is usable after trimming or cooking is called the edible yield percentage.
- ***Invoice cost per unit***—The purchase cost and unit of purchase of each ingredient in the formula is called the invoice cost per unit. This price can be obtained from purveyors or from the most recent *invoice*, or bill.

Determining the Portion Cost

Calculating the Ingredient Cost

The first step in calculating a formula's portion cost is to record all ingredients and quantities on a costing form. *See Figure 4-3.*

INDIVIDUAL INGREDIENT COST It is important to determine the cost of each ingredient in the formula to identify the excessive cost of an ingredient that may be replaced by a less-expensive substitute without affecting quality. To achieve the standard yield, the quantity of ingredients must be measured in edible-portion form. If the ingredient is not purchased in the edible-portion form, an edible yield percentage must be applied to that ingredient. It is very important that pastry chefs understand edible yield percentages.

AS-PURCHASED QUANTITY Standardized formulas list the quantity of ingredients in the edible-portion form. The individual costing the formula must be able to determine how much of the product is needed to yield a desired quantity. As seen in Figure 4-3, ingredients such as cinnamon and ginger are already in their usable form, and therefore have an edible yield percentage of 100%.

Figure 4-3 Formula costing helps determine both costs and selling prices.

Standard Formula Cost Card

Classification: Dessert
Formula Name: Sweet Potato Pie
Standard Portion: 8 slices per pie
Standard Yield: 16

Formula Quantity	Unit	Edible Yield %	As Purchased Quantity	Unit	Ingredient	Invoice Cost	Unit	Formula Cost	Unit	Individual Ingredient Cost
5.00	oz	100.0%	5.00	oz	Eggs	$24.60	30 doz	$0.0342	oz	$0.17
6.00	oz	100.0%	6.00	oz	Sugar	$20.50	50 lb	$0.0256	oz	$0.15
1.00	oz	100.0%	1.00	oz	Molasses	$36.27	4 gal	$0.0708	oz	$0.07
12.00	oz	100.0%	12.00	oz	Milk	$0.51	1 pt	$0.0319	oz	$0.38
5.00	oz	100.0%	5.00	oz	Heavy cream	$2.13	1 qt	$0.0666	oz	$0.33
1.00	oz	100.0%	1.00	oz	Pastry flour	$10.50	50 lb	$0.0131	oz	$0.01
0.00	to taste	100.0%	0.00	to taste	Cinnamon, ground	$0.00	0		to taste	$0.00
0.00	to taste	100.0%	0.00	to taste	Ginger, ground	$0.00	0		to taste	$0.00
42.00	oz	80.0%	52.50	oz	Sweet potatoes, cooked and puréed or canned sweet potatoes, puréed	$23.00	50 lb	$0.0288	oz	$1.51
2.00	each	100.0%	2.00	each	Uncooked pie shells	$0.60	each	$0.6000	each	$1.20

Total Ingredient Cost:	$3.82
Q Factor %:	5.0%
	$0.19
Formula Cost:	$4.01
Portion Cost:	$0.25
Garnish Cost:	$0.10
Cost:	
Cost:	
Total Plate Cost:	$0.35
Desired Cost %:	20.0%
Preliminary Selling Price:	$1.75
Actual Selling Price:	$3.95
Actual Cost %:	8.9%

Ingredients that are not purchased in the edible form must be evaluated for their edible yield percentage to determine the amount of product needed to yield the quantity listed. If the formula for sweet potato pie calls for 42 ounces of washed and peeled sweet potatoes, a yield test must be performed to arrive at the edible yield percentage. The 42 ounces of edible weight would be divided by the edible yield percentage (previously determined as 80%) to arrive at the as-purchased weight. This as-purchased quantity is the weight of the unprepared product needed to yield the quantity of edible product.

$$42 \text{ oz.} \div 80\% \ (0.8) = 52.5 \text{ oz. of as-purchased sweet potatoes}$$

The formula indicates that 52.5 ounces of sweet potatoes in their as-purchased form would need to be used to yield 42 ounces of usable sweet potatoes. The 52.5-ounce weight of as-purchased sweet potatoes must be used to calculate the ingredient cost.

CHANGING THE INVOICE COST PER UNIT TO THE FORMULA COST PER UNIT The *invoice cost per unit* is the cost of an ingredient in the specified unit in which it is purchased. The *formula cost per unit* is the cost of one formula unit of the ingredient.

To calculate the ingredient cost of sweet potatoes, a pastry chef would first need to arrive at the formula unit cost. If sweet potatoes are purchased at $23.00 per 50-pound bag and the formula is in ounces, the conversion from pounds to ounces is first necessary. There are 800 ounces in a 50-pound bag (50 lbs. × 16 oz.); then divide the invoice cost per unit by that number.

$$\text{Invoice cost per unit (\$23.00)} \div \text{Ounce in purchase unit (800 oz.)} =$$
$$\text{Formula cost per unit (\$0.0288 / oz.)}$$

It was previously determined that 52.5 ounces of as-purchased sweet potatoes would be needed to yield 42 ounces of edible sweet potatoes. To complete the individual ingredient cost for sweet potatoes use this equation:

$$\text{As-purchased quantity (52.5 oz.)} \times \text{Formula cost per unit (\$0.0288 / oz.)} =$$
$$\text{Individual ingredient cost (\$1.51)}$$

Total Ingredient Cost

After all individual ingredient costs have been determined (illustrated in the formula above), they are totaled. The total ingredient cost for the sweet potato pie is $3.82.

Q Factor

A *Q factor* (questionable cost) is assigned to ingredients that are immeasurable due to the small quantity used (spices and water). A Q factor is usually applied as a percentage of the total ingredient cost to achieve the formula cost. In Figure 4-3, a 5% Q factor is applied, and would be calculated as follows:

$$\text{Total ingredient cost (\$3.82)} \times \text{Q factor \% (5\%)} = \text{Q factor amount (\$0.19)}$$

Formula Cost

The *formula cost* is the total cost of measurable ingredients added to the Q factor of immeasurable ingredients.

$$\text{Total ingredient cost (\$3.82)} + \text{Q factor amount (\$0.19)} = \text{Formula cost (\$4.01)}$$

Portion Cost

The *portion cost* is the cost of one portion of the standardized formula. The formula cost is divided by the standard yield (number of portions) to determine the cost per portion as follows:

$$\text{Formula cost (\$4.01)} \div \text{Standard yield (16)} = \text{Portion cost (\$0.25)}$$

Plate Cost

Often the portion cost of a formula needs to be added with the cost of accoutrements on the plate to determine the *plate cost* of a dessert item. If the sweet potato pie is garnished or served with 2 ounces of whipped cream ($0.10 per 2-oz. portion), the portion cost of the cream must be added to the dessert cost to determine the cost of the plate:

Sweet potato pie ($0.25) + Whipped cream ($0.10) = Plate cost ($0.35)

Preliminary Selling Price

The *preliminary selling price* is the lowest suggested selling price to be listed on the dessert menu. The two basic mathematical approaches to setting the preliminary selling price are the desired cost percentage and the pricing factor.

DESIRED COST PERCENTAGE One way to determine the preliminary selling price is to divide the plate cost by the food cost percentage that the bakeshop or pastry shop is trying to achieve. If the goal is a 20% food cost, for example, the formula would be:

Plate cost ($0.35) ÷ Food cost % (20%) = Preliminary selling price ($1.75)

PRICING FACTOR A second method of calculating the preliminary selling price uses the pricing factor. Use the food-cost percentage goal of 20%. The formula to calculate the pricing factor, thus the preliminary selling price, is as follows:

100% ÷ Food cost % (20%) = Pricing factor (5)
Plate cost ($0.35) × Pricing factor (5) = Preliminary selling price ($1.75)

Both processes provide the same preliminary selling price. It is rare that the preliminary selling price is the selling price found on the menu. An adjustment (markup) to the preliminary selling is usually made by taking into consideration the location, concept, clientele, and competition. When setting the menu selling price, consider three important factors: all costs should be covered, the price must be fair and reasonable for the business in terms of profit, and fair and reasonable for the customer.

Controlling Costs

After the dessert menu has been developed and the selling prices are determined, control tools must be used to realize a profit. The cycle begins with care in purchasing and continues through the production and sales processes.

Purchasing

The size and type of foodservice establishment often determines who controls purchasing. This individual is responsible for assuring quantity control, consistent ingredient quality, and appropriate food cost. It is important to set purchasing procedures to define who has purchasing authority, which purveyors will be used, how often orders should be made, and what should be ordered. Large multiunit operations usually have employees who are dedicated to purchasing activities and are able to leverage *economies of scale,* or to negotiate price breaks, for purchasing in larger quantities.

Product Specifications

Decisions related to food quality are based on menu offerings and customer expectations. The most efficient way to ensure consistency and quality is by using standardized formulas and standard product specifications. *Product specifications* describe the quality and quantity of each standardized formula ingredient. A well-written standardized formula specifies the exact quality, such as the grade of fruits and vegetables, and the quantity of the ingredients to be used. Although bakers and pastry chefs are responsible for setting parameters, the control of costs and the quality of products is a shared responsibility of all personnel.

Types of Products Purchased

Commercial bakeshops and pastry shops do not limit purchases to food products. Bakers or pastry chefs purchase four types of products: perishable, semiperishable, nonperishable, and nonedible.

PERISHABLE   Perishable products are generally fresh foods that have a relatively short shelf life. The quality of items such as fresh fruits and vegetables tends to diminish in a relatively short period of time. These items must be ordered frequently and in quantities that can be used quickly so that the ingredients are at their peak freshness and waste is minimized.

SEMIPERISHABLE   Semiperishable products have a longer shelf life than perishable products and include unsweetened nuts and seeds, peanut butter, yeast, and fresh produce such as winter squash.

NONPERISHABLE   Nonperishable items have a long shelf life—they do not decompose rapidly when stored in their original packages at room temperature. The quality of these items remains intact when stored for up to one year. Examples of nonperishable items include canned fruits and vegetables, sugar, and flour.

NONEDIBLES   Nonedibles are nonfood products including aluminum foil, plastic wrap, paper goods, and cleaning supplies.

Determining Purchase Quantities

The quantity of items purchased depends on several factors.

- The type of product ordered
- Storage constraints
- Sales history of a dessert menu item
- Discounts on volume purchases
- Available seasonal specials

Purveyor Relationships

The relationship between bakeshop or pastry shop establishments and purveyors should be based on mutual trust and performance. Bakeshop or pastry shop operations should show integrity when interacting with purveyors, and they should be able to trust purveyors to consistently deliver products of a desired quality at a reasonable cost. Proper cost-control measures add integrity to the relationship of trust between the buyer and the purveyor.

Methods of Purchasing

There are two primary methods of purchasing products for a bakeshop or pastry shop establishment: open market and single-source buying. The ***open market*** method allows bakers or pastry chefs to obtain bids from various purveyors. Purveyors compete to offer the specified quality and quantity of product at the lowest price and best service.

At times, the effort required to purchase items on a competitive basis may not be worth the cost benefit received. If hours must be spent seeking and responding to bids for only a small cost savings, it is wiser to use single-source buying. In ***single-source buying,*** bakers or pastry chefs purchase products from a single purveyor, and negotiate a price for volume-based discounts. Single-source buying is attractive for items that are consistently purchased in large volumes.

Figure 4-4 A purchase order lists the items to be purchased.

Once the purchasing method is determined, a purchase order is prepared. A *purchase order* is a form that communicates to the purveyor the items to be purchased. *See Figure 4-4.* A purchase order lists the products ordered and specifies the type and the quantity of products. A formal purchase order also includes the unit price of the product and the total cost of the items to be purchased. Purchasing is the first opportunity to control costs.

Receiving Goods

Receiving is the process by which the product is accepted from the purveyor. An invoice is the bill that normally accompanies the delivery. Receiving standards ensure the receipt of the correct quantity as well as the quality of product ordered. A good receiving system guarantees that the correct items and the proper quantity and quality of items have been received and handled properly during shipping. Receiving is the second opportunity to control costs.

Receiving Tools and Equipment

The following tools and equipment make receiving more efficient:

- Heavy-duty gloves with nonslip fingertips
- Scales that are of the proper size to weigh food
- A calculator to check total costs or weights
- Cutting devices for opening containers, packages, and boxes
- Hand trucks to move cases from the receiving to the storage area
- Thermometers to make sure that refrigerated and frozen goods are received at the right temperature
- Business forms, such as purchase orders and invoices

Physical Inspection of Goods

All products received should be visually inspected to ensure that they are of good quality; that is, fresh, undamaged, and free from product tampering or mishandling, improper storage, and pest or rodent infestation. Weigh the products to make sure that their weights match those ordered. Check the products against the invoice. Report any discrepancies immediately.

Figure 4-5 Bar codes help track items through the inventory system.

Checking Purchase Order to Invoice

After the physical inspection of products is complete and the invoice quantities are confirmed, make sure that the items received are the ones ordered, by checking the invoice against the purchase order. The purveyor supplies an invoice, and it should match the purchase order.

Storing and Issuing Controls

When food items are received, they are moved to the proper storage area until needed for production. If a product's bar code sticker can be scanned, do so on receipt to help track the item through the inventory system. *See Figure 4-5*.

Storeroom Controls

When storing goods, keep them in the proper environment to maximize their shelf life and to ensure continued quality. Moving received products through proper inventory rotation is an important component of food storage, accomplished by using the First In, First Out (FIFO) method.

Inventory Control

Inventory control is an essential component for business success. When a customer orders an item from the dessert menu, it should be available. Running out of a popular dessert item can alienate a loyal patron. Food inventory is a large expense for a bakeshop or pastry shop operation and it is not cost efficient to hold excessive inventory. Managing inventory helps to manage capital assets efficiently.

Physical Inventory

A physical inventory is an actual count of items on hand. Physical inventories may be performed daily or weekly (for high-cost ingredients) to control theft and pilferage and to help determine how much to order. Performing periodic physical inventories at specific intervals allows the calculation of food and beverage costs (cost of sales) for a defined period of time.

Parstock

The amount of a product kept on hand from one delivery date to the next is called the *parstock*. Product shortages, bad weather, or delivery delays can affect delivery intervals. Planning for sufficient parstock to account for such unforeseen events is very important. Having a parstock quantity for each product helps to determine purchases. The purchaser simply compares the difference between the inventory on hand and the parstock quantity to determine the quantity to be ordered. This system is often referred to as the *periodic ordering method*. The amount of the product on hand is reviewed to evaluate what is needed and how much parstock of the product should be available. The formula used to calculate the amount to order is:

$$\text{Parstock} + \text{Production needs} - \text{Stock on hand} = \text{Order amount}$$

Perpetual Inventory

Perpetual inventory is a continuously updated, documented record of food items on hand. A perpetual inventory is frequently stored on computer, so that at preestablished levels, products are automatically reordered. Perpetual inventory systems are often difficult to use in bakeshops or pastry shops because thousands of different items must be tracked and many of them are perishable.

The use of computers in inventory control can help in the purchasing process to establish desired minimum and maximum inventory levels. When the inventory reaches the preestablished minimum level, an order is generated to increase the inventory to the maximum level. Computers can be programmed with ordering parameters, and tied in to point-of-sale cash register transactions to adjust inventory whenever a specific menu item is sold. When inventory levels reach parstock levels, the computer automatically generates a purchase order for the item needed. When the item is received, it is added to inventory by scanning the product bar code in the receiving area.

Issuing Controls

Owners can reduce and prevent pilferage by creating internal controls. Such methods include the use of requisitions and/or security.

A *requisition* is an internal invoice that allows management to track the release of inventory from storage. It controls the internal flow of items by allowing only authorized staff to sign requisitions.

Physical security of inventory can be accomplished by locking storage areas to restrict access to unauthorized employees. Procedures should be established and followed to monitor sales patterns and inventory levels, and to uncover any aberrations that might indicate pilferage. Video cameras can also be used to monitor storage areas.

Production Controls

When products leave the storage areas, they move to the bakeshop or pastry shop for preparation, production, and distribution. Bakers and pastry chefs often use *daily production report* forms to help control and manage costs, and to show the quantity of product prepared, sold, or not used. Not all leftover products can be reused and minimizing them is important to cost savings. By tracking the number of items remaining daily the operation may adjust production to better meet sales needs.

Sales Controls

Sales is the final information-gathering area of control that is necessary to complete the cost-control system. The recording of sales information begins when the customer places an order and the items are entered on a guest check or in a cash register system. The customer is served the product, and then pays the guest check. The daily sales information is tallied within the cash register system and is printed out for analysis. With this information, bakers or pastry chefs can determine how well the establishment performed and adjustments may be made as needed.

CHAPTER 5

Equipment, Hand Tools, Smallwares & Knives

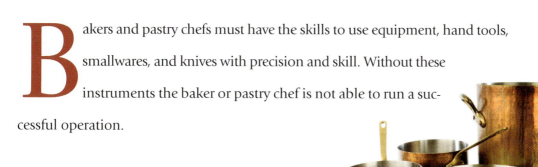

B akers and pastry chefs must have the skills to use equipment, hand tools, smallwares, and knives with precision and skill. Without these instruments the baker or pastry chef is not able to run a successful operation.

KEY TERMS

- hand tools
- smallwares
- russe
- sautoir
- tang
- rivet
- bolster
- mincing
- chiffonade
- julienne
- batonnet
- paysanne
- brunoise
- macédoine
- tomato concassée
- whetstone
- trueing

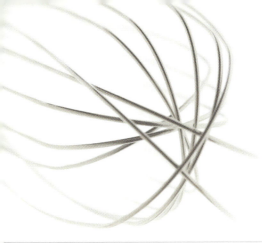

Equipment

Bakeshop equipment is constantly exposed to wet, sticky ingredients and extreme changes in temperature. It is important that it is durable and of a good quality to withstand the demanding workload of a commercial operation.

Large Bakeshop and Pastry Shop Equipment

Large bakeshop or pastry shop equipment is basic equipment that is found in a specialized foodservice operation and used in a variety of ways to prepare baked goods and pastries.

Mixers

Mixers are used to mix, knead, blend, or whip.

STATIONARY MIXER

Stationary mixers can be equipped with a variety of optional attachments, which may include the standard flat paddle, dough hooks, and wire whisks. These machines are extremely versatile and are essential to the baker and pastry chef. Stationary mixer sizes are based on the bowl size in quart measure. The standard sizes of most professional mixers are 12, 20, 30, 60, or 80 quarts. Larger models are also available up to 140 quarts. Some 80-quart mixers can actually accommodate 40- and 60-quart bowls and attachments by adding adaptors. A 12-quart bowl and attachments can also be adapted for use on some 20-quart machines.

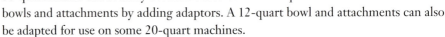

BREAD SPIRAL MIXER

The spiral mixer is particularly useful in mixing and kneading bread doughs. The movement and rotation of the spiral dough hook enables the dough to be worked against the side of the bowl and to minimize heat generated in the mixing process.

Sheeter

A sheeter is an item used primarily to efficiently roll out smooth dough with little effort. The baker controls the amount of force and pressure applied to the dough. Many sheeters are designed for laminated doughs such as puff pastry, croissants, and Danish doughs. Sheeters can also mold bread, and cut and evenly divide doughs using special cutters that attach to the sheeter.

Proofing cabinet

The proofing cabinet is designed to control the ideal conditions of proofing (the final stage of the fermentation process before the product is baked). Both the temperature and the humidity in the proofing cabinet can be adjusted.

Retarder/Proofer

A retarder/proofer is both a refrigerating unit used to help slow down, or retard, the fermentation process, and a proofer used to achieve the appropriate temperature and humidity needed for the final proofing of yeast items. This unit is particularly important in the production of artisan-style breads and in high-production bakeshops.

Bakery Ovens

Commercial ovens are used to produce a vast array of baked goods and pastries and are invaluable in the bakeshop and pastry shop. There are both gas and electric models that are equipped with a variety of accessories such as convection fans that circulate the oven's heated air; and steam injection systems that produce proper oven spring and crust development for breads. There are even old-world-type ovens lined with bricks and fueled by wood for baking specialty breads.

DECK OVEN

A deck oven is also known as a stack oven because many individual ovens can be "piggybacked" one on top of the other to save space. Each one has separate baking controls. Deck ovens are used for many production tasks in the bakeshop and pastry shop, including pizza production, artisan bread work, and fine pastry baking. These ovens are either gas or electric and can be equipped with steam injection and have automatic loaders attached as well.

RACK OVEN

The rack oven, also called a rotary rack oven, is a high-volume production unit that comes either as a single-rack (18–20 shelves) or double-rack (36–40 shelves) oven. Racks are rolled into the oven with the product on them and may be either hooked to the top of the oven or set to ride on a carousel to rotate for even baking. These ovens are usually made of stainless steel inside and out and come with a steam injection system for crusty breads. Most rack ovens use convection heat.

Convection oven

A gas or electric convection oven is equipped with a fan that provides continuous circulation of hot air around the food. Baking occurs more evenly, and up to 25% faster at temperatures approximately 50°F (10°C) lower than those of a conventional oven.

Reel oven

A reel oven contains shelves that rotate like a Ferris wheel. The standard time for the reel to complete a revolution is about two minutes. A reel oven can bake a large quantity of products evenly, at a consistent temperature and humidity. These ovens are not common in today's industry because they take up a large amount of valuable bakery space, and are time consuming because each pan must be handled separately.

Miscellaneous Items

In addition to large bakeshop equipment and bakery ovens, there are many other helpful bakeshop or pastry shop tools.

Dough dividers

Dough dividers are used to evenly flatten and divide dough into standard 18–36 equal pieces. Some units are also equipped with a rounding plate that rounds the items with the simple move of a lever.

Electric ice cream machine

This machine churns, incorporates air, and freezes ice cream or sorbet.

Deep-fat fryer

A standard deep-fat fryer is used to fry foods in hot oil at a thermostatically controlled temperature, which allows for even cooking and quick heat recovery after each batch. Automatic or computerized fryers control the movement of food baskets in and out of fat to eliminate over- or undercooking. Both gas and electric deep-fat fryers are available.

Food processor

A food processor grinds, purées, emulsifies, crushes, and kneads foods. It has a removable work bowl and an S-shaped blade and can be used to slice, julienne, and shred food by inserting specialty disks.

Work tables

Work tables can be made of stainless steel, butcher block, or marble. Most food preparation tasks in commercial kitchens are well-suited for the stainless steel work table, while butcher-block tables are more commonly used in bakeshops and pastry shops. Marble-top tables are preferred for chocolate and sugar work.

Hand Tools

Handheld tools used in the preparation, cooking, baking, and service of food are known as *hand tools*. Most hand tools are made from stainless steel, aluminum, or plastic. The price of these tools is usually determined by durability, ease of use and cleaning, and heat transfer characteristics.

Selecting Appropriate Tools

Foodservice tools must meet high standards for quality and durability to withstand the heavy use of a busy bakeshop or pastry shop. Choose tools that are well-constructed, safe to operate, and comfortable to hold. NSF International, formerly known as the National Sanitation Foundation, performs rigorous tests on tools and equipment used in the foodservice industry. Many states require that all tools and equipment used in foodservice operations be NSF-certified.

Apple corer/slicer

An apple or fruit corer quickly and cleanly removes a fruit's core in a single long, round piece. A serrated ring forms the tip and is inserted through the fruit's center. A deep, open groove under the ringed tip holds the core as it is pulled. An apple corer/slicer removes the core and slices the apple in one motion.

Bench scraper

This rectangular utensil, also called a dough cutter, has a stainless steel blade and a handle made of slip-resistant plastic or wood. The bench scraper is used to clean and scrape surfaces and to cut and portion dough.

Blowtorch

A kitchen blowtorch quickly browns items, such as lemon meringue topped pies and baked Alaska, and is practical for caramelizing crème brûlée.

Bowl scraper

A bowl scraper is a plastic card-shaped item that is curved and beveled on one side. The baker or pastry chef uses it to scrape or clean products from bowls. This tool can be used to spread icing on cakes or to hand-mix batters that contain an egg mixture that should not be deflated. Bowl scrapers may also be used to "cut" fat into dry ingredients such as pie and biscuit dough.

Mandoline

The mandoline is used for slicing fruits and vegetables into thin slices, strips, or waffle cuts.

Pastry bags and tips

Pastry bags are cone shaped bags that come in a variety of sizes. They may be made of nylon, plastic-lined cotton, polyester, or disposable plastic. Pastry bags are used with or without pastry tips. A tip is used to produce dough patterns or to decorate a finished product such as a cake. Without a tip, the bag can be used to deposit dough or batters.

Pastry brushes

These flat-edged brushes vary in shape and size. Some are used to brush on liquids and oils, and others are employed to dust off excess flour. Brushes are useful before, during, or after baking.

Pastry/cookie cutters

Pastry cutters have straight or fluted edges to cut dough into specific shapes.

Pastry wheels

Pastry wheels are small, sharp wheels on handles used for cutting pastry into strips or for decorating the tops of pies and tarts. Some are fluted for decorating pastry dough strips.

Peel

A peel is a flat wooden or aluminum board with a long handle used to move breads in and out of the oven when the product is baked directly on the hearth. The peel can also be employed to load and unload cake pans. The size of the board and the length of the handle vary depending on the depth of the oven.

Rubber, straight, and offset spatulas

A rubber spatula has a wide, flexible rubber or plastic tip for scraping food from bowls, pots, and pans or for folding whipped cream or egg whites into batters. A straight spatula, also called a palette knife, has a long, flexible metal blade with a rounded end. It is useful for spreading icing, soft cheese, or butter. An offset spatula, also referred to as a turner, has a wide, bent stainless steel blade and is designed to lift and turn food that must be cooked on both sides.

Rings

Sometimes referred to as cake rings, these rings come in various shapes and are used to mold, shape, and hold cakes, fillings, and pastries.

Rolling pins

Rolling pins are used to roll and stretch the dough used in baking and pastry. Rolling pins made from hard, tightly grained wood do not absorb fats and flavors. There are two styles of rolling pins. A long, cylindrical piece of wood is characteristic of the French-style rolling pin. The palms of the hand move the pin to roll over the dough. A second style of rolling pin is the rod-and-bearing pin. A metal rod containing bearings runs lengthwise through a wooden shaft to make rolling the pin easy.

Sieve

A sieve is a utensil with a round mesh screen used to sift dry ingredients. It is also sometimes referred to as a drum sieve.

Turntable

A turntable is a round disc set on a heavy pedestal base that can be easily rotated while frosting or decorating a cake.

Whip/Whisk

Whips or whisks are looped wires made from stainless steel that are held together by a handle. They are used to whip cream, eggs, and sauces.

Zester

The zester is used to remove the outer skin of citrus fruits, such as lemons, limes, and oranges.

Smallwares

Small, nonmechanical foodservice equipment, known as *smallwares*, are used for food preparation, and baking and pastry tasks in commercial bakeshops and pastry shops. Measuring equipment, pots, and pans are all examples of smallwares.

Measuring Equipment

Accurate measurements are essential for success in a commercial bakeshop or pastry shop. Consistency in measuring is key to controlling cost and portion sizes and creating tasty and reliable finished products. Measurements are determined in a variety of ways depending on the types of ingredients and the measurement system specified in the formula. Measuring equipment is employed to measure weight, volume, and temperature. The commercial bakeshop or pastry shop must have tools suitable for measuring large quantities of both liquid and dry ingredients and may also need specialized equipment to evaluate volume and weight in both U.S. or metric measures as well as Fahrenheit and Celsius temperatures.

Balance scale

A balance scale is most often used for weighing flour, sugar, and other baking and pastry ingredients. Ingredients are placed in a scoop on one side of the scale and standard weights are loaded on the other side. The weight of the ingredients is determined when both sides of the scale are in balance.

Electronic scale

An electronic, or digital, scale weighs items placed on its tray. The weight is displayed as a digital readout, so the electronic scale makes more precise measurements than does a portion scale.

Liquid measures

A volume of liquid is measured in either a plastic or a metal container. Glass measures are not recommended for bakeshops or pastry shops because of the risk of breakage. A lip or spout on the liquid measure helps to prevent spills and to simplify pouring. Liquid measures come in a variety of sizes, including pints, quarts, half-gallons, and gallons.

Dry measures

Dry ingredients can be measured in measuring cups or containers without pour spouts. This allows dry ingredients to be scooped into the dry measure and leveled with a knife or metal spatula for accurate volume measurement. Dry measures often come in sets with ¼-cup, ⅓-cup, ½-cup, and 1-cup measures.

Measuring spoons

Measuring spoons are used to measure small amounts of either liquid or dry ingredients. They usually come in sets of ¼-teaspoon, ½-teaspoon, 1-teaspoon, and 1-tablespoon measures. Stainless steel measuring spoons resist warping and hold up well under heavy use.

Ladle

A ladle is employed to divide liquids such as sauces and batters into portions. Its long handle simplifies reaching to the bottom of a deep pot or pan. A ladle's capacity is marked on the handle.

Scoops

Scoops are for portioning foods such as ice cream, muffin batter, or cookie dough. The bowls of scoops come in a variety of sizes coded by numbers. Color coding can help identify scoop size and use.

Thermometer

Instant-read thermometers are used to determine the internal temperature of cooked or refrigerated foods and foods that are being held for service.

Candy thermometer

This thermometer is also referred to as a sugar thermometer and is specifically designed to measure the temperature of boiling sugar. It consists of a glass tube housed in a protective cage. Digital candy thermometers used today are very accurate and easy to read.

Hydrometer

Known as a hydrometer, saccharometer, or Baumé hygrometer, this instrument measures the density of sugar in a liquid.

Selecting Cookware

Commercial bakeshops and pastry shops use a wide array of cookware. Cookware includes pots, pans, and baking dishes made from a variety of heat-conducting materials, such as stainless steel, aluminum, copper, cast iron, or ceramic.

Saucepan

The saucepan, or *russe*, has a long handle and straight sides. It comes in a variety of sizes in either shallow- or high-sided versions. This pan can be used for poaching, simmering, boiling, and blanching.

Saucepot

The saucepot is meant for range cooking and is well-suited for poaching, simmering, boiling, and blanching.

Sauté pan

The sauté pan, or *sautoir*, is a straight-sided shallow pan with a long handle, employed for sautéing or frying.

Crêpe pan

The crêpe pan is a very shallow skillet with short, slightly angled sides.

Sheet pans

Sheet pans come in standard full (18 × 26 in./45 × 65 cm) or half (18 × 13 in./45 × 33 cm) sizes as do perforated baking sheet pans used for baking breads. Perforated pans allow air to flow completely around the product to produce crispy crusts.

Nonstick bake sheets

Nonstick bake sheets are flexible silicon mats often referred to as the brand name Silpats.® These mats are full or half sheet pan size and can withstand temperatures up to 500°F (260°C). They are used to replace pan liners and can be used indefinitely.

Flexipans

Fleximolds are rubber-like molds that come in a variety of sizes and shapes, for use in baking or freezing. They can withstand temperatures of −10°F to 450°F (−23°C to 232° C).

Hotel pans

Hotel pans are rectangular pans used for baking, steaming, and also for storing foods under refrigeration. Hotel pans come in a variety of sizes. The standard-size pan is 12 in. (30.5 cm) wide, 20 in. (51 cm) long, and 2.5 in. (6.5 cm) deep.

Loaf pans

Loaf pans are rectangular in shape and usually made of metal. They come in a variety of sizes. The standard size is 5 in. (12.5 cm) wide, 9 in. (23 cm) long, and 3 in. (7.5 cm) deep. The loaf pan is used primarily for baking loaf bread, but accommodates quick breads and pound cake, as well.

Tube pans

Tube pans are used for baking angel food cakes and specialty cakes. These pans have a hollow tubular center that promotes even baking.

Springform pans

These pans contain a clamp that is used to release the pan's bottom from its circular wall. They are commonly used for cheesecakes.

Tart pans

Tart pans are usually round, but are available in various shapes and sizes. The sides of these pans are straight and usually fluted. These pans are shallow in depth and may have a removable bottom.

Madeleine pans

Madeleine pans have small indentations that resemble a shell and are used to make madeleines.

Baguette screens

Baguette screens are metal perforated pans that are curved to create a cylindrical finished product similar to the French baguette.

Brioche tins

Brioche tins are small individual fluted cups with sloped sides. They come in a variety of sizes and shapes.

Muffin and cupcake tins

These are pans consisting of 6–24 cups arranged for even baking. The cups vary in ounce capacities.

Cooling racks

Cooling racks are slotted shelves that allow air to pass through them to rapidly cool products removed from the oven.

Knives

Knives are some of the most versatile tools used in bakeshops and pastry shops. They are utilized to perform numerous tasks such as peeling fruit for pie fillings or dicing vegetables for a quiche.

Knife Construction

The knife is one of the most valuable tools in the bakeshop or pastry shop. An understanding of the construction of cutlery aids in the selection of appropriate knives and the care required to maintain them. *See Figure 5-1.*

Blade

The blade of a professional knife is a single piece of metal that has been cut, stamped, or forged into shape. Most professional knife blades are made from high-carbon stainless steel—an alloy of iron, carbon, chromium, and other metals. The metal combines the best features of stainless steel and carbon steel, in a blade that can be easily sharpened but is resistant to rust and discoloration.

Before the development of high-carbon stainless steel, most professional knives were made of carbon steel. Although carbon steel knives are still available today, they are not as desirable as those made from high-carbon stainless steel. Carbon steel blades rust easily and can impart a metallic flavor to food. The edge of these knives also wears down quickly.

Some knives are made from stainless steel, which is an alloy of chromium and carbon steel. Stainless steel knives are extremely durable and do not rust, discolor, or contribute off-flavors to food. They also hold their edge longer than carbon steel knives. However, because the metal is very hard, these knives are difficult to sharpen.

Tang

The *tang* is the portion of the blade that extends into the knife's handle. A full tang that runs through the entire length of the knife handle contributes strength and durability to the knife. Knives meant for heavy use, such as French knives and cleavers, should have a full tang. Paring knives, utility knives, or other knives used for lighter work may have a partial tang.

Figure 5-1 A knife is constructed of the following elements.

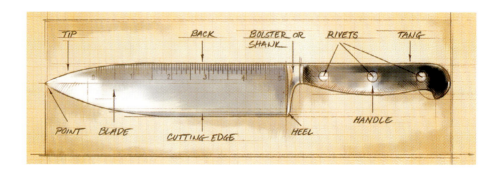

Handle

The knife handle can be made from hardwoods, such as walnut or rosewood, or other materials, such as plastic or vinyl. When choosing a knife, consider the feel and fit of the handle. Knives are held for long periods of time, so the handle must fit comfortably in the hand. A handle that is either too small or too large can be uncomfortable to hold.

Rivets

Metal *rivets* hold the tang in the handle. Rivets should lie smooth and flat against the handle to prevent rubbing and irritation against the hand. They should also be flush with the surface so that there are no crevices in which dirt or microorganisms can collect.

Bolster

The shank, or *bolster,* is the metal area on the knife where the blade and handle meet. Although not all knives have a bolster, those that do are stronger and more durable. The bolster also helps block food particles from entering the space between the tang and the knife handle.

Types of Knives

The type of knife required depends on the specific cutting or chopping task. Although there are a number of specialty knives, the following list summarizes the basic types used by bakers and pastry chefs.

French Knife

The French knife, or chef's knife, is the most common knife choice. This all-purpose knife comes in lengths from 8 to 14 in.(200 to 350 mm). The longest knives are best suited for heavy cutting and chopping. The blade of the French knife is wide at the heel and tapered at the point.

Paring Knife

The paring knife has a pointed, rigid, 2- to 4-in.(50- to 100-mm) blade. It is used for paring or trimming fruits and vegetables.

Slicer

The slicer has a long, thin blade with either a rounded or pointed tip. Slicers come in rigid or flexible blade styles, and may have serrated, or wavy, edges. The serrated slicer smoothly slices coarse foods, such as bread or pastry items.

Knife Skills

The proper use of a knife is one of a chef's most important skills. Good techniques improve the appearance of food items, speed preparation times, and reduce fatigue.

Grip

A good grip provides control over the knife, increases cutting efficiency, minimizes hand fatigue, and lessens the chance of an accident. The size of the knife, the task at hand, and personal comfort determine how best to grip the knife. Regardless of the gripping style used, avoid placing the index finger on the top of the blade. *See Figure 5-2 and Figure 5-3.*

Figure 5-2 Grip the knife by placing four fingers on the bottom of the handle and the thumb against the other side of the handle.

Figure 5-3 Grip the knife by placing three fingers on the bottom of the handle, the index finger flat against the blade on one side, and the thumb on the opposite side. Although this grip may be uncomfortable for some, it offers maximum control and stability.

Control

Knife movement must be controlled in order to make safe, even cuts. Guide the knife with one hand while holding the food firmly in place with the other hand. Allow the sharp edge of the blade to do the work, rather than forcing the blade through the food. A sharp knife provides the surest cuts and is the safest to use. Smooth, even strokes work best. The hand holding the food is referred to as the guiding hand. To protect the guiding hand from cuts, curl the fingertips back. The knife blade should rest against the second knuckle of the index finger, to act as a guide for the cut. See *Figure 5-4*.

Knife Cuts

A knife is used to cut food into desired shapes or sizes. Food that has been cut into equal pieces looks more attractive and cooks more evenly.

Basic Peeling Techniques

Many types of fruits and vegetables must be peeled before cuts are made. Before peeling, thoroughly wash fruits and vegetables to remove dirt and other contaminants. Wash and peel fruits and vegetables as close as possible to the time they are used to retain taste and quality.

Using a Vegetable or Swivel Peeler

A vegetable or swivel peeler effectively removes the skin from fruits and vegetables. It shaves off a thin layer of skin without removing too much of the edible flesh. Pears and apples are examples of thin-skinned foods. See *Figure 5-5* and *Figure 5-6*.

Using a Paring Knife

A paring knife can also be used to remove the peel from fruits and vegetables. Hold the paring knife at a 20 degree angle to the food surface, and cut the skin from the fruit or vegetable. Thick skinned fruits and vegetables should be peeled with a paring knife rather than shaved with a peeler. See *Figure 5-7*.

Using a French Knife

Use a French knife to remove the peel from large fruits and vegetables or from those with very thick skins. The French knife works well for removing the peel, core, stems, and

Figure 5-4 To make slices of equal width, adjust the index finger while working, moving the thumb and fingertips down the length of the food.

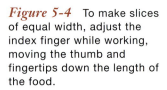

Figure 5-5 Peeling an apple with a vegetable peeler.

Figure 5-6 A swivel peeler is ideal for removing thin peels and minimizing waste.

seeds from fruits and vegetables such as pineapple, grapefruit, and winter squash. Before beginning to peel, cut off the top and bottom of the fruit or vegetable to create a flat, stable end. When peeling, follow the contours of the fruit or vegetable. *See Figure 5-8*.

Coarse Chopping

Coarse chopping is the process of cutting food into imprecise but relatively uniform pieces. Coarse chopping is especially appropriate when cut fruits are to be puréed as for a coulis.

Mincing

Mincing is the process of cutting food into very fine pieces. Herbs and some vegetables, such as fresh mint, shallots, and garlic, lend themselves to mincing. To mince, use a French knife and coarsely chop the food. Then use the guiding hand to hold the tip of the knife against the cutting board. Quickly rock the knife up and down, keeping the blade tip stationary. Gradually move the knife blade back and forth across the food. Continue mincing until the food reaches the desired fineness. *See Figure 5-9*.

Fruit and Vegetable Cuts

Fruit and vegetable cuts affect the appearance and the ability of foods to cook evenly. Cuts can enhance the natural shape of a fruit or vegetable and create an impressive presentation.

CHIFFONADE The *chiffonade* cut creates fine ribbons or strips and can be used for cutting herbs, such as fresh mint, oregano, or rosemary.

Figure 5-7 A paring knife is ideal for removing the peel from citrus fruit.

Figure 5-8 Use a French knife to peel extremely thick-skinned vegetables and fruits, such as a pineapple.

Figure 5-9 Mincing creates a very fine cut.

Julienne A *julienne* is a 1/8-in.(4-mm)-thick, matchstick-shaped cut about 1–2 in.(25–50 mm) long.

Batonnet A *batonnet* is a long, rectangular cut that is similar to, but slightly wider and longer than the julienne cut (1/4 × 1/4 × 2–2 1/2 in./6 × 6 × 50–60 mm). To make a batonnet cut, follow the same procedure as for julienne, but increase the dimensions.

Dicing Dicing produces cubes between 1/8- and 3/4-in.(4- to 20-mm) square. To make the cubes, begin by cutting the foods into julienne or batonnet.

- The *paysanne* cut is similar to a dice, but it produces a square slice that is 1/2 × 1/2 × 1/8 in.(12 × 12 × 4 mm) rather than a true cube. To cut paysanne, cut 1/2-in. (12-mm) sticks into 1/8-in.(4-mm) slices.
- The *brunoise* is a very fine dice cut, yielding a 1/8-in. square (4-mm) cube.
- The *macédoine* is a 1/4-in. square (6-mm) dice cut.

Additional Preparation Techniques

A variety of special techniques are used to cut tomatoes, onions, and other vegetables.

Tomato concassée *Tomato concassée* describes a tomato that has been peeled, seeded, and diced or chopped. The French verb *concasser* refers to the act of chopping or pounding. Tomatoes are not the only food that the French prepare concassée, but today, the term is used mostly in reference to this technique of tomato preparation. In baking, tomato concassée is added to focaccia, pizza, and quiche.

Fanning Fanning is a decorative cut for raw foods, such as apples, pears, and strawberries. The cut is used to spread fruits into a fan shape.

Knife Safety and Care

Good knives are an investment. With proper use and care, quality knives can last for many years. Maintain knives by keeping them clean and sharp.

Knife Use Guidelines

Knives can be dangerous if mishandled or used improperly. Follow the safety guidelines in *Figure 5-10*.

Sharpening Knives

Use a sharpening stone, or *whetstone,* to sharpen knives. The stone can be dry, or wet with water or mineral oil. As the edge of the blade is passed over the whetstone, the grit in the stone sharpens the cutting edge. A whetstone can be made of either silicon carbide or stone and may have up to three sides, which range from coarse to fine grain. Begin by sharpening against the coarsest stone and end with the finest stone, taking only about 10 strokes against the coarsest stone before moving on to the next.

Trueing Knives

In a process called *trueing,* a steel is used to keep the knife blade straight and to smooth out any irregularities. Trueing does not sharpen the blade, but it does help maintain the edge between sharpenings.

Sanitizing Knives

Wash, rinse, and sanitize knives after every cutting task to avoid cross-contamination and to destroy harmful microorganisms. Do not leave a knife in a water-filled sink where it cannot be seen.

Figure 5-10

Safety Guidelines

Always use the knife appropriate for the cutting task.

Never use a knife for a task for which it was not designed. Opening cans and prying open lids are not tasks meant for knives.

Always use a sharp knife. Dull knives require more force, and may cause the knife to slip and cause an injury.

Always use a cutting board with a knife. Marble and metal surfaces dull the blade and may damage the knife.

Never let the knife blade or its handle hang over the edge of a cutting board or work table. Someone might be injured by bumping into the knife, or the knife might fall and be damaged.

When carrying a knife, hold it by the handle with the point of the blade pointed straight down. See Figure 5-11.

Never try to catch a falling knife. Step away from the knife, and let it fall.

To hand a knife to someone else, lay the knife down on the work surface, or hold the knife by the dull side of the blade while carefully extending the handle toward the other person.

Do not leave a knife in a water-filled sink as someone may reach into the sink without seeing the knife and get cut.

Always wash, rinse, sanitize, and air-dry knives before putting them away.

Do not clean knives in the dishwasher. They pose a risk to the person loading and unloading the dishwasher, and the blades could be dented or damaged through contact with other utensils.

Wooden handles on knives cannot withstand intense heat and prolonged exposure to water and should be dried by carefully wiping from the dull side out.

Figure 5-11 Always carry a knife by the handle, blade pointed down.

Storing Knives

Storing knives properly will protect both the knives and the individuals who work with them. Two convenient storage solutions are slotted knife holders or magnetized bars hung on the wall. A custom-built drawer with a slot for each knife is another storage option. A knife kit with individual slots to hold knives is another convenient and portable storage unit. Vinyl cases are easy to clean and sanitize.

STEP PROCESS

Sharpening a Knife

A whetstone may be used to sharpen a knife, but it must be oiled to be effective and avoid damaging the stone.

To use a whetstone to sharpen a knife:

1 Moisten the stone by applying honing oil or by dipping it into the oil reservoir. Lock into place. Start using the coarsest grade and then proceed to the finest grade.

2 With the cutting hand, hold the knife at a 20 degree angle with the heel of the blade against the stone. Use three fingers on the guiding hand to apply pressure.

3 Maintain a 20 degree angle, and gently push the blade from heel to tip across the stone. The guiding hand should continue to apply even pressure on the blade.

4 Continue to move the knife smoothly and evenly across the stone, all the way to the tip of the blade.

5 Gently bring the knife off the stone. Repeat the process 3–10 times as necessary.

6 Turn the knife over, and repeat the process, pulling the knife toward you as you move it across the stone.

7 Rotate the stones to the finger grits, maintaining the same 20 degree angle and pressure throughout. Finish with a sharpening steel and then wash, sanitize, and air-dry the knife before using.

STEP PROCESS

Trueing a Knife

Trueing a knife with a steel maintains the edge between sharpenings.
Follow these steps to use a steel:

1 Rest the heel of the blade at a 20 degree angle against the inner side of the steel at its tip. The steel should be held slightly above waist height, a comfortable distance away from the body, on a 0 degree to 45 degree angle.

2 Draw the blade downward along the entire length of the steel, maintaining the knife's 20 degree angle.

3 Complete the movement with the tip of the blade just above the base of the steel.

4 True the opposite side of the blade by repeating steps 1–3, but this time pass the blade over the outer side of the steel.

5 Repeat these steps 3–5 times on each side of the blade.

6 Wipe the blade to remove any metal particles.

89

CHAPTER 6

Fundamentals of Baking & Pastry

Bakers and pastry chefs use formulas and ingredients in measured and precise amounts. These measurements, often called baker's percentages, include the percentage of each ingredient in relation to the weight of flour in the final baked product. The distinction in baked products often lies in the ratio of ingredients used. Most baked products are made from the same seven basic ingredients— flour, water, sugar, salt, egg, fat, and a leavening agent.

KEY TERMS

weight	glutenin	shortening	folding
mass	gliadin	high-ratio shortening	kneading
volume	bran	leavening agent	rubbing
balance scale	germ	dough	sifting
scaling	endosperm	batter	stirring
tare weight	diastase	blending	whipping
gluten	triglyceride	creaming	

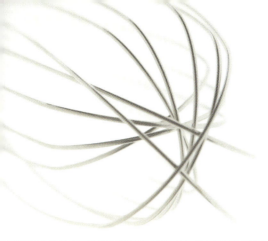

Bakeshop Measurements

Bakeshop ingredients are generally calculated by weight or volume. **Weight** is a measure of the force of gravitational attraction between two objects, and is defined in many baking books as the mass or the heaviness of an ingredient. **Mass** is sometimes defined as the amount of matter, or the weight of an object.

Scaling and Volume Measurement

Volume measures the amount of space that an ingredient takes up, while weight measures mass or heaviness. These two methods of measurement often produce very different results. In baking, volume measures are generally used for lighter liquids, such as water, milk, oils, egg whites, etc. *See Figure 6-1.* Syrups such as honey, corn syrup, and molasses, on the other hand, must be weighed on a baker's scale because they are much heavier.

Exact and constant quantity is significant in baking, so bakers tend to weigh most ingredients on a digital electronic scale or on a **balance scale** that has two platforms. The left platform holds the item being weighed and the right platform supports the weights. Bakers refer to weighing as *scaling*. Most of the dry ingredients used in baking, such as flour and sugar, are easily and accurately weighed. Liquid ingredients, such as water and milk, can also be scaled but are often measured by their volume.

Figure 6-1

Volume Measures

HALF PINT

Cups	Ounces	Pounds
1	8	1/2

PINT

Cups	Ounces	Pounds
2	16	1

QUART

Cups	Ounces	Pounds
4	32	2

HALF GALLON

Cups	Ounces	Pounds
8	64	4

GALLON

Cups	Ounces	Pounds
16	128	8

Baker's Percentage

Bakers frequently convert formulas to baker's percentages to produce the desired number of servings. A baker's percentage assigns each ingredient a certain percentage, by weight, in relation to the amount of flour in the formula. The weight of flour is important because it is the central ingredient in baked goods. The weight of flour in baking formulas is always 100%. In the United States flour is measured in pounds and ounces, while in countries using the metric system, it is measured in kilograms and grams.

Using the Baker's Percentage

The key to using the baker's percentage is understanding that all ingredients are in ratio, or in relationship, to the total flour weight. Once the total amount of flour needed to produce the desired quantity of baked goods has been established, the weight of all the other ingredients can be calculated, assuming that the ratio of those ingredients to the total flour weight is known. The ratio of every ingredient is always expressed as a percentage of the total flour weight. The following examples include basic math equations used to determine either the weight or the percentage of ingredients in a formula.

To convert the weight of an ingredient into a percentage using the *Baker's Math System*, use the following equation:

$$\text{Ingredient Weight (IW)} \div \text{Total Flour Weight (TFW)} \times 100 = \text{Ingredient Percentage (IP)}$$

For example, if yeast is 8 ounces and the total flour weight is 25 pounds, first calculate the yeast percentage by converting both figures to pounds or both figures to ounces.

1. *Yeast = 8 ounces = .5 pound*
   (8 ounces = 1/2 pound = .5 pound)
2. *Flour = 400 ounces = 25 pounds*
   (25 lbs. × 16 = 400 ounces because there are 16 ounces in a pound)
3. *Then, calculate the yeast percentage:*
   (8 ounces ÷ 400 ounces) × 100 = 2(%)

   OR

   .5 lb ÷ 25 lbs. × 100 = 2(%)

In both calculations, the answer is 2%.

For the sake of clarity, conversion examples are in U.S. Standard measurements, but these examples may also apply to metric measurements. *(See Appendix for metric conversions.)*

If a formula is expressed in percentages, the weight of every ingredient can be found regardless of batch size, by using two equations:

First, find the total flour weight (TFW) using this equation—

$$\text{Total Weight (TW)} \div \text{Total Percentage (TP)} = \text{Total Flour Weight (TFW)}$$
(expressed as a decimal Ex: 90% = .90)

In the following formula for Lean Bread Dough, the total weight (TW) is 42 pounds and the total percentage (TP) of the formula is 168%.

FORMULA FOR LEAN BREAD DOUGH

16 pounds	Water	64%
8 ounces or .5 pound	Comp. yeast	2%
25 pounds	Bread flour	100%
8 ounces or .5 pound	Salt	2%
42 pounds total weight		168%

Thus the total flour weight for this batch is 25 pounds.

$$42 \text{ lbs.} \div 1.68 \text{ (168\%)} = 25$$

Next, calculate the weight of the other ingredients.

$$\text{Total Flour Weight (TFW)} \times \text{Ingredient Percentage (IP) (expressed as a decimal)} = \text{Ingredient (individual) Weight (IW)}$$

In this formula for Lean Bread Dough, water is listed as 64% (.64). To find the water weight multiply the total flour weight times the water percentage.

$$25 \text{ lbs.} \times .64 \text{ (64\%)} = 16 \text{ lbs.}$$

Thus, the water weight is 16 pounds.

The total flour weight can also be calculated from any other ingredient in the formula, by using this equation:

$$\text{Ingredient Weight (IW) (expressed in pounds)} \div \text{Ingredient Percentage (IP) (expressed as a decimal)} = \text{Total Flour Weight (TFW)}$$

Using the same percentages and ingredients as those listed in the Lean Bread Formula, now change the 8 ounces of yeast to 4 ounces, and calculate the total flour weight.

$$4 \text{ oz.} = .25 \text{ lb.}$$

(See Figure 6-2 in which ounces are divided by 16 to convert them into percentage fractions of a pound.)

$$.25 \text{ (IW)} \div .02 \text{ (2\%)} = 12.5 \text{ lbs. or 12 lbs., 8 oz. (TFW)}$$

STEP PROCESS

Using a Balance Scale

Professional bakers and pastry chefs use a twin platform/single-beam balance scale or a digital electronic scale for measuring.

To use a balance scale, follow these procedures:

1 Set the weight on the horizontal bar to zero and make sure the scale is in balance. Position the scoop on the left side of the scale. Add a counterweight, or **tare weight**, to adjust for the weight of the scoop.

2 To obtain the exact amount of an ingredient, add weights to the right side of the scale that equal the desired weight of the ingredient. Move the weight on the horizontal bar to make adjustments.

3 Add the ingredient to the scoop on the left side of the scale until the scale is balanced.

Figure 6-2

Ounce Weight	Decimal Pound Weight
1	.0625
2	.1250
3	.1875
4	.25
5	.3125
6	.375
7	.4375
8	.5
9	.5625
10	.625
11	.6875
12	.75
13	.8125
14	.875
15	.9375
16	1.00

Conversion Chart

Essential Bakeshop Ingredients

Most baked goods are made from the same basic ingredients: flour, fats, sweeteners, leaveners, thickeners, a liquid, and various flavorings.

Wheat Flour

Wheat and wheat flour are the most essential ingredients used in baking. Wheat flour gives structure and strength to most baked goods. It is classified by the season in which it is sown—spring or winter, and by the kernel—hard or soft.

Hard wheat is high in protein content and is used mostly in the production of bread items. Soft wheat is lower in protein and is utilized in the production of pastry items such as cakes, cookies, and pie dough.

The protein found in wheat flour is called *gluten.* Gluten is ample in wheat flour and forms a tough, rubbery, and elastic substance when mixed or kneaded with water or other liquids. Gluten affects the texture of baked goods. Yeast dough items need an extensive amount of gluten to help hold the gases during the fermentation and baking processes. Pastry items such as pie dough, biscuits, cakes, and cookies need less gluten because of their tender finished quality.

Gluten is made up of two other proteins: glutenin and gliadin. *Glutenin* gives a baked good strength and structure and helps retain gases during leavening and baking. *Gliadin* gives dough its elastic or stretching ability, allowing it to expand during the leavening and baking processes. Wheat flour is the only grain and flour that has an abundant amount of these vital proteins. Rye, corn, rice, and oats lack an appropriate amount of glutenin and gliadin.

The Wheat Kernel

Wheat kernels contain three sections: the bran, germ, and endosperm. See Figure 6-3.

Bran

The *bran* is the outer protective kernel coating. It is high in fiber and mineral content and is generally removed during the milling process.

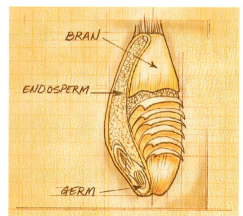

Figure 6-3 Wheat kernel

Germ

The *germ* contains high levels of fat and oils (vitamin E) and gives wheat its nutty taste. In this section the embryo is formed and later removed during the milling process. Wheat flour with a high amount of wheat germ has a tendency to turn rancid if not stored properly.

Endosperm

The *endosperm* is the innermost portion of the kernel. It contains mostly carbohydrates, starches, and proteins. When the bran and germ are removed during the milling process, the starchy endosperm remains. When only the endosperm is milled, 100% "straight flour" is produced. Straight flour is the most commonly used flour in today's bakeshop or pastry shop.

Milling Process

After the wheat has been harvested, it is sent to the mill for processing. There the beard and shaft are removed by threshing the wheat. This process leaves only the edible wheat kernel, which goes through a washing and tempering process before it is ground and sifted into flour.

When 100% of the whole kernel is ground and sifted, it is called whole-wheat flour. Whole-wheat flour is an excellent food source as it contains substantial amounts of proteins, vitamins, minerals, fats, and carbohydrates. Whole-wheat flour contains the bran and germ that affect the gluten production in baked goods. Baked goods that have a high amount of whole-wheat flour tend to be flat and less leavened.

To produce baked goods that have a lighter and less dense texture, the bran and germ are removed to expose only the creamy, starchy endosperm. Millers have been removing the outer bran and germ for thousands of years. In ancient times, aristocrats preferred white flour because the finished baked goods were lighter in texture and easier to digest. The whole-grain flour was left for the peasants.

Straight Flour

Straight flour is 100% milled endosperm with all the bran and germ having been removed during the milling process. Straight flour contains about 70% starch. The first grinding and sifting of straight flour is called first clear or clear flour.

The measure of ash content indicates the quality of milled flour. High ash content—0.75%–1.0%—indicates a lower-quality milled wheat flour while low ash content—0.7% to less than 0.5%—indicates a higher quality. The ash content is determined by flash burning a small amount of milled straight flour and finding out how much mineral or ash content remains. The percentage of the remaining minerals or ash determines the quality of the milled wheat flour.

First-Clear or Clear Flour

This type of flour is milled from the outer portion of the endosperm. It is high in mineral or ash content because it contains some residue of bran, which gives clear flour its gray to tan color. Baked goods that use clear flour leave the crumb darker and less creamy or white in color. First clear is generally used in multigrain and rye breads.

Patent Flour

Patent flour is milled from the center of the endosperm. It is considered pure flour because of the low mineral or ash content. Patent flour has a more consistent amount of protein, and is used in high-end production. Most flours used today are said to be of patent quality.

Wheat Flour Treatments and Additives

When wheat flour is milled into straight flour by removing the bran and germ, some of the vital minerals and vitamins are removed. Most millers are required by law to enrich their flour with the vital minerals and vitamins that have been removed, before the flour can be sold to the general public.

Flour used for bread production sometimes has other added ingredients, such as malted barley and ascorbic acid. The miller may also add enzymes if the wheat is found to be lacking in a group of enzymes known as *diastase,* which convert starches into sugars and are essential in yeast breads.

Aging and Bleaching

When wheat is first milled, it does not make very good bread. The gluten is rather inelastic and the flour has a yellow color. Over time the oxygen in the air matures the proteins to affect the bonding properties and increase the flour's strength and elasticity. Flour can be bleached naturally or chemically. Chemical aging takes merely three to four days. When wheat is milled, oxygen is slowly mixed into the flour to increase the length of the aging process. The high cost of storage and the great demand for high-quality baked goods sometimes requires chemical bleaching using chlorine, benzoyl peroxide, and bromates. Bromates increase the oxygen levels in wheat, which are essential in the aging process. Many millers in the United States have stopped using chemically aged or bleached flour and have gone back to a more natural aging process. Naturally aged flour is not usually bleached. Unbleached flour gives the finished crumb a more creamy color. Some states, including California, have banned the use of chemically treated and bleached flour.

Common Additives

The USDA requires that millers list all additives and the type of aging or bleaching process performed on the flour. Common additives in flour inlcude:

- malted barley or wheat, which helps in the fermentation process
- ascorbic acid, a dough conditioner that aids in the mixing process and helps strengthen the flour so it can perform better with modern mixers
- bleaching agents that help to keep the flour white in color; chlorine gas is used in cake flour processing
- chemical aging agents, such as potassium bromates or benzoyl peroxide
- enrichments of vitamins, minerals, folic acid, and iron

Hard-Wheat Flour

Hard wheat makes up about 75% of U.S. wheat production. The difference between hard and soft wheat is the amount of protein in the kernel. The total protein in the kernel runs from about 7%–14%. Hard-wheat kernels have a higher percentage of protein. When milled, hard-wheat kernels break into larger fragments because of a reduced number of free starch granules. Hard wheats also form strong bonds in the baking process. Common hard-wheat flours include hard red, white spring wheat, and hard red winter wheat. *See Figure 6-4.*

High-Gluten Flour

High-gluten flour is wheat flour that has a protein content of 13% or greater. Generally, the gluten protein is added at the mill. High-gluten flour is used for making hard, crusty products such as pizza dough, bagels, and multigrain breads.

Figure 6-4

HARD-WHEAT FLOUR

Characteristics of Hard-Wheat
- *coarse to the feel*
- *absorbs large amounts of water (50%–75% of its weight in water)*
- *flows evenly when used for dusting the work bench*
- *will not pack when squeezed*
- *excellent for yeast dough items*

Patent or Bread Flour

Patent or bread flour is the most commonly used flour in the bakeshop because of its high quality. This type of flour has a protein content of 11%–12% and an ash content of 0.5%–0.7%. It is excellent for French- and Italian-style breads, laminated doughs, and most high-quality bread items.

Clear Flour

Clear flour has a dark color that makes it unsuitable for use in white bread. It is appropriate for mixing with other grains that have their own color and do not require the purity of patent or bread flour.

Whole-Wheat Flour

As the name implies, whole-wheat flour consists of the entire wheat kernel, including the bran, endosperm, and germ.

Soft-Wheat Flour

Soft-wheat flour has a lower protein content—6%–11%—than hard-wheat flour. Soft winter wheat is the most common soft-wheat flour in the United States. *See Figure 6-5.*

Pastry Flour

Pastry flour, like clear flour, has more color, a creamy texture, and a slightly higher gluten content than other soft flours. Pastry flour is of medium strength, silky and smooth, and can be squeezed into a ball. It is used in pie crusts, cookies, brownies, cake doughnuts, and muffins, where low gluten is desired and the color and extraction rate are not as important.

Cake Flour

Cake flour is much finer than the coarser hard flours, which allows it to carry maximum quantities of sugar and liquids. Cake flour is very high in starch content. Finely ground

Figure 6-5

SOFT-WHEAT FLOUR

Characteristics of Soft-Wheat
- *soft and silky to the feel*
- *absorbs less water than does hard wheat (25%–50% of its weight)*
- *will not flow evenly if used to dust the work bench*
- *will pack if squeezed*
- *excellent for cakes, cookies, and fine pastries*

cake flour also provides excellent volume and crumb color to products such as high-ratio cakes, layer cakes, foam cakes, sheet cakes, cupcakes, and loaf cakes. Cake flour is generally bleached to keep it snowy white during and after the baking process.

Flour Blends

Flour blends are flours of varying strengths blended together to produce products that are better suited to meet the specific needs of the baker.

All-Purpose Flour

All-purpose flour is a blend of hard and soft wheats with a moderate gluten content. This flour is the most popular in homes and supermarkets, and is commonly called for in cookbook recipes. Bakers, on the other hand, generally blend their own flours to suit their needs.

Fats

Fats can be used to add moisture and tenderness to baked goods. They also add shelf life to products because they help retain moisture. The flakiness of many products, such as Danish, puff pastries, croissants, and pies, is not possible without fats. Some fats, including butter and lard, add flavor to baked goods and pastries as well. *See Figure 6-6.*

Fats and oils are components of the same class of chemicals known as triglycerides. *Triglycerides* are made up of glycerol and fatty acids, called fats, oils, and shortenings. The difference between fats and oils is generally their melting point. Oils are liquid at room temperature, and fats are solid or waxy and plastic in consistency at room temperature.

Fats used in baking and pastry are called *shortenings* because of their function in the product. They shorten the gluten strands and help lubricate the gluten during the mixing and kneading processes. The high melting point of most shortenings gives a tender mouthfeel.

Fats in solid form are also creamed. Mixing fats with sugar increases the amount of fat-coated air cells and aids in the leavening process. The fats most commonly used by the baker or pastry chef are shortening, butter, margarine, and sometimes lard. The oils selected are usually made from vegetable or nut oils, such as soybean, canola, peanut, and corn. Olive oils are rarely used in the bakeshop because of their strong taste and aroma.

To gain optimal use when working with fats, it is important to keep them cool—approximately 60°F (16°C). Fats are less effective in warm environments because they shorten the gluten strands too much and capture air needed for leavening.

Figure 6-6 Baking fats and oils add moisture to baked goods.

Shortening

Shortening is usually produced by injecting hydrogen gas into heated oils. The amount of gas used dictates the firmness of the shortening. Shortening is 100% fat, while butter and margarine consist of about 80% fat. All shortenings are white in color, tasteless, and have a waxy to plastic consistency at room temperature. Shortenings have a wide variety of uses in the bakeshop and pastry shop. There are two basic kinds of shortenings: all-purpose and high-ratio.

All-purpose shortening This type of shortening is made from 100% vegetable oils. All-purpose shortening is used in a variety of baked goods, in pie dough and biscuits, and for creaming. All-purpose shortening can also be used for frying doughnuts.

High-ratio shortening   *High-ratio shortening* is made from 100% hydrogenated vegetable oils added to mono- and diglycerides. These emulsifying agents increase the absorption and retention of moisture and allow high-ratio cakes to retain greater amounts of liquids and sugars. High-ratio shortening can be used in the production of sweet dough items and icings. Even though all-purpose and high-ratio shortenings are both made from 100% vegetable oils, they are not interchangeable as they do not perform identically in different applications.

High-ratio liquid shortening   Liquid shortenings are oils derived primarily from plants that have been combined with emulsifiers and solid fats. They remain liquid at room temperature and are used in place of solid shortenings because they blend faster and more easily with other ingredients.

Puff-pastry shortening   Puff-pastry shortening is 80%–100% fat and has a mild to salty flavor. This shortening is designed to have a texture that is soft enough to roll out evenly and thinly between layers of a laminated dough. Regular shortenings have no flavor and are harder, breaking up in colder dough and giving finished products uneven lift. Puff-pastry shortening has an extremely high melting point for better steam leavening. A disadvantage of this type of fat is the waxy feel it leaves on the palate.

Butter

Butter is a dairy product made by churning milkfat after it has ripened. Its unique flavor is the reason for its use when flavor is most important in a product. It also adds a creamy mouthfeel to baked products because of its low melting point. The United States produces primarily sweet cream butter that is lightly salted, unsalted, or whipped. The USDA requires that butter have a fat content of at least 80%. Cultured butter, a rich butter made from cultured cream, is found mainly in Europe. This butter has a fat content of 82% or higher.

All butters are hard and brittle when cold. They work best in baked goods when they are close to room temperature, so that they are smoother and creamier. Butter creams better when it is soft. It is recommended for laminated dough items such as croissants, Danish, and puff pastry. Butter gives baked products an excellent taste and aroma. Products made with butter have a richer crust and crumb color. Although butter is preferred by most bakers and pastry chefs, it is very expensive to use and has a low melting point. It must be handled with extreme care.

Margarine

Margarine was invented in France in the 1800s and is used as a butter substitute. It is made from either animal fats or vegetable oils, and has always been considered a product that is inferior to butter. Many bakers and pastry chefs blend margarine with butter to reduce the cost of baked goods and pastries.

Federal guidelines for margarine state that it must consist of 80% oil, have at least 15,000 added IUs of vitamin A, and be an aqueous solution (one made of milk products and water). Ingredients such as salt, vitamins, and those that enhance taste, texture, or stability may also be added. As it is made from vegetable oils, margarine contains vitamin E. Margarine products that have less than 80% oil are frequently called "spreads" or lowfat, reduced-fat, or fat-free margarine. These products are rarely used in a production bakeshop or pastry shop.

Lard

The quality of lard (rendered and clarified pork fat) depends on two variables:

- the section or interior organ from which it is derived
- and the method of rendering.

Lard rendered from the outer parts of the carcass has a lower melting point than that taken from the body cavity (leaf lard). Lard can be processed in many ways, including prime steam, dry render, open kettle, or continuous. It is then normally filtered and bleached. Typically, processed lard has roughly the consistency of vegetable shortening and a nutlike flavor. Lard is richer than many other fats because it is harder and has a higher melting point. Consequently, it is used in extremely tender and flaky products such as biscuits and pie dough. Lard is used mostly in regional and ethnic baking.

Oils

Oils are derived primarily from plants that have been purified of fibrous materials. Many oils have been deodorized, and their color ranges from clear to slightly yellow. They are used primarily in the production of muffins and quick-bread batters.

Considerations When Selecting Fats

When selecting fats, consider the following:

- their purpose and function in the product
- the cost of the fat
- storage and temperature conditions to which the fat and final product will be subjected
- the appearance of the final product
- the desired flavor and texture

Each fat has a variety of characteristics that distinguish it from other products. If flavor is an important factor, choose butter.

Storing Fats and Oils

The type of storage facilities, the holding time, and temperature must be seriously evaluated when selecting fats and oils.

Sweeteners

Sugars and sweeteners have many purposes in baking. *See Figure 6.7.* These include:

- providing flavor and sweetness
- adding color to brown a crust
- serving as food for yeast
- aiding in the retention of moisture
- breaking up the gluten to tenderize products
- serving as a base for icings
- helping stabilize egg foams

Figure 6-7 Common sugars and sweeteners in baking.

Sugar

The word *sugar* usually refers to table sugars derived from sugarcane or sugar beets. Although the chemical classification for regular sugars is sucrose, other sugars of numerous chemical structures also exist.

Types of Sugars

Sugars belong to a group of chemical compounds called carbohydrates, a group that also includes starches. The following is a list of the types of sugars:

- sucrose, made from sugarcane or sugar beets
- carbohydrates, contained in wheat and other grains
- maltose, or malt sugar
- fructose, found in fruits and vegetables
- lactose, found in milk products
- glucose, contained in fruits and vegetables

BROWN SUGAR   Brown sugar was originally an unrefined or partially refined sugar that retained some molasses. Molasses, a cane sugar derivative, is the brownish liquid residue present after heating up raw cane sugar/sucrose and using a centrifuge to refine, or drain off the liquid. Molasses contains uncrystallizable sugars, some residual sucrose, and other nutrients.

Today "brown sugar" is generally white or refined sugar to which molasses has been added. Brown sugar comes in varying stages of refinement that range from dark brown to light brown. The lesser the amount of molasses added, the lighter the color. Brown sugar helps retain moisture in baked goods and gives cookies a soft, chewy texture.

GRANULATED SUGAR   Standard granulated sugar, also called table sugar, is the most commonly used type of sugar. It is derived from refined sugarcane or sugar beets and is white in color because all the molasses and impurities have been removed.

COARSE SUGAR   Coarse, or sanding sugars are larger particles used primarily for decorations on cookies, cakes, and other bakery items.

FINE AND SUPERFINE SUGAR   Very fine and ultrafine sugars (also known as caster sugar to the British) have smaller crystals than granulated sugar but are still not as fine as powdered sugar. They are used in mixes that require rapidly dissolving sugar for a more uniform batter.

CONFECTIONERS' OR POWDERED SUGAR   These granulated sugars are produced by grinding the crystals finer and mixing them with a small amount of starch (about 3%), to prevent caking. They are grouped by the finished crystal size, which is indicated by Xs. The most commonly sold form is 6X, but 10X is the finest. These sugars are referred to as icing sugars because they are used primarily for cold work, such as icings and whipped cream.

Inverts and Syrups

Inverts and syrups are liquid forms of sugar. They are hygroscopic, which means they have the ability to hold and retain moisture. When used in baked goods and pastries, inverts and syrups help to retain moisture and keep the product softer. They also increase the shelf life of the product.

INVERT SUGARS   Invert sugars are liquids produced from ordinary sugar. The sugar is boiled with a solution of hydrochloric acid and water that changes the disaccharide sucrose to monosaccharides. Invert sugar is somewhat sweeter than table sugar. It can aid in moisture retention and is sometimes used as a fat substitute in fat-reduced products.

CORN SYRUP   Corn syrup is a natural product made by converting cornstarch to a liquid through the use of enzymes. It is very economical and is the most common syrup used in the foodservice industry. Corn syrup is added to products to prevent recrystallization of sugars in candy making or sugar work and can also help to retain moisture in baked goods.

MAPLE SYRUP   Maple syrup is natural sugar harvested from the sugar maple tree. Depending on how sweet the sap is, 30 to 40 gallons of raw sap may be required to make

1 gallon of maple syrup. Sugar maple sap is about 2% sugar, making it a very expensive sweetener that is not used often in commercial bakeshops or pastry shops except as a flavoring.

HONEY   Honey was most likely the first sugar ever used by humans. It is produced by bees from the nectar of flowers. Honeybees are essential in the pollination of fruit, flowers, and nut trees. The most common honeys are clover honey and orange blossom honey.

Clover honey is used for its robust flavor and is amber to light tan in color. Orange blossom honey is similar in color to clover honey but has a mellower taste. Other types of honey are also produced from a variety of flowers or blossoms, including apple blossom, rose hips, and sage.

Flower nectar is about 80% water and 20% sugar. In the hive it is converted to an invert sugar and dried to about 70% sugar. The remaining product consists of oils, gums, and water. The oils give each honey its unique characteristics. Honey's high acid content and bitter taste can be counteracted in baked goods and pastries by adding baking soda.

GLUCOSE   Glucose is similar to corn syrup. It is prized by bakers and pastry chefs because of its purity. Glucose is produced from corn, rice, or potatoes. It is used in the same way as corn syrup to prevent the recrystallization of sugar in candy making and sugar work.

MALT SYRUP   Malt syrup or malt extract is a concentrated water extract from barley grain or corn that has been allowed to sprout. Sprouting makes the grain relatively high in vitamins and a valuable nutritional additive to bakery goods. Malt syrup contains diastatic enzymes that help break down starches into sugars. These enzymes are excellent for yeast fermentation. Malt syrups are used in the production of beer and other alcoholic beverages. Nondiastatic malt that is without enzymes can also be used as a flavoring agent.

Leaveners

A *leavening agent* is a substance that causes a baked good to rise by adding carbon dioxide (CO_2), steam (water vapor), or air into the mix. The gases in the leavening agent increase in molecular size because of the heat of the oven and stretch the gluten strands or protein cell walls of the product. The end result is a baked good that is light, fine textured, and increased in volume. Leavening action occurs in baked goods either physically (air and steam), chemically (by using baking soda or baking powder), or biologically (through the fermentation process). See Figure 6-8.

Physical Leavening Agents

Air and steam are important physical leavening agents.

AIR   Air is classified as a physical leavening agent because it occurs as a result of a physical action, such as sifting, creaming, or whipping. In the making of Genoise sponge cake, for example, the flour is double sifted to increase the air, and the egg mixture is whipped to produce foam. In some cookie- and cake-making methods, fats and sugars are creamed, to incorporate air that aids in the leavening process.

STEAM   When water is heated to 212°F (100°C) steam is created. This action swells the cell structure, causing the product to rise. All baked products contain liquids with some amount of water content that converts to steam. In the production of pie dough, puff pastry, and pâte à choux (cream puffs or éclairs), steam is the primary leavening agent.

Figure 6-8   Leaveners cause baked goods to rise.

Chemical Leaveners

Baking powder and baking soda are the two main chemical leaveners.

BAKING POWDER Baking powder is composed of an alkaline (such as baking soda), an acid/salt (such as cream of tartar or aluminum sulfate), and a moisture absorber (such as cornstarch). When mixed with a liquid, baking powder releases carbon dioxide gas (CO_2). Baking powders serve as leaveners in cakes, muffins, cookies, and quick breads. They come in three basic types:

- Slow-acting baking powders that react to the heat of the oven.
- Fast-acting baking powders that react to the liquid in the formula.
- Double-acting baking powder is a mixture of both fast- and slow-acting baking powders that react at different rates known as action rates. Two-thirds of the powder react in the presence of a liquid and one-third reacts when it comes in contact with the heat from the oven (2/3 moisture and 1/3 heat). Though slow- and fast-acting baking powder can be purchased for the baking industry, double-acting baking powder is most commonly used by bakers today.

BAKING SODA Baking soda, or sodium bicarbonate, is a chemical leavener that requires the presence of moisture and an acid to release CO_2 gas into the baked good. Baking soda neutralizes or changes the pH levels in products and helps to release CO_2. Buttermilk, chocolate or cocoa powder, yogurt, fruits, honey, molasses, and sour cream are common sources of acid. Baking soda has the ability to weaken gluten strands that help to tenderize the baked good during baking. Baking soda also has the ability to darken the color of a baked good. Products such as devil's food cake have a richer color when the proper amount of baking soda is used. If baking soda is not mixed completely or is out of ratio with the acid, it may leave a soapy aftertaste that can weaken the gluten strands and produce an undesirable texture. Too much baking soda can cause the product to rise and fall before it can set in the oven, making the product inedible.

Biological Leaveners

Yeast is a living organism, a single-celled fungus, that multiplies quickly in a warm, moist environment. The cell division that occurs in yeast is called "budding." During the process of fermentation, yeast converts carbohydrates into carbon dioxide gas and alcohol. The carbon dioxide expands and leavens the product. Yeast products get their characteristic aroma and taste from this fermentation process. Yeast needs food (sugars and carbohydrates), warmth (proper water temperature, room environment, and mixer speed), and moisture for proper fermentation to occur. The ideal temperature for yeast to thrive is 78°F–82°F (26°C–28°C). Yeast cell reproduction ceases if the temperature of the dough reaches the thermal death point, or TDP, at 138°F (59°C).

Commercial baker's yeast was first discovered in the 1800s by the French chemist Louis Pasteur, who is responsible for discovering how yeast cells reproduce. He was able to collect and grow yeast cells in the lab and then introduce them into bread. This important discovery gave bakers a fast and reliable way to produce bread. The yeast cells that Pasteur discovered are called *Saccharomyces cerevisiae*, or baker's yeast. Three types of baker's yeast are most frequently used in bakeshops: compressed, dry active, and instant dry yeast.

COMPRESSED YEAST Compressed yeast, sometimes called fresh or wet compressed yeast, is partially dried to about a 70% moisture level. Compressed yeast is the standard in most formulas unless otherwise stated. It should be tan in color, have a crumbly texture, and smell like freshly baked bread. Compressed yeast quickly deteriorates at room temperature, so it must be refrigerated when not in use. To use compressed fresh yeast, crumble it into water to create a slurry, or crumble it directly into the dough as it is being mixed. Compressed yeast has a shelf life of about 24–30 days if properly stored.

DRY ACTIVE YEAST Dry active yeast retains only 8% of its moisture. It must be rehydrated in a portion of liquid in a formula at approximately 110°F (43°C) before it is added to other ingredients. This type of yeast is normally sold in vacuum-sealed bags of 1- or 2-pound quantities. Unopened packages can be stored in a cool, dry place for several months. Once opened, containers of dry active yeast should be kept frozen. When substituting this type of yeast for compressed yeast, cut the amount of yeast in half. That is, if a formula calls for 2 ounces of compressed yeast, use just 1 ounce of dry active yeast.

INSTANT DRY YEAST Instant dry yeast is similar in appearance to dry active yeast but consists of a different strain of yeast cells. Its leavening action is much more rapid because carbon dioxide is created earlier and in greater quantities to speed up the fermentation of the dough. When substituting instant dry yeast for compressed yeast, use one-third of the amount. If a formula calls for 3 ounces of compressed yeast, use only 1 ounce of instant dry yeast. To use instant dry yeast, blend it with the dry ingredients and then add the results to the water or liquid ingredients. Instant dry yeast lasts at least 24 months in unopened packages or when stored frozen. Once opened, it is good for about 6 months. Instant dry yeast is very popular in today's bakeshops because of the small amount necessary in a formula and its longer shelf life. It is also very reliable and produces a consistent finished product.

LIQUID YEAST Liquid yeast is primarily used in large commercial bakeshops and in brewing. Various grains are wetted and inoculated with yeast spores to begin the process of cultivation. The end product is a yeast slurry that can be added to the mix in a continuous stream without slowing down production. This type of yeast is rarely used in small commercial bakeries.

SOURDOUGH STARTERS Sourdough starters, the forerunners of modern yeast, were accidentally introduced into grain mashes by wild yeast in the air. Today, starters are used for the unique flavor they bring to the final product, such as San Francisco sourdough bread. Sourdough starters are often fortified with yeast for reliability and consistency.

Storing Leavening Agents

Dry yeast products, baking soda, and baking powder must always be kept in properly sealed containers to protect them from moisture and heat. Moisture can reduce their leavening ability, and exposure to heat causes deterioration. Yeast should be kept in the refrigerator and purchased just before use.

Thickeners

Most thickening agents are forms of starch derived from the roots or seeds of certain plants. When used in conjunction with proteins, starches perform many valuable functions. These include:

- quickly thickening fillings and sauces
- producing glossy fillings and glazes
- offsetting the action of certain acids
- maintaining a filler's thickness and consistency upon cooling
- maintaining the flavor and color of fruit fillings and glazes

See Figure 6-9.

Figure 6-9 Thickening and gelling agents.

Arrowroot

Arrowroot is a root starch. It is similar to cornstarch but more expensive. A sauce made from arrowroot has a very clear appearance. Arrowroot is often used in frozen food sauces because it does not break down when frozen.

Cornstarch

Cornstarch is a grain starch like flour. A powdery dense flour with almost twice the thickening power of flour, cornstarch creates sauces and fillings that are glossy and almost clear.

Flour

When flour is used in baking, the starch interacts with the proteins and forms a dough. Flour also absorbs water and gelatinizes during baking to give products a firm structure. When used as a thickener, flour leaves a pasty aftertaste, unless it is cooked with fat to make a roux.

Tapioca

A starchy substance extracted from the root of the cassava plant, tapioca is used like cornstarch as a thickening agent for fruit fillings and glazes. Although sold in various forms, the most common is pearled tapioca.

Eggs

Eggs are used for thickening when flavor and leavening are also important. The protein structure of the egg captures air. As heat is applied, the egg begins to dehydrate and coagulate, creating a soft gel. (**See Chapter 7: Dairy Products and Eggs for more information on eggs.**)

Modified Food Starch

Modified food starch helps in the reduction of "set-back"—the process in which the starch weeps and begins to thin, generally in the presence of an acid. Modified food starches do not rupture or lose viscosity as do unmodified starches.

Pregelatinized Starches

Pregelatinized starches are precooked and dried and need no further cooking to absorb water. They are used in products such as instant puddings. When using pregelatinized starches, mix them with sugar (about one part starch to four parts sugar) before adding the liquid. Use pregelatinized starches to quickly thicken sauces and to thicken coulis-type sauces.

Gelling Agents

Gelling agents come from both plants and animals. These agents are fluid at high temperatures and set as they cool. Three commonly used gelling agents are gelatin, agar-agar, and pectin.

GELATIN Gelatin is the generic term used to identify a group of protein-related materials found in the collagen of animals, particularly bones and pork skins. It has fairly strong gelling power, that allows molds to hold their shape after they have been unmolded.

AGAR-AGAR Agar-agar is a water-soluble plant gum derived from seaweed. It sets up a gel that is stiff, clear, and reversible. Agar-agar is used in some pies, jellies, and icings.

PECTIN Pectin is a fiber that occurs naturally in many fruits and becomes soft in water. It is used to thicken fruit-based items, such as jams, jellies, and glazes. Pectin can be purchased in liquid or powdered form.

Liquids

Liquid is necessary in baking to combine all the ingredients into a homogeneous mass. Liquids also dissolve and readily disperse all the ingredients, allowing them to bind together. Some of the more common liquids in baking and pastry are water, milk, cream, eggs, butter, and oil. (*See Chapter 7: Dairy Products and Eggs for more information on other liquids.*)

Water

Water is a solvent used for the redistribution of salt, sugar, and other water-soluble ingredients. It is used to hydrate flour to form gluten and is added during the mixing stage of all bakery products. Water may be added either directly to the mix or through the use of eggs, milk, juices, or other liquid products.

Milk and Cream Products

Milk and cream products add moisture, sweetness, flavor, and fat to desserts such as custards and ice creams. Milk also contains important minerals and proteins that improve the flavor and texture of certain baked goods, and extends their shelf life. Buttermilk, sour cream, yogurt, heavy cream, and dried milk solids such as nonfat dry milk are also used in bakeshops and pastry shops.

Nondairy Toppings

Rich's® Whip Topping®, nondairy topping, and Cool Whip nondairy whipped topping are examples of nondairy products available on the market. These products have become increasingly popular as more and more people become lactose-intolerant. Imitation toppings are made with soy and gums that mimic milk and cream.

Flavorings

Although flavorings do not usually impact the baking method, they do add to the unique and identifying flavor, aroma, and color of the finished baked good. Many items are available to enhance products, such as extracts, fruits, spices, and nuts. *See Figure 6-10.*

Extracts and Essences

Extracts and essences are concentrated flavorings derived from various foods or plants. Extracts are liquid flavorings that contain alcohol and are usually derived through evaporation or distillation. Vanilla extract, the most commonly used extract in the bakeshop, is an exception as it is made by merely soaking the vanilla bean in alcohol and water to extract flavor. Essences are concentrates, usually oily substances extracted from food.

Figure 6-10 Flavorings enhance the final baked product.

Fruit

Fruits add color, flavor, and nutritional value to baked goods. They may also supply texture and visual appeal to the finished product and can improve moisture retention and sweetness. Juices, purées, skins, zests, and pulps may all be used. (*See Chapter 8: Fruit for more information.*)

Compounds

Compounds are mixtures of flavors bound to sugars or other dry ingredients to enhance the taste of products. They may also be blends of flavors in a dry form such as butter, lemon, orange, and vanilla.

Seasonings/Spices and Herbs

Seasonings and herbs are the fragrant leaves and seeds of plants. Seasonings are sometimes referred to as spices. They can be made from the bark, roots, flower buds, or seeds of a plant. Herbs are the tender leaves without woody stems or branches.

Spices come in ground or whole form. Ground spices release their flavor quickly and keep for about three months. The flavor of whole spices comes out during the baking process. (*See Chapter 9: Herbs and Spices / Nuts and Seeds for more information.*)

Nuts

Nuts provide flavor, color, and texture to baked goods. They are used in a variety of forms, including whole, halved, broken, chopped, chipped, or ground. Nuts should be stored in a cool, dry place to prevent them from turning rancid. Nuts can be used as a topping as well as an ingredient.

Salt

Although many salts are known, the one most significant to the baker is sodium chloride, or table salt. Salt is naturally present in many foods and is necessary for human life. It is obtained from several sources, but the two most important are salt deposits from the ground and from oceans and lakes.

Salt that is mined from the ground consists of large, irregular pieces with many impurities. It is generally referred to as rock salt and used mostly for home freezing. Rock salt is sometimes purified and used for koshering, or meat curing.

Table salt is mined by pumping water into underground salt deposits. The brine, or salt-laden water, leaves behind crystals as it evaporates. To keep the salt from absorbing moisture and clumping together, chemicals are often added. In the United States, iodine is an important nutrient added to salt.

Evaporation is another processing method. Water from lakes, ponds, or seas is evaporated and the impurities are removed until the salt has dried and crystallized. Then it is further purified, ground, and graded for use.

An additional method is the vacuum process, in which the brine is placed in vacuum pans and boiled off, leaving lump salt that is then purified, ground, and graded.

Functions of Salt

Salt has many functions in bakery and pastry shop products.

Flavor Salt enhances the flavors of ingredients in products by adding a unique contrasting flavor.

Controlling yeast activity By lowering the yeast's ability to produce gas (CO_2), salt slows down the fermentation process.

Strength and structure Salt interacts with gluten in flour-based products to toughen the gluten giving items more strength and structure.

Stretch and texture As salt interacts with gluten, it creates stretching ability, which produces a finer texture in the final product.

Preventing bacterial growth Salt changes the pH level of doughs and helps reduce water activity to prevent bacterial growth.

Bleaching Salt can be used as a bleaching agent to create a less creamy color in the crumb of the bread if desired.

Aiding in digestion Salt helps make food more digestible.

Aerating properly Salt's ability to control fermentation also helps to properly aerate doughs.

The Effects of Too Much or Too Little Salt

Although salt has many benefits in the bakeshop, it must be measured accurately. Too much salt causes the dough to ferment slowly and have a strong salty flavor. If the product contains too little salt, it is tasteless and rises uncontrollably, with poor structure and texture.

Chocolate

Figure 6-11 Chocolate comes from the bean pods of the cacao tree.

Chocolate, a popular bakeshop ingredient, adds body, bulk, flavor, and color to cookies, cakes, candies, and pastries. It comes from the bean pods of the tropical cacao tree. Chocolate is grown in Mexico, Ecuador, Brazil, and in some parts of Africa. *See Figure 6-11.*

The cacao is a delicate tree that requires six years to grow to maturity. When the fruit ripens, it is collected. The berry is extracted from the pod, which contains 30–40 cacao berries. They are removed, washed, fermented, and allowed to dry naturally in the sun or are blown dry mechanically. Once it is dry, cacao is transferred to a processing center where it is roasted, cracked, and separated into the nibs. The nibs are then shipped to manufacturers, where they are roasted twice and processed into a variety of chocolate or cocoa products, which include cocoa butter, a natural vegetable fat, and chocolate liquor, a thick, dark brown semifluid.

If other ingredients are added (milk powder, sugar, etc.), the chocolate is refined again. The final step for most chocolate is conching, a process by which huge granite rollers slowly combine the heated chocolate liquor, eliminating some remaining water and unstable acids. The conching continues for 12–72 hours, during which other ingredients may be added, such as small amounts of cocoa butter and lecithin. Lecithin reduces the fat requirement and the viscosity, and emulsifies the sugar and fat. Soy lecithin, a mixture of fatty substances that are derived from the processing of soybeans, may also be added. These ingredients give chocolate its silky texture. *See Figure 6-12.*

Couverture Chocolate

Couverture, the French word for "coating," is a term used for high-quality dark, milk, and white chocolates. Couverture is made from chocolate mass (chocolate and cocoa butter) and extra cocoa butter. If couverture chocolate is used to coat candies or other products, it must be tempered, carefully melted, and cooled off to just the right temperature.

Figure 6-12 Chocolate in many forms.

Compound Chocolate

Compound chocolate, also called coating, is made of hydrogenated fat and lecithin, rather than cocoa butter. It is therefore more stable and does not have to be tempered like other chocolates for dipping, coating, or molding. Despite its convenience, it does not have the mouthfeel of couverture chocolate.

Dark Chocolate

Pure chocolate or dark chocolate is marketed as unsweetened chocolate and is referred to as baking or bitter chocolate. U.S. standards require that unsweetened chocolate contain 50%–58% cocoa butter. It has no added sugar or milk solids, giving it a bitter taste. Dark chocolate is intended as a flavoring for products such as fillings, mousses, and marzipan. Since it contains all the cocoa butter from the bean, it gives products a very rich flavor.

The amount of sugar, lecithin, and vanilla added creates bittersweet, semisweet, or sweet chocolate. Bittersweet chocolate must contain at least

35% chocolate liquor; while semisweet and sweet chocolate require 15%–35% chocolate liquor.

Milk Chocolate

Milk chocolate is made from cocoa paste that has been finely ground and conched and combined with additional cocoa butter, sugar, and vanilla. Milk chocolate must contain at least 12% milk solids and 10% chocolate liquor. It is rarely melted and added to batters because of its somewhat low chocolate liquor content. Instead, it is added to a variety of confections or as a coating chocolate.

White Chocolate

White chocolate is not really chocolate at all but a mixture of sugar, cocoa butter, lecithin, vanilla, and dried or condensed milk. This product cannot be officially classified as chocolate because it contains no chocolate liquor. It is used as a confectionery or may be eaten alone.

Cocoa Powder

Cocoa powder is the dry, brown powder that remains after cocoa butter has been removed from the chocolate liquor. Cocoa powder can be either alkalized (known as Dutch-process cocoa powder) or nonalkalized (regular) cocoa.

Dutch-process cocoa is treated with an alkali solution of potassium carbonate to make it milder, less acidic, and darker. It is less likely to lump than regular cocoa and can be substituted for unsweetened chocolate by adjusting the amounts of cocoa and shortening.

Regular cocoa has no added sweeteners or flavorings. It is commonly used in baking and is able to absorb moisture and provide structure.

Combining Doughs and Batters in the Bakeshop and Pastry Shop

Doughs and batters are produced by mixing dry and liquid ingredients. ***Doughs*** have a low moisture content and a thick, pliable consistency that is generally firm enough to handle or to work with. Breads and rolls are products made of dough. ***Batter*** can either be thin or thick. Waffle and pancake batters are examples of thin batters, whereas muffins and loaf breads have a thicker batter. The consistency of the batter determines the mixing method. Blending and creaming are often utilized to produce batters.

Blending

Blending *is incorporating ingredients until they are evenly combined. A spoon, rubber spatula, wire whisk, or a bench mixer with a paddle attachment may be used to blend.*

Creaming

Creaming *refers to beating the ingredients, such as fat or butter with sugar, and gradually adding the remaining ingredients until light and smooth. An electric mixer on medium speed is employed to cream.*

Folding

Gently mixing a lighter ingredient, such as whipped cream or whipped eggs, into a heavier one (such as batter or a cream) is known as **folding**. Folding can be accomplished by utilizing a spatula or a balloon whisk.

Kneading

Folding and pressing a dough with the hands to develop gluten is known as **kneading**.

Rubbing

Rubbing is employed when solid fat is rubbed or cut into dry ingredients such as flour to create a mixture of flour and large fat flakes.

Sifting

Sifting is the transference of dry ingredients such as flour or sugar through a wire mesh in order to aerate, separate, and remove the lumps from a fine mixture. A drum or rotary sifter or mesh strainer may be employed to execute sifting.

Stirring

Blending ingredients gently by hand until they are equally combined is known as **stirring**. A spoon, wire whisk, or rubber spatula may be used for stirring.

Whipping

Whipping is performed by beating ingredients for the sole purpose of aerating. A whisk, or a mixer with a whip attachment can be used for whipping.

The Mixing Process and Gluten Development

As mentioned earlier, gluten is the part of the wheat protein that, when hydrated, mixed, and kneaded properly, develops into a tough, rubbery, and elastic substance. Wheat flour is the only grain to naturally contain both the proteins gliadin and glutenin. Gliadin gives dough its stretching ability and allows it to expand during leavening and baking. Glutenin contributes strength and structure to dough and helps it retain gases during leavening and baking. Together, gliadin and glutenin create a cell structure that gives bread products form and structure and enables them to swell with gas during fermentation and baking.

Controlling Gluten

The presence of certain ingredients affects gluten formation.

Flour

Different types of flour have varying degrees of gluten-forming proteins. Bread flour, for example, has a high gluten content, while pastry flour and cake flour have lower gluten contents.

Liquid

Liquids such as water must be mixed or kneaded for the gluten structure in flour to form.

Shortenings and Fats

Shortenings and fats lubricate the gluten and shorten the gluten strand, to tenderize baked goods. Shortening also prevents water from hydrating too quickly, giving the dough a longer mixing time. In the production of pie dough and biscuits, the gluten-developed flour envelops the fat, melts it, and creates steam so that the product is flaky and tender.

Salt

Salt strengthens the gluten, conditioning it and giving it more rigidity and stability. Too much salt in a dough prevents gluten development.

Sugar

Sugar has hygroscopic tendencies that slow gluten development by competing with gluten for water. If a dough requires a high percentage of sugar, it is best to add the sugar in stages to help the development of the gluten strands.

Mixing Method

Mixing, or mechanical manipulation, is very important to the distribution of ingredients. It also aids in the full and proper hydration of the gluten. Yeast dough items need an ample amount of gluten development for consistent finished products. Pastry items including cookies, cakes, pie dough, and biscuits need less gluten because they require a more tender finished quality. If a yeast dough is underdeveloped, the bread will be flat and less aerated. Pastry items with too much gluten development will have a tough and less desirable finished texture.

The Baking Process

All doughs and batters go through the same changes during the baking process. The eight stages of baking follow.

Gas Formation and Expansion

The first stage of baking, gas formation and expansion, produces a boost in volume known as oven spring. Carbon dioxide (CO_2), air, and steam are the main gases that cause the leavening of baked goods. Carbon dioxide is released from yeast, baking soda, or baking powder. Air is added to batters and doughs during the mixing process. Steam forms during baking, as the dough's moisture is heated. The rise continues until the product's internal temperature reaches 140°F (60°C).

Gases Trapped in Air Cells

Proteins in dough, such as gluten and egg, trap the gases that are formed during baking. Without these proteins, the gases would escape, and little leavening would occur in the baked good.

Gelatinization of Starches

After the heat has permeated the dough and the product has reached a temperature of 140°F (60°C), the gelatinization of starches begins. During this stage, the starches gain structure as they absorb moisture, expand, and become firm.

Coagulation of Proteins

As the baking process continues, the proteins begin to coagulate until the structure becomes set at a temperature of 165°F (74°C). The interior temperature of the dough will not exceed 212°F (100°C) because of the evaporation of the water and alcohol. The baking temperature is crucial because of its effect on coagulation. Temperatures that are too high create premature coagulation that splits the crust, or a low volume in the product. Low temperatures do not allow the proteins to coagulate soon enough, causing the product to collapse.

Water Evaporation

The evaporation of water occurs throughout the baking process. Bakers must consider this moisture loss when scaling dough. Although factors such as baking time and proportion of surface area to volume affect the percentage of weight loss, all dough loses about 10% of its original mass as a result of evaporation.

Melting of Shortenings and Fats

This stage of baking may occur at a variety of temperatures, depending on the type of shortening or fat used. The melting of shortening and fat break gluten and emulsify the product to make the item tender.

Crust Formation and Browning

As water evaporates and leaves the surface dry, crust is formed. Sugar and starch on the surface of the product begin to brown, or caramelize, at 266°F (130°C) as they react to chemical changes from the heat. The baker generally washes the product with water, milk, or egg wash to increase browning and to add moisture to the bakery item. This moisture helps to give the final product good form and shape.

Carryover Baking

Baked products continue to bake for a short time after they have been removed from a hot oven because of a process called carryover baking. This process occurs because the chemical and physical reactions that take place during baking do not stop immediately. Bakers must keep this in mind to not end up with overbaked products.

chapter 6 Fundamentals of Baking & Pastry

CHAPTER 7

Dairy Products & Eggs

Bakers and pastry chefs use a wide variety of dairy products and eggs to produce baked goods and desserts. The most popular dairy products are butter, cream, and milk.

KEY TERMS

pasteurization
ultrapasteurization
ultra high temperature (UHT) processing

homogenization
certified milk
crème fraîche
curds

whey
ripening
clarified butter
lecithin

albumen
chalaza
candling

Milk

Milk is a dairy product that is used to add moisture, sweetness, texture, and volume to baked goods. It is also the foundation of other dairy products, including butter, cheese, and yogurt.

The composition of milk varies, depending on the type of animal, the breed, the animal's diet, the season, and even the time of day the animal is milked. Milk, in this book, refers to cow's milk. Cow's milk is composed primarily of water (88%). It contains about 3.5% milkfat and about 8.5% milk solids in the form of proteins (primarily casein), lactose (milk sugar), minerals, and vitamins. Milk provides a rich source of calcium, as well as vitamins A, D, E, K, C, and B. Most milk is fortified with additional vitamin D. Low-fat and skim milk products are fortified with vitamin A to replace the vitamin A drawn off with the removal of milkfat. A government grading system classifies milk as Grade A, B, or C on the basis of the milk's bacterial count. Grade A milk is the highest grade.

Casein is the primary protein in milk. It causes milk to clump and curdle in the presence of acid, large amounts of salt, or high heat. When milk is heated, a skin can form on top. The skin is actually coagulated protein and fat. Milk can also coagulate, or become firm, and then burn on the bottom of a pan. Continuous stirring, heavy cookware, moderate heat, and minimal cook times help limit coagulation.

Milk Processing Methods

Milk provides a favorable host for the growth of bacteria, many of which can cause serious illness. Almost all milk is treated to kill bacteria and to increase its keeping qualities and shelf life.

Pasteurization

In the 1860s, French chemist Louis Pasteur discovered a way to use heat to prevent spoilage of beer and wine without destroying the flavor. A similar process known as *pasteurization* is now used for the treatment of milk. Law requires the pasteurization of all Grade A milk. In fact, nearly all milk sold commercially in Western countries is pasteurized. The most common pasteurization technique heats milk to 161°F (72°C) for 15 seconds and then rapidly cools it to 45°F (7°C). Pasteurization kills most bacteria and extends the keeping qualities of milk by killing naturally occurring enzymes that cause milk to spoil. Pasteurized milk, properly refrigerated, stays fresh for about one week.

Ultrapasteurization

Often used to extend the shelf life of cream, *ultrapasteurization* subjects dairy products to very high temperatures for shorter periods, to destroy nearly all bacteria. Unopened, refrigerated ultrapasteurized milk and cream stays fresh for 60–90 days. Once opened, the product stays fresh for the same amount of time as conventionally pasteurized milk products.

Ultra High Temperature Processing

Ultra high temperature (UHT) processing combines the process of ultrapasteurization with specialized packaging. Holding the milk for 2–6 seconds at 280°F–300°F (138°C–149°C) sterilizes it. Hermetically sealed sterile containers block out bacteria, gases, and light. Unopened UHT milk can safely be stored without refrigeration for up to 3 months. After it is opened, handle UHT milk in the same way as conventionally pasteurized milk products.

Homogenization

In untreated milk, milkfat particles float to the top and create a layer of cream. This occurs because the fat globules, which are lighter than the surrounding water, are too large to stay suspended in the emulsion. To prevent the separation of milkfat, most commercially sold milk is homogenized. ***Homogenization*** breaks down fat globules by forcing the warm milk through a very fine nozzle. The fat particles are reduced to one-tenth their original size, allowing the milkfat to remain suspended.

Certification

Certified milk has met strict sanitary conditions. Veterinarians examine the herd to make sure that all the cows are disease-free. Doctors review the health of dairy workers and rigid inspections ensure that milking equipment is sanitary. Some states allow certification of raw milk as well as pasteurized products.

Removal of Milkfat

Removing milkfat fills the demand for variety in milk and cream products. To remove milkfat, milk is processed in a separator, a type of centrifuge. The milk spins at high speeds, causing the lighter milk to collect at the outer wall of the separator, while the denser cream collects in the center, where it is piped off. Modern separators can produce a range of milk products, including whole milk, lowfat, reduced-fat, or skim. *See Figure 7-1* for a

Figure 7-1

Liquid Whole Milk
Contains no less than 3.25% milkfat and 8.25% milk solids. Vitamins A and D are optional additions, as are flavoring ingredients, such as chocolate.

Liquid Lowfat or Reduced-fat Milk
Milk with some fat removed. Contains between 0.5% and 2% milkfat and no less than 8.25% milk solids. Common lowfat milk includes 1% milk and reduced-fat milk has 2% milk. Lowfat and reduced-fat milks must contain added vitamin A. Vitamin D is optional.

Liquid Skim Milk
Also called fat-free or nonfat milk and has all or most fat removed. Must have less than 0.5% milkfat and no less than 8.25% milk solids. Vitamin A must be added. Vitamin D is optional.

Evaporated Milk
A canned concentrated milk formed by evaporating about 60% of the water from whole or skim milk. It has a cooked flavor and darker color than regular milk.

Sweetened Condensed Milk
Whole milk with 60% of its water removed and a large quantity of sugar added. Pasteurized and usually sold in cans it is evaporated milk that has been sweetened. Sweetened condensed milk is not an acceptable substitute for whole or evaporated milk.

Dry Milk Powder
Milk with nearly all moisture removed. Often made from skim milk, although whole milk powder is also available. Dry milk can be used to enrich baked goods or can be reconstituted in water.

Figure 7-2

Varieties of Cream

Half-and-Half
A mixture of milk and cream that contains at least 10.5% milkfat but no more than 18% milkfat. Used in baking and as an enrichment.

Light Cream
Also called coffee cream or table cream, light cream contains between 18% and 30% milkfat and is used in baking and as an enrichment.

Light Whipping Cream
Also called whipping cream, contains between 30% and 36% milkfat. It can be whipped or used to make sauces, ice cream, and other desserts.

Heavy Whipping Cream
Also called heavy cream, contains at least 36% milkfat, whips well and serves as a topping, a thickener, and as an enrichment for sauces and desserts.

Figure 7-3 Soft peaks of heavy cream form when the cream is whipped to between one-half and three-quarter volume.

description of milk products. *See Figure 7-2* for information about cream varieties and *Figures 7-3, 7-4, and 7-5* for examples of whipping cream.

Receiving and Storing Milk, Cream, and Cultured Dairy Products

Proper temperature is key to maintaining the wholesomeness of dairy products. Receive fresh milk, cream, and cultured dairy products, such as yogurt and sour cream, at 41°F (5°C) or lower. Inspect canned milk products, such as evaporated milk and sweetened condensed milk, to look for dents, rust, bulges, and signs of leakage. Examine packages of dry milk for leakage, lumps, evidence of exposure to excess moisture, and signs of insects and rodents.

Label and date fluid milk, cream, and cultured dairy products, and then refrigerate them at 41°F (5°C) or lower. To minimize absorption of odors, store fresh dairy products in closed containers, away from other foods. Stored in this way, fresh milk products will keep for about one week. Freezing is not recommended.

Store canned milk products at least 6 inches off the floor and 6 inches away from the wall at 50°F–70°F (10°C–21°C). Inspect cans regularly for signs of spoilage. Unopened canned milk products keep for about one year at room temperature. Dry milk powder can be stored for several months, provided that it is kept in tightly closed containers in cool, dry storage.

Figure 7-4 Firm peaks of heavy cream are whipped to maximum volume.

Figure 7-5 Overwhipped cream loses its sheen, becomes grainy, and starts to curdle.

Foodborne Illness and Dairy Products

Bacteria grow well in dairy foods because of the high protein and moisture content. Unpasteurized dairy products are particularly vulnerable. Pasteurization destroys *Mycobacterium tuberculosis* and many other disease-causing microorganisms found in milk. However, pasteurized milk is not sterile and contains a small number of bacteria. Other organisms can infect milk after pasteurization. Improperly stored and handled milk products can harbor *Staphylococcus aureus*, *Salmonella* spp., *Escherichia coli*, and *Listeria monocytogenes*.

Fermented Dairy Products

Fermented dairy products, also called cultured dairy products, result from the addition of a starter bacterial culture to fluid dairy products. Under the right temperature conditions, these lactic-acid-producing bacteria reproduce rapidly, causing milk or cream to ferment. The fermentation process gives dairy products a tangy flavor and thicker consistency.

Cultured Products

Buttermilk, yogurt, sour cream, and crème fraîche are the most common cultured dairy products.

Buttermilk

Traditionally, buttermilk was a by-product of the butter-making process. Today cultured buttermilk is manufactured by introducing starter cultures from bacterial strains into skim or lowfat milk and then holding the milk at a controlled temperature for 12–14 hours. Buttermilk adds a distinctive, tart taste to baked goods and other dishes.

Yogurt

Yogurt is made from either whole, lowfat, or nonfat milk. It adds a tangy flavor to sauces and dishes and serves as a lowfat substitute for sour cream. Yogurt is made by the same process used to make buttermilk, incorporating different bacterial strains. These bacterial strains cause the milk to ferment and coagulate into a custard-like consistency. Some commercial yogurt contains live active bacteria. Medical research shows that yogurt with live active cultures helps the body to produce lactase, an enzyme that breaks down lactose. Yogurt with live active cultures aids digestion, especially in individuals who have difficulty digesting milk products because of lactose intolerance. Research also suggests that yogurt containing live active cultures speeds recovery from some forms of intestinal illness.

Sour Cream

Sour cream is made by using the same process and bacteria used for buttermilk, with a different base of cream with 18% milkfat. Sour cream is smooth, thick, and tangy.

Crème Fraîche

Commercial *crème fraîche* is cultured, heavy cream that resembles a thinner, richer version of sour cream. Widely available in Europe, crème fraîche is expensive in the United States. Crème fraîche is made by heating 1 quart (1 L) of heavy cream to 100°F (38°C), adding 3 ounces (100 ml) of buttermilk, and putting the mixture in a warm place until thickened, about 6–24 hours.

Cheeses

Cheese is a product derived from the interaction of milk and bacteria or an enzyme called rennet. The milk proteins (casein) coagulate, forming solid *curds*, which are then drained from the liquid *whey*. The main ingredient in all cheeses is milk. The milk most frequently used is often cow's milk, but goat and sheep milks are also employed. The quality and type of milk determines the flavor and texture of the cheese. The federal grades for American-produced cheeses are U.S. Grades AA, A, B, and C. Foreign-produced cheeses are not graded by the USDA.

Common Bakeshop and Pastry Shop Cheeses

Cheeses are divided into several categories: hard, firm, semisoft, ripened soft, and fresh soft. The hardness of the cheese is affected by the degree of ripening and other factors, such as aging. Ripening is also a way of classifying cheese. Some cheeses are ripened by the addition of bacteria or mold. Other cheeses—for example, Roquefort and Swiss—are ripened from the inside out. Still others, such as Camembert, Brie, and Limburger, are ripened from the outside in. The outside-in varieties soften with the ripening.

Hard Cheeses

Hard cheeses have been aged to reduce moisture content to about 30%. Hard cheeses are often used for grating. Maximum flavor comes from freshly grated cheese.

Parmigiano-Reggiano (Parmesan)
True Parmigiano-Reggiano is a cow's milk cheese from an area in Italy near Parma. It has a sharp, spicy taste, a very hard, dry texture, and is nearly always used for grating or shaving. Parmesan produced in the United States and elsewhere does not match the flavor of the original.

Pecorino Romano
Made in central and southern Italy from sheep's milk, Pecorino Romano has a robust and piquant flavor and is noticeably salty.

Manchego
Known as the cheese of Don Quixote, Manchego comes from the Spanish region of La Mancha. It is a sheep's milk cheese with a slightly mild, nutty flavor. Manchego is often served with dried fruit, nuts, figs, and fruit preserves.

Firm Cheeses

Firm cheeses have a moisture content of 30%–40%, for a firm solid texture. Their flavors can range from mild to quite sharp, depending on age.

Cheddar
With origins in Great Britain, cheddar is now the most popular cheese in the world. This cow's milk cheese ranges from mild to sharp in flavor and has a dense texture. Orange cheddars owe their color to vegetable dye. Uncolored cheddars are pale yellow. Colby is a popular mild American cheddar used in baking and cooking.

Emmental
Emmental is the original cow's milk Swiss cheese with very large holes made by gases that form during ripening. It has a mild, nutty taste and comes in 200-pound wheels enclosed in rind. Emmental is served on dessert trays.

Gruyère
Gruyère is a Swiss cow's milk cheese. It has a mild, nutty taste, moist texture, and small holes. Gruyère is classically used in quiche but can also be served as a dessert cheese or appetizer.

Provolone
Provolone is a cow's milk cheese from southern Italy. It has a pale yellow color and a flavor that ranges from mild to sharp, depending on age. Provolone is used in calzones and on pizza.

Swiss
Swiss cheese is a domestically produced product made in the style of the classic Swiss cheese, Emmental.

Semisoft Cheese

Semisoft cheeses have a moisture content of 40%–50%. Their texture is smooth and sliceable, but not spreadable. Semisoft cheeses can be classified into two groups: the smooth, buttery cheeses and the veined cheeses, which owe their distinctive appearance and taste to the veins of mold running through them.

Fontina
Fontina is a nutty, rich cow's milk cheese from Italy. It has a slightly elastic touch and a few small holes. The French make their own version, called Fontal. Use Fontina in cooking and as a dessert cheese.

Gorgonzola
Gorgonzola is a blue-veined cow's milk cheese from Italy. Aged Gorgonzola piccante is sharp in flavor, while younger Gorgonzola dolce is softer and sweeter. Its texture is smoother than that of many other blue-veined cheeses and it is often blended into milder cheeses to make cheesecakes and tarts. Gorgonzola is an excellent dessert cheese and often used on cheese trays.

Gouda

Gouda is a Dutch cow's milk cheese with a pale yellow color and a mellow, buttery flavor. Mature Gouda has a firm texture and a more pronounced flavor. Gouda is often packaged in red or yellow wax-covered wheels. It is used in cooking and served as an appetizer, with fruit, and as a dessert cheese.

Roquefort

Made from sheep's milk, Roquefort is a crumbly blue-veined cheese with a pungent taste and strong aroma. Roquefort is aged for three months in the limestone caves of Mount Combalou, where the humid air helps to develop the cheese's characteristic blue veins. Roquefort is a dessert cheese.

Stilton

Stilton is an English cow's milk blue-veined cheese. It has a crumbly texture, edible rind, and pungent tang. Traditional accompaniments to Stilton are fruit, walnuts, and port.

Ripened Soft Cheeses

Ripened soft cheeses have rich flavors and a buttery smoothness. They are characterized by thin rinds and soft, creamy centers.

Brie

Brie is a French cow's milk cheese with a white crusty rind and a buttery texture that oozes when the cheese is fully ripe. Brie has little flavor before it is ripe and stops ripening once cut. Overripe Brie develops a strong ammonia odor. Serve Brie when its center begins to bulge slightly. Brie is suitable on dessert trays and in pastry. Brie should be served at room temperature.

Camembert

Similar to Brie, Camembert is a cow's milk cheese that originated in the French village of Camembert. It has a slight tang but is generally milder than Brie. Its uses mirror those of Brie.

St. André

St. André is a French triple-crème cheese with a white downy rind and a slightly sweet, buttery taste. It is most often served as a dessert cheese.

Fresh Soft Cheese

Fresh soft cheeses are unripened cheeses with mild flavors and a moisture content of 40%–80%. The high moisture content gives these cheeses their soft texture and short shelf life.

CHÈVRE FRAIS

Chèvre frais is fresh goat cheese. It is soft and spreadable with a mild but characteristic goat cheese tang. Use chèvre frais on dessert boards.

FARMER'S CHEESE

Farmer's cheese, also known as baker's cheese, is a form of cottage cheese from which most of the water has been drained. Made from cow's milk, farmer's cheese has a mild taste with a slight tang. Farmer's cheese can be dry and crumbly or firm enough to be sliced. It is served as is or used in baking and cooking.

MASCARPONE

Mascarpone is an Italian cow's milk cream cheese with a rich, creamy taste and a silky, smooth texture. Prized for its flavor, Mascarpone is used in tiramisu.

MOZZARELLA

Mozzarella is the firmest of the fresh soft cheeses. Traditionally, mozzarella is a small oval cheese made with water buffalo's milk, although cow's milk is now a common substitute. Fresh mozzarella is white and quite mild. Commercial mozzarella has a much firmer texture and a blander flavor. Use mozzarella on dessert trays or shredded on pizza.

NEUFCHÂTEL

Neufchâtel is a cow's milk cheese, similar to cream cheese, from the Neufchâtel region of Normandy. Neufchâtel has a soft rind, creamy texture, and slightly tart flavor that builds as the cheese ripens. It can be used as a substitute for cream cheese in cheesecakes, fillings, and icings.

RICOTTA

Ricotta is an Italian cheese made from the whey left from other cheeses, such as mozzarella and provolone. Ricotta has a smooth, slightly grainy texture and is a bit sweeter than cottage cheese. Ricotta is used in variations of cheesecake and as a filling in the Sicilian specialty, cannoli.

CREAM CHEESE

Cream cheese is a soft, unaged cheese that is made from cow's milk. It is mainly used in cheesecakes, fillings, and icings.

Artisan Cheeses

Cheese making is an art that has long been practiced. Scholars believe that cheese production began in the Middle East or Central Asia and later spread to the European continent. Artisan cheese making is an art that is often passed down through generations of families. Throughout the centuries European

Figure 7-6 For maximum flavor, it is important to serve cheese at room temperature.

restaurants, hotels, and inns have served cheese as a dessert course. Traditionally, an artisan cheese course was served with port, sherry, or Madeira. Today the cheese course might also include wines, dried fruits, nuts, olives, spreads, crackers, or breads. The art of cheese making has recently gained popularity in the United States due to customer demand. *See Figure 7-6.*

Handling Cheese

Cheese requires special care to develop and maintain its distinctive characteristics and wholesomeness. Recommendations for ripening, storing, and serving cheese vary depending upon the type of cheese.

Ripening Cheese

Ripening is the stage in the cheese-making process in which added bacteria or molds begin to work on the fresh curds. Ripening develops the unique texture, aroma, and flavor that define a cheese. The length of the ripening phase and the conditions under which ripening occurs also contribute to the characteristics of a cheese. Uncut cheeses can continue to ripen after they leave the controlled cheese-making environment.

Receiving and Storing Cheese

Receive cheese at 41°F (5°C) or lower. Inspect packaged cheeses for soiled wrappers, leakage, broken tubs or cartons, or an ammonia odor. Check expiration dates, and return damaged products for credit.

Refrigerate cheeses at 35°F–41°F (2°C–5°C). Store cheese in its original wrapping until it is ready for use. After opening, tightly rewrap cheese in plastic wrap to keep it from absorbing odors, developing mold, and drying out.

Hard and firm cheeses generally keep for several weeks. Whole cheeses and large wedges maintain their freshness longer than shredded cheese. Soft and unripened cheeses stay fresh for 7–10 days. Although freezing is a safe way to extend the life of cheese, it changes the texture.

Butter and Margarine

Butter is made from cream that has been churned, or agitated, until the fat solids separate from the liquid, forming a solid mass. Butter is a water-in-oil emulsion with at least 80% milkfat. Butter also contains about 16% water and 2%–4% nonfat milk solids.

Butter is prized for the unmistakable flavor it contributes to a variety of baking and pastry products. Butter has a relatively low melting point, which contributes to its smooth, melt-in-your-mouth texture. Butter softens at 80°F (27°C), begins to melt at 88°F (31°C), and has a final melting point of 94°F (34°C). Because of butter's relatively low smoking point, which ranges from 250°F–260°F (121°C–127°C), oil is sometimes added to butter to prevent burning when vegetables are sautéed for a quiche or savory meat pie.

Butter Grades

Government regulations do not require the grading of butter. Even so, most manufacturers choose to pay for this service because graded butter carries the USDA shield on the package. Federal graders judge butter by its flavor, texture, keeping qualities, color, and aroma.

USDA Grade AA

The highest grade, Grade AA butter is made from high-quality fresh, sweet cream. It has a smooth, creamy texture, a pleasant aroma, and a subtle, sweet flavor.

USDA Grade A

Grade A butter is made from fresh cream and has a fairly smooth texture and a pleasant flavor.

USDA Grade B

Grade B butter is wholesome, although its flavor is judged to be slightly acidic. It is used most often in manufactured foods.

Types of Butter

Sweet cream butter and cultured cream butter are two main types of butter. Most of the butter made in the United States is made with sweet cream and sold in salted and unsalted forms. Many European butters, or European-style butters made in the United States, are made from cream to which a lactic acid starter has been added. European-style butters usually have a slightly higher fat content than domestic butters, ranging from 82%–88% milkfat. Both U.S. and European butters are usually made from cow's milk, although they can be made from other milks, such as goat and sheep milk.

Salted Butter

Salted butter contains up to 2.5% salt. Salting extends butter's shelf life but can also mask undesirable flavors or signs of rancidity. Some people prefer salted butter, both on the table and for cooking. Adjust the quantity of added salt when preparing dishes containing salted butter.

Unsalted Butter

Unsalted butter has a pure, clean taste. It is usually preferred for baking and cooking.

Whipped Butter

The light, fluffy texture of whipped butter results from the air that has been whipped into it. Whipped butter is mainly used on the table.

Clarified Butter

Clarified butter, also called drawn butter or ghee, is pure butterfat that has all water and milk solids removed. One pound of butter yields three-quarters of a pound of clarified butter. Clarified butter does not burn as easily as regular butter, so it is preferred for making sauces, frying, and other baking and cooking needs.

STEP PROCESS

Clarifying Butter

Because clarified butter has a higher smoking point than regular butter, it is frequently used for baking and cooking.

Follow these steps to make clarified butter:

1 Melt the butter in a heavy saucepan over medium heat. Cook until the foam rises to the surface.

2 Let the butter rest about 5 minutes, and then skim the foam from the surface.

3 Ladle the clarified butter into a container, being careful not to pick up the milk solids and water that have sunk to the bottom of the saucepan.

4 Clarified butter should be golden yellow and perfectly clear.

Margarine

Margarine may contain vegetable oil or animal fat, with flavorings, vitamins, salt, milk, whey, and emulsifiers. It was developed as a butter substitute and is divided into two main types: oleomargarine (made chiefly of beef fat and a blend of other oils), and vegetable margarine (made from a blend of vegetable oils, usually corn or soybean). The oils are pressed from the source, purified, hydrogenated, and then colored and fortified with vitamins. Vegetable margarine, with or without added milk products, is most commonly found in the United States. Margarine contains approximately 80% fat; 15%–20% water; and 2%–5% solids, salt, preservatives, and emulsifiers.

A variety of margarines is available to the baker and pastry chef. Baker's margarine, which is cakelike and soft, is excellent for creaming. Roll-in margarine, sometimes referred to as puff pastry margarine, is firmer than baker's margarine and may be used in laminated doughs to produce products such as puff-pastry, croissants, and Danish. Margarine blends are also available to bakers and pastry chefs. These blends are made mostly of vegetable margarine, but contain small amounts of butter to improve flavor and assist in melting.

Receiving Butter and Margarine

Receive butter and margarine at 41°F (5°C) or lower. Look for the USDA grading stamp on butter and the expiration date on both products. Check for off-odors and examine packaging for tears or signs of leakage.

Storing Butter and Margarine

Store butter and margarine tightly wrapped and away from foods with strong odors at 41°F (5°C) or lower. Freeze any product not intended for use within a few weeks. Frozen butter and margarine keep for two months.

Eggs

Eggs are among the most versatile and nutritious of foods. They contribute high-quality protein at a relatively low cost. In addition, eggs possess many qualities that make them indispensable in the bakeshop and pastry shop. They add color, flavor, richness, and moisture to baked items and cooked dishes.

Eggs are a very important and costly ingredient in baked goods and pastries. They are second only to flour as a structural component and can represent nearly half the cost incurred in cake production. Although most U.S. bakeshops and pastry shops use hen eggs, other types of eggs, such as duck or quail eggs, are in fact commonly used in other countries. Brown eggs and white eggs are virtually the same, the difference lies in the breed of hen. Information that follows is limited to that of hen eggs.

Eggs serve many purposes in baking and pastry:

- *Leavening:* Whipped or beaten eggs entrap air that expands when heated.
- *Creaming:* Eggs increase the number of air cells formed and coat these cells with fat that allows expansion.
- *Color:* The rich yellow color of the yolk provides a distinct color to baked goods.
- *Flavor:* Eggs impart odor and flavor to baked goods.
- *Nutritional value:* The egg white is a valuable source of protein.
- *Richness:* The yolk is 30% fat. Combined with other solids, eggs provide shortness to a mix and also act as a tenderizer.
- *Freshness:* Because eggs contain almost 75% moisture, they can bind and retain moisture to improve the shelf life of a product.
- *Structure:* The high moisture content of eggs also allows them to act as a binding agent.
- *Emulsification and thickening:* The *lecithin*, an emulsifier found in the yolk of the egg, allows fillings and sauces to come together, bind, and thicken.

The Composition of Eggs

Eggs are composed of three main parts: the shell, yolk, and white. *See Figure 7-7.*

Shell

The shell acts as a protective vessel for its contents. Shells are composed of calcium carbonate and can be white or brown in color, depending on the chicken's breed. The color

Figure 7-7 Composition of an egg.

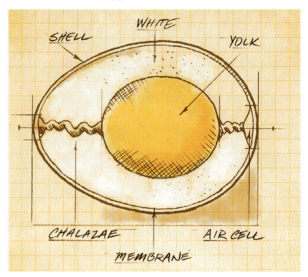

has no relation to the quality or taste of the egg. A thin membrane lines the shell and forms an air pocket at the large end of the egg. Because eggshells are porous, the air pocket enlarges. Over time, the eggs lose moisture and absorb flavor and odors. Egg processors coat shells with a thin film of mineral oil to extend the egg's shelf life.

White

The egg white is composed of water and a protein called **albumen**. The white also contains minerals, including sulfur. The thinnest part of the white is close to the shell. The thickest part surrounds the yolk. When raw, the white is clear. The white coagulates when it is heated, turning white in color and firm. When a raw egg white is whipped, the protein and water molecules bond together to create a stable foam.

Yolk

The yolk contains fat, including lecithin, as well as protein, vitamins, and iron. The yolk makes up about one-third of the total weight of the egg, but contains all the fat. It also encompasses all of the egg's cholesterol and most of its calories. The yolk is yellow, but the diet of the chicken determines the intensity of the color. The **chalaza** is a twisted white cord that holds the yolk in place. Yolks improve creaming and emulsification and act as thickeners in sauces. Because of their ability to hold air cells, whipped yolks create volume.

Forms of Eggs

Eggs come in fresh, frozen, and dried forms. Understanding the differences in egg products enables bakers and pastry chefs to choose the right one for the job.

Fresh Eggs

Fresh eggs, also known as shell eggs, are the most common form of egg used. Eggs begin to deteriorate almost immediately upon cracking. Great care must be taken to properly handle them both for sanitary reasons and to retain their valuable qualities. Fresh eggs should be stored at 41°F (5°C) or lower when not in use. Eggs generally have a shelf life of up to four weeks. Fresh eggs come in a variety of sizes, from peewee to jumbo. Most standard formulas assume the use of large eggs. *See Figure 7-8*.

Figure 7-8 Egg sizes are based on the net weight per dozen eggs.

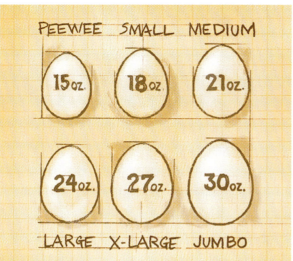

Frozen Eggs

Frozen eggs are broken, strained, mixed, and most often pasteurized. Frozen eggs are best thawed by ordering them in advance so that thawing may take place under refrigeration. Convenient to use, frozen eggs can be substituted for fresh eggs in many items, although their use in cooked fillings and custards should be avoided to prevent the formation of yellow spots in the final product.

Pasteurization Pasteurization destroys harmful bacteria in eggs. Pasteurized egg products safely replace raw eggs in formulas for unbaked meringue pie topping and mousses. Frozen, fluid, and dried eggs are pasteurized. Pasteurized fresh eggs are also available.

Whole eggs Frozen cracked and blended whole eggs can be thawed and used for baking and in cooked egg dishes.

Egg yolks In the bakeshop and pastry shop, pasteurized egg yolks and sugared egg yolks are used to produce a variety of items. Sugared egg yolks are pasteurized with 15% added sucrose. They are used in custards and ice cream.

Egg whites Frozen egg products that contain only egg whites can be thawed and used in baking or cooking. Like frozen egg yolks, they offer convenience and minimal waste.

Dried Eggs

Dried eggs are made from fresh eggs that have been dehydrated to remove moisture. Dried egg products include whole-egg, whites, and yolks. Whole dried eggs are primarily utilized for cake mixes and candy manufacturing. Dried egg whites are used for meringue and dried egg yolks to make egg wash. Bakers and pastry chefs must follow manufacturer's instructions when calculating the ratio of egg to water, which will vary based on the product used.

Egg Substitutes

Egg substitutes can accommodate the needs of individuals on low-cholesterol diets. Some egg substitutes replace egg yolks with vegetable or milk products but leave the white. Other egg substitutes replace both the yolk and white with soy or milk proteins. When cooked, egg substitutes taste and react differently than real eggs. Not all egg substitutes are interchangeable with real eggs in baking and cooking.

Testing for Freshness

Candling is a process used to determine the quality and freshness of an egg. While holding an egg up to a light, check the shell for cracks and tiny holes. Determine the location of the yolk. If the yolk is in the center and the air pocket is small, the egg is fresh. Turn the egg. If the white holds the yolk in place, the egg is of good quality.

Water can also be used to test egg freshness. Dissolve 3.5 ounces (100 g) of salt in a quart (1 L) of water. Place the egg in the solution, and observe what happens. Fresh eggs sink to the bottom, but old eggs float. Although old, an egg that floats may still be safe to eat.

After the egg is cracked, check the odor. Old eggs smell slightly of sulfur. Look at the contents. Fresh eggs have round, firm yolks surrounded by a firm egg white with a more liquid white around the outside. Old eggs have flat yolks and thin, runny whites. A cloudy white indicates that the egg is very fresh. Do NOT use an egg if the white appears pink or iridescent, as this is an indication of bacterial spoilage. Blood spots result from the rupture of a blood vessel in the yolk and do NOT indicate that the egg is unsafe.

The U.S. Department of Agriculture grades the quality of eggs on the basis of the appearance and characteristics of the yolk, white, and shell.

Receiving Eggs

When fresh eggs are delivered, the air temperature of the truck must be 45°F (7°C) or lower. Eggs should be refrigerated immediately upon delivery. Frozen eggs should be received frozen. Examine eggs carefully. Check packaging for the USDA stamp to verify that the eggs came from a USDA-approved processing plant. Inspect the packing date on fresh eggs. Eggs should be received within a few days from the packing date. Fresh eggs should be clean, dry, and uncracked.

Storing Eggs

Store fresh eggs in their original containers or in covered containers at 41°F (5°C) or lower. Keep frozen eggs between 0°F–10°F (–18°C–12°C) and dried eggs tightly sealed in the refrigerator or freezer. Store eggs away from foods with strong odors and separated from raw foods. Thaw frozen egg products in the refrigerator. Refrigerated eggs stay fresh for several weeks. Remove eggs from the refrigerator immediately before use and take out only as many eggs as needed.

Preparing and Cooking Eggs Safely

Eggs can harbor salmonella bacteria. Improperly stored, cooked, and served egg dishes can cause salmonella poisoning. Raw and undercooked eggs are especially risky. Follow these procedures to minimize the risk of foodborne disease.

- Keep fresh eggs refrigerated until immediately before use.
- Use only pasteurized eggs in formulas calling for uncooked eggs in ready-to-eat products such as meringues and mousses.
- When using fresh eggs for a sauce such as sabayon or a crème anglaise/vanilla custard sauce, cook to an internal temperature of 185°F (85°C).
- Discard eggs with broken or cracked shells.
- Carefully open eggs to keep shell from falling into liquid eggs.
- Wash hands, utensils, equipment, and work areas with hot, soapy water before and after contact with eggs.

STEP PROCESS

Separating Eggs

Follow these steps to separate eggs:

1 Crack the egg in half, and allow the white to fall into a bowl.

2 Transfer the egg yolk back and forth between shells until all of the egg white has been collected in the bowl.

3 Alternatively, eggs may be separated by cracking them in half and allowing the yolk to remain cradled within your fingers as the white drops into the bowl.

CHAPTER 8

Fruit

California, Florida, and Washington are the top three fruit-producing states in the United States. California is the leading supplier of grapes, strawberries, peaches, nectarines, and kiwi fruit and a major producer of citrus fruit, apples, pears, plums, and sweet cherries.

KEY TERMS

phytonutrients
supremes

achenes
fraise des bois

drupe
bloom

ethylene gas

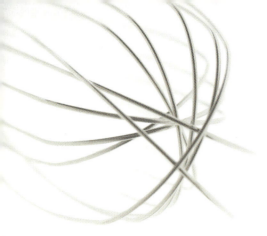

Fruit

Fruits are an essential ingredient in baked goods and pastries and are utilized in cakes, tortes, pies, crisps, cobblers, and sauces. A fruit dessert such as a poached pear is a healthy alternative for the customer who wants to follow a balanced diet. The natural sugars in fruit provide quick energy, and fruit acids aid digestion. Fruit is also an important source of fiber. Fruits are rich in antioxidants called **phytonutrients**, which appear to be active agents in fighting heart disease, cancers, and other age-related illnesses.

Types of Fresh Fruit

Fresh fruits are those that have not been canned, frozen, or dried. They may be ripe or unripe, depending on the type of fruit, how it was harvested, and the storage conditions used.

Fruits can be divided into eight categories: citrus fruits, melons, berries, drupes, pomes, grapes, tropical fruits, and exotic fruits.

Citrus Fruits

Citrus fruits grow in tropical and subtropical climates. Their origin has been traced to Southeast Asia, particularly Malaysia, more than 6,000 years ago. The classification includes lemons, limes, grapefruit, kumquats, tangerines, oranges, and several hybrids. Citrus fruits are identified by their thick rind, composed of a bitter white pith covered by a thin layer of brightly colored skin, or zest. Citrus fruits are acidic and have a strong aroma. The flesh is segmented and juicy, and varies in flavor from bitter, to tart, to sweet. Citrus fruits are a source of vitamin C, folic acid, dietary fiber, and potassium. The pulp, skin, and juice of citrus fruits are all edible.

Lemons

Lemons are the most commonly used citrus fruits. Their strongly acidic flavor makes them nearly unpalatable raw, but they are well-suited for flavoring desserts and confections. The zest adds flavor to pastries and savory dishes.

Eureka, Lisbon, Bearss, Avon, Harney, Villafranco, Meyer

Description and Purchasing Specifications Oval shape with a thin yellow to orange skin. Firm, heavy, fine-textured skin.
Packaging and Receiving 75, 95, 115, 140, 165, 200, and 235 count cartons
Grade and Size U.S. No. 1, Combination, No. 2
Storage and Usage 45°F–48°F (7°C–9°C), 89%–91% relative humidity. Garnish, juice, beverage enhancer, desserts.
Flavor Very tart, bitter. Meyer lemons are sweet in comparison to other lemons.
Yield 49%

STEP PROCESS

Zesting a Lemon

The thin strips of the rind, or zest, of a lemon add intense flavor to foods. When zesting, it is important to avoid the pith because it is very bitter.

To zest a lemon:

1 Wash the lemon very well before cutting.

2 Draw a five-hole zester across the skin to remove the zest, being careful to avoid cutting into the pith.

3 Use a French knife to chop the zest into tiny pieces.

4 The peel includes the bitter white pith. The zest is only the outer yellow skin.

LIMES

Small, thin-skinned limes can often be used interchangeably with lemons. Limes are too tart to eat raw, but desserts often include their juice. The zest adds color and flavor to a variety of desserts.

Key, Mexican, Persian, Tahiti

> **Description and Purchasing Specifications** Round or oval shape with a bright green skin. Fully ripe, heavy, smooth skin.
> **Packaging and Receiving** 10-lb. (4.4-k), 20-lb. (8.8-k), and 38-lb. (16.8-k) cartons (27, 48, 54, and 63 count)
> **Grade and Size** U.S. No. 1, Combination, No. 2
> **Storage and Usage** 58°F–60°F (14°C–16°C), 90% relative humidity. Used like lemons; can also be used in salad dressings.
> **Flavor** Tart, slightly sweeter than lemons.
> **Yield** N/A

GRAPEFRUITS

The grapefruit is a large, round hybrid of the orange and the pomelo. Grapefruits have yellow-pink skin, thick rind, and tart flesh. They are available in white-fleshed and the sweeter pink- or ruby-fleshed varieties. Grapefruit segments are available canned in syrup.

Duncan, Marsh, White, Pink, Ruby Red

Description and Purchasing Specifications Large, round, yellow to pink skin, acidic flesh. Firm, plump, heavy, well-shaped, smooth outer peel.
Packaging and Receiving 18, 24, 32, 40, 48, and 56 count boxes; 5-lb. (2-k, 225-g) and 8-lb. (4-k) bags
Grade and Size Fresh: U.S. Fancy, No. 1, No. 2, Combination, No. 3. Canned: Grade B
Storage and Usage 40°F–45°F (4°C–7°C), 85% relative humidity. Eaten raw, juiced, or in marmalade.
Flavor Juicy, acidic; sweetness varies.
Yield 47%

KUMQUATS Kumquats are very small, oval-shaped, orange fruits with a soft, sweet skin and slightly bitter flesh. They can be eaten whole, either raw or preserved in syrup, and used in jams and preserves.

Description and Purchasing Specifications Small, oval fruit whose flesh contains tiny, edible white seeds. Firm, unblemished skin, not shriveled.
Packaging and Receiving 1-lb. (450-g) and 2-lb. (900-g) baskets
Grade and Size None
Storage and Usage 40°F (4°C), 90%–95% relative humidity. Garnish, salad ingredient, raw or cooked, jams and preserves.
Flavor Sweet peel, sour pulp, tart flavor.
Yield 94%

MANDARINS, TANGELOS, AND TANGERINES Mandarins, a hybrid of the orange and the tangerine, have loose, smooth, deep orange skin that can be easily peeled to reveal sweet flesh. Mandarins can be eaten raw or they can be juiced.

Tangelos are a hybrid of the mandarin and the pomelo, with a smooth, deep orange skin that is easily peeled. Tangelos are large, though smaller than the pomelo, and slightly elongated on one end, resulting in a bell shape. Their juicy, seedless flesh can be an ingredient in cooked desserts or can be eaten raw.

Tangerines are small, dark oranges with loose, easily removed rinds. They are most often eaten fresh, but they are also available canned like mandarin oranges.

Description and Purchasing Specifications Small, very sweet oranges. Soft, heavy for size, loose skin with deep orange color.
Packaging and Receiving 54, 66, 80, 100, 120, and 156 count bushels or boxes
Grade and Size U.S. Fancy, No. 1, No. 2, No. 3
Storage and Usage 38°F–40°F (3°C–4°C), 90%–95% relative humidity. Raw, juice, salad, garnish.
Flavor Juicy and sweet.
Yield 72%

ORANGES
Oranges are round fruits with a juicy orange flesh and a thin orange skin. They may be either sweet or bitter. Oranges can be juiced for beverages or sauces, eaten raw, cooked in desserts, or used as a garnish.

Navel, Ruby Red, Temple, Valencia, Blood

Description and Purchasing Specifications Round, thick-skinned fruit. Firm, heavy, good color from pale orange-yellow to deep orange-red, unblemished skin.
Packaging and Receiving 40-lb. (17.7-k) case; California and Arizona: 48, 56, 72, 88, 113, 138, and 168 count. Florida: 100, 125, 163, 200, 252, and 324 count; 5-lb. (2.2-k) and 8-lb. (4-k) bags
Grade and Size U.S. Fancy, No. 1, Combination, No. 2, No. 3
Storage and Usage 45°F–48°F (7°C–9°C), 90% relative humidity. Eaten raw, juiced, or as a garnish.
Flavor Sweet, juicy, with an acidic bite.
Yield 70%

STEP PROCESS

Segmenting an Orange

Citrus segments are known as **supremes**. Orange segments may be added to salads, cooked in desserts, or used as a garnish.

To segment an orange:

1. Cut off both ends of the orange to create a flat surface and expose the flesh.

2. Lay the cut side flat on a cutting board, and remove the peel. Take caution to waste as little of the orange as possible while trimming away the entire pith. Follow the rounded shape of the orange.

4. Remove the orange segment from the membrane.

3. Cut alongside each membrane, cutting down to the center of the orange.

Melons

Melons are divided into two categories: sweet (or dessert) melons and watermelons. They vary in shape, size, color, and taste. Melons can be oval, oblong, or long and slender. They may have smooth, grooved, or netted surfaces. The color of a melon's flesh can be white, pink, green, orange, or yellow, and its flavor ranges from sweet to spicy.

Because melons are almost 90% water, cooking gives the flesh a mushy consistency. Most melons are served chilled and sliced, sometimes with a bit of lime or lemon juice. Melons blend well in fruit salads. Melons should always be vine-ripened, yield slightly to pressure at the blossom end, and give off a pleasant aroma.

CANTALOUPES (MUSKMELONS) Cantaloupes, or muskmelons, have a sweet, moist, orange flesh and a strong aroma. Peak season for cantaloupes is summer, but Mexican imports provide a year-round supply.

Unlike other sweet or dessert melons, muskmelons are further divided into the net-skinned varieties, including cantaloupes and Persian melons, and the smooth-skinned varieties, including casabas and honeydews.

Description and Purchasing Specifications Round-to-oval-shaped melon with a netted skin. Pleasant aroma, heavy, firm, good skin, sweet salmon-colored flesh.
Packaging and Receiving 12-lb. (5.3-k) to 15-lb. (6.6-k) crate; 18, 23, 27, 36, and 46 count. 38-lb. to 41-lb. half-cartons; 9, 12, 18, and 23 count. 53-lb. to 55-lb. two-thirds cartons; 12, 14, 18, 24, and 30 count. 80-lb. to 85-lb. jumbo crates; 18 to 45 count
Grade and Size Frozen: Grade A. Fresh: U.S. Fancy, No. 1
Storage and Usage 38°F–40°F (3°C–4°C), 90% relative humidity. Store in a warm, dark space if melon is hard. Eaten raw or in salads, as a dessert, or side dish.
Flavor Moist, mildly sweet.
Yield 58%

CASABA MELONS These teardrop-shaped melons have coarse yellow skin and a thick, ridged rind. The sweet flesh is creamy white to yellow with no aroma. Purchase melons with a deep skin color and no dark spots or moist patches. Peak season is September and October.

Description and Purchasing Specifications Large, round melon, yellow rind. Creamy white flesh.
Packaging and Receiving 4 to 8 count cartons
Grade and Size U.S. No. 1
Storage and Usage 40°F (4°C), 95% relative humidity. Eaten raw, as a side dish, or in salads, beverages, and desserts.
Flavor Sweet, juicy.
Yield 59%

CRANSHAW OR CRENSHAW MELONS Cranshaws are large, pear-shaped sweet melons with a strong aroma. The orange-pink flesh is very flavorful with a strong, spicy accent. The skin is mottled green and yellow with a ridged rind. Cranshaws are available from July through October, with peak season occurring in August and September.

Description and Purchasing Specifications Light yellow-red skin, soft pink flesh.
Packaging and Receiving 4 to 8 count carton
Grade and Size U.S. No. 1
Storage and Usage 40°F (4°C), 95% relative humidity. Eaten raw as a dessert, as a side dish, or in fruit salad.
Flavor Sweet, spicy flavor.
Yield 66%

HONEYDEWS Honeydews are large, oval, sweet melons with a smooth rind that ranges from white to pale green. The flesh is usually pale green, but it can also be pink or gold. Honeydews have a mild, sweet flavor. They are available almost year-round, with peak season from June through October.

Description and Purchasing Specifications Large, heavy melon with a waxy, creamy white to yellow skin and thick green flesh.
Packaging and Receiving 4, 5, 6, 8, 9, and 12 count crates
Grade and Size Fresh: U.S. No. 1. Frozen: Grade A
Storage and Usage 45°F (7°C), 90%–95% relative humidity. Store in a warm, dark space if melon is hard. Eaten raw as a side dish or dessert or in a salad.
Flavor Firm, moist, slightly sweet flavor.
Yield 58%

JUAN CANARY MELONS These rather large melons have a bright yellow skin that has a waxy feel when ripe. The flesh has a unique, sweet flavor that makes it suitable for sorbet when puréed or as an ingredient in fruit salad.

Description and Purchasing Specifications Oval melon, pale green to white flesh. Uniform canary yellow skin, fairly large and firm.
Packaging and Receiving 4 to 8 count carton
Grade and Size U.S. No. 1, Commercial, No. 2

Storage and Usage 50°F (10°C), 90%–95% relative humidity. Eaten raw, as a side dish, or in fruit salad.
Flavor Distinctive sweet flavor.
Yield N/A

PERSIAN MELONS   Persian melons are larger than cantaloupes, with dark green skin covered with yellow netting.

Description and Purchasing Specifications Finely netted, dark green rind with slight tan cracks. Firm orange/pink flesh.
Packaging and Receiving 4 to 8 count carton
Grade and Size U.S. No. 1, Commercial, No. 2
Storage and Usage 40°F (4°C), 95% relative humidity. Eaten raw as a side dish, dessert, or in fruit salad.
Flavor Very sweet and buttery.
Yield N/A

SANTA CLAUS MELONS   Santa Claus melons, or Christmas melons, are large, elongated sweet melons with a smooth, green-and-yellow-striped rind. The flesh is creamy white or yellow, with a taste similar to that of casaba melons. Santa Claus melon is a winter variety, with peak availability during December.

Description and Purchasing Specifications Oblong with a mottled green and yellow rind and light netting.
Packaging and Receiving 8 to 10 count carton
Grade and Size U.S. No. 1, Commercial, No. 2
Storage and Usage Ripen: 65°F–72°F (18°C–22°C), 95% relative humidity. Store: 50°F (10°C), 90%–95% relative humidity. Eaten raw as a side dish, dessert, or in fruit salad.
Flavor Sweet, light green flesh.
Yield N/A

WATERMELONS

Watermelons belong to a different genus than sweet melons. Native to tropical Africa, they are grown commercially in Texas and other Southern states. Watermelons are large, weighing up to 30 pounds (13.5 kilograms). They are round or oval-shaped with a thick rind. The skin may be green, green-and-white striped, or mottled with white. The crisp, juicy flesh may be pale pink to deep red. Most watermelons are flecked with small, hard, dark seeds, though more expensive seedless hybrids are available.

Charleston Gray, Carmon Ball, Seedless, Picnic, Icebox

Description and Purchasing Specifications Symmetrical shape, heavy for its size.
Packaging and Receiving 8 to 15 count cartons. Carmon Ball: 20 to 25 lb. is best
Grade and Size U.S. Fancy, No. 1, No. 2
Storage and Usage 50°F–60°F (10°C–16°C), 90% relative humidity. Eaten raw in salads, as a side dish, or dessert.
Flavor Sweet, moist flesh, usually pale pink to dark red.
Yield 49%

Berries

A good source of potassium, fiber, and vitamin C, berries grow on bushes and vines in many parts of the world. Berries are juicy with thin skins and numerous tiny seeds. Berries can be added to many dishes or used to make jams, jellies, and preserves. Berries are eaten raw, featured in desserts, and baked.

When purchasing fresh berries, avoid fruit that is soft or moldy. Berries do not ripen further after picking, so choose plump, fully colored ones. Highly perishable, berries can be loosely packed and refrigerated for two to four days or frozen to store them for longer periods.

Blackberries Blackberries resemble raspberries, but their color ranges from deep purple to black. They are larger and heavier with shiny skin. Wild blackberries are rather common, but there is only limited commercial production. Mid-June through August is their peak season.

Blueberries Blueberries are small, firm berries with deep blue skin and lighter blue flesh that originate in North America. They are cultivated from Maine to Oregon, with peak season mid-June to mid-August. Frozen blueberries are available year-round.

Boysenberries Boysenberries are large, juicy, and very sweet. They are often used for preserves and syrups as well as in baked goods and desserts.

Cranberries Cranberries are firm berries with blotchy red skin and a tart flavor. They grow in bogs in Massachusetts, Wisconsin, and New Jersey. Fresh cranberries are widely available from September through October, but they are seldom served raw. They are usually frozen, canned, dried, or made into jelly, and added to sauces, pies, pastries, and breads for their unique flavor.

Currants Currants are tart berries that are extremely small and grow in clusters on bushes. Most currants are a bright, jewel-like red, but varieties with a black or golden color are also grown. Sauces, jams, and jellies use all types of currants. Black currants, however, are the basis of *crème de cassis*, a liqueur. Peak season is late summer.

Raspberries Raspberries are extremely delicate berries that are available year-round. They are usually marketed frozen. A cool climate crop, raspberries are cultivated from Washington state to western New York, but they are also imported from New Zealand and South America. Domestically, peak season is May through November.

Gooseberries Fresh gooseberries, in both sweet and tart varieties, are available only for a short period during the summer. Sweet gooseberries can be eaten raw, and tart varieties are used in preserves and baked desserts.

Blackberries, Blueberries, Boysenberries, Cranberries, Currants, Raspberries, Gooseberries

> ***Description and Purchasing Specifications*** No mold or insect damage. Firm, plump, well-colored.
> ***Packaging and Receiving*** 12 pints per flat tray
> ***Grade and Size*** Fresh: U.S. No. 1. Canned: Grade B. Frozen: Grade A
> ***Storage and Usage*** 40°F (4°C), 90%–95% relative humidity. Eaten raw, in salads, jams, jellies, syrups, and baked desserts.
> ***Flavor*** Berries vary in flavor, from rather tart (cranberries) to very sweet and juicy (blackberries).
> ***Yield*** 92%–97%

STRAWBERRIES Strawberries are botanically classified as a perennial herb, though they are commonly considered berries. The skin is a vivid red color and is covered with minute, black seeds called *achenes.* Look for berries with good color that still have the green leafy hulls. Avoid soft or brown-spotted berries. The most flavorful and aromatic strawberries are the small alpine berries called *fraise des bois.* These are seldom available domestically.

> *Description and Purchasing Specifications* Fresh strawberries should have an intact green hull.
> *Packaging and Receiving* 5-lb. (2.2-k) to 10-lb. (4.4-k) boxes, or 24-qt. (22.8-l) crates
> *Grade and Size* U.S. No. 1, Combination, No. 2
> *Storage and Usage* 40°F (4°C), 90%–95% relative humidity. Eaten raw, in jams, preserves, and baked desserts.
> *Flavor* Quite sweet, succulent.
> *Yield* 87%

Drupes

A *drupe,* or stone fruit, is a type of fruit with one large, hard stone, or pit, often containing toxic acids. This category includes apricots, cherries, peaches, nectarines, plums, and olives. With their soft flesh and thin skin, drupes are delicate and bruise easily. Care must be taken when packing and transporting drupes because they are also highly perishable. They should not be washed until just before use. Drupes are delicious fresh or dried, and are used as flavoring for many brandies and liqueurs.

In North America, stone fruits are in season during the summer, but they need varying periods of cold temperatures in the winter in order to flower in the spring. Peaches grow best in Georgia, where winters are mild. Cherries grow better in the upper Midwest and the Pacific Northwest, where winters are cold. Apricots thrive in hot, dry regions such as Central California.

APRICOTS

Apricots are small, round fruits with succulent flesh. Their color ranges from yellow to bright orange. They can be eaten raw, or they can be poached, baked, stewed, candied, dried, or processed into jams and preserves. Apricot juice, or nectar, is widely available.

Apricots are often used in sweet and savory sauces, compotes, quick breads, pastries, custards, and mousses. Early summer is their peak season. Select firm apricots that are well shaped and plump. Avoid mushy or greenish-yellow ones.

Moorpack, Royal, Tilton, Blenheim

> *Description and Purchasing Specifications* Plump, firm, no worms or insect damage, bright golden yellow to orange color.
> *Packaging and Receiving* Fresh: 24-lb. (10.6-k) to 28-lb. (12.4-k) lugs, 12-lb. (5.3-k) flats. Dried: 5-lb. (2.2-k) boxes
> *Grade and Size* Fresh: U.S. No. 1, No. 2. Canned: Grade B. Dried: Grade A
> *Storage and Usage* Refrigerate at 40°F–45°F (4°C–7°C), 80%–90% relative humidity. Eaten raw or cooked in sauces, jams, or desserts.
> *Flavor* Very sweet, chewy fruit.
> *Yield* 94%

CHERRIES

Cherries are available in sweet and sour varieties. Northern states, especially Washington, Oregon, Michigan, and New York, are the largest suppliers. Sweet cherries are round or heart-shaped, and their color varies from yellow to deep red to almost black. The flesh is sweet and juicy. Deep red Bing cherries are the most popular variety, although yellow-red Royal Ann and Rainier cherries are also grown in some regions. Select plump, firm cherries with the green stems still attached. Refrigerate fresh cherries, and avoid washing them until ready for use.

Sour cherries are seldom eaten raw. The most common varieties are Montmorency and Morello. They are usually available canned, frozen, or as prepared pie and pastry filling that has been cooked with sugar and starch. All varieties of cherries are available dried.

Bing, Lambert, Rainier, Burlat, Brooks, Tulare, Lapins, Chelan, Sweetheart

> ***Description and Purchasing Specifications*** Bright skin, firm, smooth, green stem still attached, matured, no mold.
> ***Packaging and Receiving*** Fresh: 18-lb. (8-k) to 20-lb. (8.8-k) lugs and baskets. Frozen: 30-lb. (13.3-k) crates
> ***Grade and Size*** Canned or frozen: Grade A. Fresh: U.S. No. 1, Commercial
> ***Storage and Usage*** 40°F–50°F (4°C–10°C), 80%–90% relative humidity. Eaten raw or prepared as garnish or pastry filling.
> ***Flavor*** Sweet, juicy, crisp.
> ***Yield*** 88%

PEACHES

Peaches and nectarines are closely related, and although their flavors are slightly different, they can be used interchangeably. Peaches have a thin, fuzzy skin; the skin of nectarines is smooth. Flesh color varies from white to pale orange.

Fresh peaches and nectarines can be eaten raw, used in pies and pastries, and prepared as jams, preserves, chutneys, and relishes. They are available as freestones, with a stone that separates easily from the flesh, or clingstones, with a stone that is firmly attached to the flesh. Freestones are best eaten raw, but clingstones hold their shape better when cooked or canned. Choose fruit that has an appealing aroma and a color ranging from creamy yellow to yellow, to yellow-orange. Avoid those with green skin, which indicates that the peaches were harvested too soon, because they will not ripen further.

Alberta, Hale, Blake, Red Top, O'Henry, June Lady, Flavor Crest

> ***Description and Purchasing Specifications*** Creamy yellow with varying degrees of red blush or fuzz, fresh looking, not too hard.
> ***Packaging and Receiving*** 40-lb. (16-k) bushels, 48 to 80 count. 22-lb. (9.7-k) lugs
> ***Grade and Size*** Dried: Grade A. Canned or Frozen: Grade B. Fresh: U.S. Fancy, Extra Fancy No. 1, No. 1, No. 2
> ***Storage and Usage*** Ripen: 65°F–72°F (18°C –22°C), 95% relative humidity. Store: 32°F (0°C), 90% relative humidity. Eaten raw or in desserts, pastries, jams, or relishes.
> ***Flavor*** Very sweet and juicy with a hint of dryness about the skin.
> ***Yield*** 76%

> NECTARINES Nectarines are related to the peach. They have a smooth skin and sweet flesh.
>
> ***Description and Purchasing Specifications*** Red with orange-yellow shading, no green, firm and plump.
> ***Packaging and Receiving*** 19-lb. to 23-lb. lugs (50 to 84 count)
> ***Grade and Size*** U.S. Fancy, Extra Fancy No. 1, No. 1, No. 2
> ***Storage and Usage*** Ripen: 65°F–72°F (18°C–22°C), 95% relative humidity. Store: 40°F (4°C), 95% relative humidity. Eaten raw or in desserts, pastries, jams, or relishes.
> ***Flavor*** Crisp, juicy, sweet with a slight tang.
> ***Yield*** 86%

PLUMS

Plums have a round to oval shape with skin color that may be yellow, green, red, or deep purple. The flesh is sweet and juicy and varies from pale yellow to dark red. There are dozens of varieties of plums, but only a few are commonly available.

Plums are used in tarts, pies, cobblers, jams, and preserves. Fresh plums are available from June to October. Select plump fruit with smooth, unblemished skin. Plums will ripen at room temperature, and then they should be refrigerated. Dried plums, called prunes, are produced from special plum varieties.

Purple, Red, Yellow, Green, Angeleno, Kelsey, French, Stanley, Fellenburg (European), Burbank (Japanese)

> ***Description and Purchasing Specifications*** Full, deep color, no bruises or blemishes, flesh just beginning to soften. Thin red to purple skin, meaty green to yellow flesh.
> ***Packaging and Receiving*** 25-lb. (11.1-k) lugs
> ***Grade and Size*** Fresh: U.S. Fancy, No. 1, Combination, No. 2. Canned: Grade B. Frozen: Grade A
> ***Storage and Usage*** Ripen: 65°F–72°F (18°C–22°C), 95% relative humidity. Store: 32°F (0°C), 90% relative humidity. Eaten raw or in jams and jellies.
> ***Flavor*** Very sweet, meaty and juicy flavor.
> ***Yield*** 90%

Pomes

Pomes are tree fruits with a central core that contains many small seeds. The skin is thin, and the flesh is firm. The most familiar pomes are apples, pears, and quinces.

APPLES

There are hundreds of varieties of apples. They vary in shape, color, texture, taste, and nutritional value. Their variety, convenience, flavor, and availability are keys to the enduring popularity of apples. Apples grow in temperate zones throughout the world, though they grow best in cool climate countries such as China, Russia, Germany, France, England, and the northern United States (Washington, Michigan, and New York).

Apples are a source of vitamin C, folic acid, and potassium. The high fiber content of apples aids in the digestive process. Apples also contain antioxidants that can improve the immune system and help prevent heart disease.

Apples can be baked, stewed, and cooked in pies and sauces. They are pressed to make juice and cider. Commercial growers harvest apples before they are fully ripe and store them until they are sold. Though the peak season is autumn, fresh apples are plentiful year-round.

Golden Delicious, Granny Smith, Jonathan, McIntosh, Newtown, Red Delicious, Yellow Delicious, Winesap, York, Gala, Fuji, Rome

> ***Description and Purchasing Specifications*** Fragrant, well-colored. Color varies from pale greenish-yellow (Golden Delicious) to deep green (Granny Smith) to deep red (Winesap). Firm, no bruises or spots.

Packaging and Receiving 40-lb. (17.7-k) boxes; 88, 100, 113, 125, 138, 150, and 163 count. 40-lb. (17.7-k) bushels. 46-oz. (1.175-k) or 10-lb. (4.4-k) canned

Grade and Size Fresh: U.S. Extra Fancy, Fancy, No. 1. Canned: Grade A

Storage and Usage 31°F–32°F (−1°C–0°C), 85%–90% relative humidity. Eaten raw, cooked in pies, sauces; used for a garnish or as juices.

Flavor Crispy, mildly sweet to tart.

Yield 76%

Pears

Washington, Oregon, and California produce more than 90% of the pears sold in the United States. Pears are also grown in New York, Michigan, and Pennsylvania. The main international producers are Chile, Canada, New Zealand, Australia, and Argentina. Pears are a source of fiber and vitamin C and are available year-round.

Pears ripen at room temperature after harvesting. Choose pears with smooth, unbroken skin and intact stems. When fully ripe, pears have a good aroma and yield slightly to pressure near the stem. Pears can be eaten raw, alone or as a complement to cheese. They are added to fruit salads, preserves and pastries; made into cider; and baked.

Bartlett, Red Bartlett, Bosc, Comice, Forelle, Seckel, Anjou

Description and Purchasing Specifications Bright color depending on variety, firm, no bruises or damage to the skin.

Packaging and Receiving Boxes packed 20 to 245 count; 110 to 135 count is recommended.

Grade and Size Washington Extra Fancy, U.S. No. 1, No. 2, Combination

Storage and Usage Ripen: 60°F–65°F (16°C–18°C), 95% relative humidity. Store: 40°F (4°C), 95% relative humidity. Eaten raw or cooked in salads, compotes, or preserves.

Flavor Tart, sweet, juicy.

Yield 78%

Quinces Quinces are large, fragrant, yellow fruits that resemble pears but have an uneven shape. Quinces are too astringent to eat raw, but they develop a sweet flavor and delicate rosy pink tone when cooked. The hard flesh is very high in pectin, encouraging gelling when added to other fruit jams and preserves.

South America and Europe are the main producers of fresh quinces, which are available from October through January. Look for firm, yellow fruits. Cut away any small blemishes, taking care to avoid damaging the fruit. Quinces can be stored up to a month under refrigeration.

Grapes

Fresh table grapes are plump, sweet, and juicy. They are classified by color, with green grapes referred to as white and red grapes referred to as black. White grapes have thinner skins, firmer flesh, and less intense flavor than black grapes.

California supplies nearly all the grapes sold in the United States. Table grapes are used as an accompaniment to cheese or desserts. They are available year-round. Select those that are unblemished and still attached to the stem. A dusty white coating, or *bloom*, indicates that the grapes have been recently harvested. Grapes are always picked ripe and do not ripen further once off the vine. All grapes should be rinsed before use.

Champagne Grapes Champagne grapes, with their reddish-purple color, are classified as the Black Corinth variety and were originally used to make raisins and wine. These sweet, juicy grapes are so tiny that they are most often eaten by the bunch rather than one by one.

Concord Grapes The Concord grape is an American original, first cultivated in 1849 in Concord, Massachusetts. Deep purple in color, Concords are intensely sweet and juicy with a nice aroma. Concords are commonly used for jelly and juice.

Emperor Grapes Emperor grapes are eaten fresh in salads or as a dessert. Select large, plump, firm grapes. Red varieties are best when the red color dominates most or all of the grapes.

Red Flame Grapes Red Flame grapes are a seedless hybrid with a variegated red color. They are large, round, and slightly tart in flavor.

Thompson Seedless Grapes Pale green Thompson seedless grapes are sweet with a crisp texture. They are the predominant commercial table variety.

Champagne, Concord, Black, Red Emperor, Red Flame, Thompson Seedless

> **Description and Purchasing Specifications** Plump, well-colored, pliable green stem.
> **Packaging and Receiving** 18-lb. (8-k) to 22-lb. (9.7-k) lugs
> **Grade and Size** U.S. Fancy, No. 1
> **Storage and Usage** 36°F–40°F (2°C–4°C), 90% relative humidity. Eaten raw alone, in salads, or as garnish; prepared as juices, jellies.
> **Flavor** Ranges from very sweet and juicy (Concords) to firm and slightly tart (Red Flame).
> **Yield** 94%

Tropical Fruits

Modern transportation and refrigeration methods make tropical fruits available year-round. Tropical fruits are native to the tropical and subtropical regions of the world. The flavors of different tropical fruits usually complement one another, and they are used in baked goods and pastries.

Bananas
The most popular tropical fruit in the United States is the banana. Ranging from 7 to 9 inches in length, bananas are actually the berries of a large tropical herb. They have soft, sweet flesh and a thick, yellow skin that peels easily. Bananas are harvested before ripen-

ing and ripen at room temperature. Brown flecks in the yellow skin signal that they are fully ripe. They are available year-round.

Cavendish, Burro, Manzano, Red, Petites, Gros Michel

> **Description and Purchasing Specifications** Firm, uniform shape and color, strong peel, bright.
> **Packaging and Receiving** In bunches, 25-lb. (11.1-k) boxes, 40-lb. cases
> **Grade and Size** U.S. No. 1, No. 2
> **Storage and Usage** 50°F–60°F (10°C–16°C). Eaten raw, in salads, or baked in breads and muffins.
> **Flavor** Sweet and sticky.
> **Yield** 68%

PLANTAINS Plantains are also called "cooking bananas" because they are used as a vegetable in tropical or ethnic dishes. Plantains are larger than bananas, but they are not as sweet.

> **Description and Purchasing Specifications** No bruises, gouges, black spots, or mold on the skin.
> **Packaging and Receiving** 25-lb. (11.1-k), 40-lb. (17.7-k), and 50-lb. (22.2-k) cartons
> **Grade and Size** U.S. No. 1, No. 2
> **Storage and Usage** 50°F–60°F (10°C–16°C). Cooked as side dish when green; in desserts when ripe.
> **Flavor** Less sweet than bananas.
> **Yield** 72%

COCONUTS Coconuts are the seeds of the tropical coconut palm tree. The shell is a thick, hard, dark brown oval covered with coarse fibers. Inside the shell is a layer of sweet, moist, white flesh. The flesh surrounds clear liquid called coconut water. This liquid is different from coconut milk or cream, which is processed from the flesh.

Shredded or flaked coconut flesh, either sweetened or unsweetened, is readily available. Coconut is often featured in desserts, candies, and Indian and Caribbean cuisines. Look for fresh coconuts that are heavy and without cracks, moisture, or mold.

> **Description and Purchasing Specifications** Heavy for their size, no cracks, no wetness around the eyes.
> **Packaging and Receiving** 40-lb. (17.7-k) or 50-lb. (22.2-k) sacks and cartons
> **Grade and Size** N/A
> **Storage and Usage** Room temperature. Refrigeration will dry the liquid. Used in fruit salad, pastries, candies.
> **Flavor** Sweet, nutty flavor.
> **Yield** 53%

DATES

Dates grow on the date palm tree in lengths of 1 to 2 inches (2.5–5 cm). Their coloring when ripe is golden to dark brown, they have a papery skin, and a single grooved seed inside the flesh. Dates have a sticky texture and sweet taste and can be served with other fruits, stuffed with cheese as an appetizer, or baked in breads, muffins, and pastries.

Deglet Noor, Zahidi, Khadrawy, Halawy, Medjool

> **Description and Purchasing Specifications** No sugar crystallization or insect damage. Plump with smooth, golden brown skin.
> **Packaging and Receiving** Bulk: 15 lbs. (6.6 k). Chopped: 30 lbs. (13.3 k)
> **Grade and Size** Fresh: None. Dried: Grade A
> **Storage and Usage** 50°F–60°F (10°C–16°C), 30%–40% relative humidity. Fresh or dried, stuffed as appetizer, in baked goods and pastries.
> **Flavor** Rich, sticky-sweet flavor.
> **Yield** 90%

STEP PROCESS

Pitting and Cutting Mangoes

The pit of the mango is quite large with a relatively flat, oval shape, about 3/4 inch (2 cm) thick. It is located in the center, and conforms to the basic shape of the mango.

To pit and cut a mango, follow these steps:

1 Hold the mango in a vertical position. Slice downward gliding the knife along the pit. Turn the mango and repeat to separate the other half from the pit.

2 Take one section of the mango, and make lengthwise and then crosswise cuts, without cutting through the skin.

3 Fold the skin back, and cut away the cubes of mango.

Kiwis

Originally called Chinese gooseberries, kiwis are small, oval fruits with fuzzy brown skin and bright green flesh surrounding a white core that is bordered with numerous tiny black seeds. Sweet but bland, kiwis are often served raw or sliced for garnishes. Puréed kiwis add flavor to sorbets or mousses.

Hayward, Golden

> ***Description and Purchasing Specifications*** Ripe kiwis have soft, furry, and unwrinkled skin that gives under gentle pressure; 2 to 3 inches long.
> ***Packaging and Receiving*** Single-layer flat cartons. New Zealand: 5 lbs. (2.2 k) to 6 lbs. (2.6 k). California: 11 lbs. (4.8 k) to 12 lbs. (5.3 k) 25 to 46 count
> ***Grade and Size*** U.S. No. 1, No. 2, Fancy
> ***Storage and Usage*** 40°F (4°C), 90%–95% relative humidity. Eaten raw in salad or as garnish, puréed for sorbet and mousse.
> ***Flavor*** Sweet, slightly bland.
> ***Yield*** 84%

Mangoes

These oval- or kidney-shaped fruits have thin, smooth skin, with color that ranges from yellow to red to green. The bright orange, juicy flesh is firmly attached to a large pit. Mangoes usually weigh between 6 ounces and 1 pound. As they ripen, they lose their green coloring.

The flavor of mangoes is a slightly acidic blend of sweet and spicy. Puréed mangoes are added to drinks and sauces, and the sliced or cubed flesh can be used in desserts. Mexico provides most of the mangoes consumed in the United States. Peak season is from May through August. When purchasing mangoes, look for firm fruit that yields to gentle pressure and which has no blemishes or wrinkles. When ripe, mangoes have a good fragrance. They will ripen at room temperature, and can then be refrigerated for as long as one week.

Tommy Atkins, Keitt, Ataulfo, Haden, Kent, Van Dyke

- ***Description and Purchasing Specifications*** Strong fruity fragrance, tight skin, full and firm shape.
- ***Packaging and Receiving*** 10-lb. (4.4-k) to 12-lb. (5.3-k) flats. 24-lb. (10.6-k) lugs (16 count)
- ***Grade and Size*** None
- ***Storage and Usage*** 50°F–55°F (10°C–13°C), 85% relative humidity. Eaten raw in salads, chutneys, or desserts; puréed in sauces or drinks.
- ***Flavor*** Smooth texture, sweet taste.
- ***Yield*** 69%

Papayas (Hawaii)

Papayas are greenish yellow with a golden to pinkish flesh and a center cavity full of edible, silver-black seeds. Ripe papayas can be served raw sprinkled with lemon or lime juice. Puréed papayas add flavor to sweet or spicy sauces, or sorbets.

Papayas are available year-round, but peak season is April through June. As papayas ripen, their skin color changes from green to yellow. Papayas will ripen at room temperature, and they can be refrigerated for up to one week.

Waimanalo, Kapoho, Sunrise

- ***Description and Purchasing Specifications*** Good yellow-green to full yellow-orange color, firm and unblemished, no dark spots on the peel.
- ***Packaging and Receiving*** 10-lb. (4.4-k) cartons, 6 to 14 count
- ***Grade and Size*** Hawaii: Fancy, No. 1, No. 2. No U.S. grade
- ***Storage and Usage*** 90%–95% relative humidity. Mature green to yellow papaya: 55°F (13°C); partially ripe yellow papaya: 50°F (10°C); ripe to full yellow papaya: 45°F (7°C). Eaten raw or puréed in soups, sauces, and drinks.
- ***Flavor*** Sweet and juicy. Seeds and pulp are edible.
- ***Yield*** 67%

Passion Fruit

A passion fruit has a firm, tough, dark purple skin and is about the size of a large egg. The orange-yellow pulp surrounds large, edible, black seeds. Passion fruit has a rich, sweet flavor reminiscent of citrus. The fruit should be heavy with dark, shriveled skin, and a strong fragrance. Bottles or frozen packs of flavorful passion fruit purée are commonly available. Fresh passion fruit will ripen at room temperature and can then be refrigerated. Peak season is February and March.

Edgehill, Black Knight, Kahuna, Purple Giant

- ***Description and Purchasing Specifications*** Heavy, large, wrinkled dark purple to black indicates ripeness.
- ***Packaging and Receiving*** Not standardized
- ***Grade and Size*** U.S. No. 1
- ***Storage and Usage*** Refrigerate 41°F–45°F (5°C–7°C), 90%–95% relative humidity. Used in sauces, custard, ice cream.
- ***Flavor*** Very aromatic, distinctive sweet flavor. Seeds and pulp are edible.
- ***Yield*** 64%

Pineapples

Pineapples resemble oversized pinecones, with their rough, brown skin and prickly eyes. The sweet, juicy, pale yellow flesh surrounds an edible but tough core that is usually removed. Raw pineapple can be served alone, grilled, or sautéed. Cooked pineapple can also be used in gelatin. Pineapple juice can be mixed into punch or cocktails.

Pineapples do not ripen after harvesting, so they must be picked when completely ripe, which makes them very perishable. They are available year-round but are most plenti-

STEP PROCESS

Peeling, Coring, and Trimming a Pineapple

Pineapples have a rough, brown skin and prickly eyes. One must first peel, core, and trim the fruit before eating.

Follow these steps to prepare pineapple for consumption:

1 Cut off both ends of the pineapple to create a flat surface and expose the flesh.

2 Use a French knife and cut downward to remove the peel.

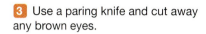
3 Use a paring knife and cut away any brown eyes.

4 Cut the pineapple in half lengthwise, and then cut it into quarters.

5 Remove the core from the quarters.

6 Slice the pineapple quarters to the desired thickness.

---

ful March through June. Look for heavy pineapples with a strong, sweet aroma, avoiding any with soft spots or dried leaves. The skin should have an even greenish brown or golden color. Use fresh pineapples immediately. Canned pineapple is available as slices, cubes, crushed, dried, or candied.

Smooth Cayenne, Gold, Queen, Red Spanish, Sugarloaf, Cabaiani

> **Description and Purchasing Specifications** Heavy for their size, well-shaped, fresh-looking, dark green crown leaves. Dry, crisp shell. Greenish yellow to golden color.
> **Packaging and Receiving** 35-lb. (15.5-k) crates, 9 to 21 count
> **Grade and Size** U.S. No. 1
> **Storage and Usage** 45°F–50°F (7°C–10°C), 85%–90% relative humidity. Eaten raw in salads or alone; also used in cakes and pies.
> **Flavor** Sweet, crisp, meaty flesh, white to yellow in color.
> **Yield** 52%

Exotic Fruits

Improved transportation and storage methods have introduced numerous exotic fruits to North American menus.

CHERIMOYAS Cherimoya, also known as a custard apple, is native to Central America. The conical cherimoya measures from 4 to 8 inches (10–20 cm) long, and weighs on average

from 5 to 18 ounces (140–510 g). Cherimoyas have two types of skin: smooth and marked with fingerprint-like swirls or covered with conical protuberances. Ripe cherimoya has sweet, juicy flesh and an appealing aroma. The glossy, dark seeds that stud the flesh are highly toxic and must be discarded.

FIGS Figs have an extremely sweet flavor, and their tiny seeds add crunchiness to the soft flesh. Figs are pear shaped with various skin colors, including black, yellow, green, and white. Sliced figs may be baked, poached, or made into jams and preserves.

Mission figs are a dark-skinned variety whose name stems from their cultivation in early mission settlements along the Pacific coast. Their thin skin surrounds sweet flesh and small seeds. They can be purchased fresh, dried, or canned. The more common Calimyrna fig is large and yellow with large seeds. These are also available dried.

Figs should be harvested when fully ripe, but ripened figs are delicate and hard to transport. Peak season for most figs is June through October, but June is the only month when fresh Calimyrna figs are available.

GUAVAS Guava is a small fruit with an oval or pear shape, an aroma that is strong, and flesh that is mild with some graininess. Guava juice is served plain or blended with the juices of other tropical fruits. Guavas are well-suited for use in jams and preserves. The fruit is best when it is fully ripened and slightly soft. It will ripen at room temperature.

FEIJOA Feijoa, or pineapple guava, is a distant relative of the guava. It has a flavor reminiscent of pineapple and guava, sometimes with a suggestion of mint. The fruit's size varies from just under an inch to about 3 inches long. Skin color is blue-green to blue or grayish green, occasionally with areas of orange or red. The waxy skin can be smooth or rough and pebbly. Feijoa has an intense and long-lasting aroma with grainy, white, thick flesh, and numerous edible seeds.

LOQUATS Loquats are about the size of an apricot, globular to pear shaped, with yellow-orange skin. The tender, juicy flesh is creamy to orange, with a flavor that ranges from sweet to tart, depending on the variety. The flavor is similar to that of apricots, cherries, and plums.

Because fresh loquats bruise easily when transported, they are usually only available in areas where they are grown. California and Florida have some commercial production, but limited imports come from Chile and Spain. Loquats ripen at room temperature. To prepare them, peel, remove the seeds, and slice into fruit salads.

LYCHEE The lychee, a native of southern China, may be round, oval, or heart shaped and measures from 1 to 1-½ inches long. This sweet, tender fruit is encased in a leathery rind that varies in color from pink to red. The white flesh is firm and juicy.

PERSIMMONS Persimmons have a shiny orange skin and are shaped like an acorn, although much larger in size. The jellylike flesh is bright orange with a mild taste similar to that of honey and plums. Ripe persimmons can be refrigerated. The fruit should be peeled and the seeds discarded before use.

Underripe persimmons are almost inedible; they have a strong tannic taste and chalky texture. When completely ripe, they are very soft, and the skin is almost translucent. Fresh persimmons are available from October through January.

POMEGRANATES Pomegranates are native to Iran. Round, bright red, and about the size of an orange, pomegranates have a hard shell encasing hundreds of small, brilliant red seeds surrounded by a juicy red pulp. An inedible yellow membrane surrounds groups of seeds and divides the fruit into segments. Pulp and seeds are tangy and sweet, and the seeds are crunchy. Seeds can be used as an attractive garnish, while the juice is popular in Mediterranean cuisines. Grenadine syrup is produced from concentrated pomegranate juice.

Select pomegranates that are heavy but not rock hard, cracked, or heavily bruised. Whole pomegranates can be refrigerated for several weeks. The fruit's peak season is in October, but pomegranates are available from September to December.

PRICKLY PEARS Prickly pears, also called cactus pears or barbary figs, are the berries of several kinds of cacti. They grow to the size of a large egg and have a barrel or pear shape. Their skin is green or purple, thick, and covered with small thorns and stinging fibers. Prickly pears have spongy, pink-red flesh with scattered, tiny black seeds. Ripe prickly pears have a sweet flavor. They can be peeled and diced and eaten raw or puréed for jams, sauces, custards, or sorbets.

Look for heavy fruits that have deep color. Avoid soft fruits or those with bruises. Ripe prickly pears can be refrigerated for about a week. Peak season is September through December.

POMELO The pomelo is native to Southeast Asia and is the largest of all the citrus fruits. An ancestor of the grapefruit, it ranges from nearly round to pear shaped and grows up to 12 inches (30.5 cm) wide. The thick skin may be greenish-yellow or pale yellow, the flesh color varies from greenish-yellow or pale yellow to pink or red, and the taste is sweeter and less acidic than that of grapefruit.

When purchasing pomelo, look for firm, unbruised fruit that has a sweet aroma and is heavy for its size. Ripe pomelos can be refrigerated for up to one week. Skinned segments can be broken apart and used in salads and desserts or made into preserves. Extracted juice makes an excellent beverage. The peel can be candied.

RHUBARB Botanically, rhubarb is a vegetable, although it is usually prepared as a fruit. Rhubarb grows abundantly in temperate and cold climates. The pinkish-red stalks are edible and do not need to be peeled. The leaves contain high amounts of toxic oxalic acid.

Rhubarb is extremely acidic and very bitter when raw. It requires large amounts of sugar to develop a characteristic sweet-sour taste. Rhubarb can be used for pies, stewing, or preserves, and it is sometimes combined with other fruits, such as strawberries. Choose crisp stalks without blemishes. Cooked rhubarb has an appealing pink color. Peak season for fresh rhubarb is February to May, but frozen rhubarb is readily available year-round.

STAR FRUITS Star fruits are also known as carambola. They have an oval shape with five raised ribs running the length of the fruit. When sliced into cross sections, the resulting shape is similar to that of a star. An edible, orange-yellow skin encases the light yellow flesh. Sweet but bland, star fruit's taste is similar to that of plums. Sliced star fruit is often used as a garnish or added to fruit salad.

Look for fruit that is deep golden yellow, with brown along the edges of the ribs. There should be a strong floral aroma. Unripe fruit will ripen at room temperature and can be refrigerated for up to two weeks. Fresh star fruit is available from August to February.

Purchasing and Storing Fresh Fruit

Fresh fruit may be purchased ripe or unripe. To get the maximum benefit from fresh fruit, it is important to consider several factors: size, the grade or quality, the stage of ripeness, and the nutritional content. All of these factors affect how the fruit can be used for cost-effectiveness, flavor, and presentation.

Grading Fresh Fruit

The USDA's voluntary fruit grading program grades on size, uniformity of shape, color, texture, and the absence of flaws. Most fruits purchased are U.S. Fancy, the highest grade. Lower grades are U.S. No. 1, U.S. No. 2, and U.S. No. 3. Fruits with grades lower than U.S. Fancy may have some cosmetic or other minor defects that do not affect the nutritional quality. They are adequate for use in sauces, jams, jellies, or preserves.

Purchasing Fresh Fruit

As fruits ripen, they change from hard and inedible to succulent and flavorful, often with an appealing aroma. Ripening fruit may change in size, color, weight, or texture. Most fruit is harvested long before it ripens to avoid damage during shipping and spoilage at the destination. *Figure 8-1* shows how various fruits ripen after harvesting.

Receiving

Fresh fruits are sold by count or by weight and are packed in flats, lugs, or cartons. Lugs are shallow wooden containers that hold various amounts of produce weighing up to 40 pounds. Flats are shallow boxes used to ship pints and quarts of produce such as berries.

Some fruits, such as melons, berries, and pineapples, may be cleaned, peeled, or cut before shipping. These prepared fruits are purchased in bulk, often with sugar and preservatives added. Prepared fruit is a labor-saving convenience that can be more costly than unprocessed fruit.

Ripening and Storing Fresh Fruit

For sensory appeal, flavor, and nutrition, fresh fruits should be used at the height of ripeness. Some fruits, such as pineapples, must be harvested when fully ripe and transported quickly to market because they only ripen on the plant. Others, such as bananas, are harvested when unripe and continue to ripen afterward.

Because the quality of fruits begins to deteriorate after they reach peak ripeness, it is important to know how long and under what conditions various fruits can be stored before spoilage sets in. As fruits ripen, they give off colorless, odorless *ethylene gas*. To speed up the ripening process, place unripened fruits near other fruits that emit high

Figure 8-1

Fruit Ripening Process

BANANAS
After harvest, change in color, texture, moisture content, and sweetness

APPLES, PEARS, CHERIMOYAS, MANGOES, KIWI, PAPAYAS
After harvest, become sweeter

APRICOTS, NECTARINES, PEACHES, MELONS (EXCEPT WATERMELON), BLUEBERRIES, FIGS, PASSION FRUIT, PERSIMMONS
After harvest, change in color, texture, and moisture content; no change in sweetness

CITRUS FRUITS, CHERRIES, GRAPES, SOFT BERRIES, PINEAPPLES, WATERMELONS, OLIVES
No changes after harvest

levels of ethylene gas, such as bananas or apples. To slow the ripening process, keep fruits chilled and separated from other fruits.

Purchasing and Storing Preserved Fruit

In addition to fresh fruit, foodservice operations often use preserved fruit that has been canned, frozen, or dried.

Canned Fruit

Most fruits can be canned successfully. Pineapple, peaches, and pears may be preserved in water, fruit juice, or heavy, medium, or light syrup. They may also be preserved in a solid pack can that contains little or no added water. The heat used in the canning process destroys microorganisms, and the sealed, airtight can eliminates oxidation. Canned fruit is usually softer than fresh fruits. Canned fruits are available in standard-sized cans. *See Figure 8-2.*

When receiving canned fruits, inspect the cans for any signs of damage such as dents or bulges. Return cans with bulges immediately without opening them because bulges can indicate the presence of the botulism toxin. Damaged cans in storage or prep areas should be discarded. Store cans in a cool, dry area. Once opened, any remaining fruit should be transferred to a clean storage container and refrigerated.

Frozen Fruit

Freezing is an effective method of preserving fruits. It inhibits the growth of the microorganisms responsible for spoilage, but it does not destroy nutrients. Freezing does alter the appearance and texture of fruits, however. This happens because ice crystals form in the water contained in the cells, causing the cell walls to burst.

Today many fruits are individually quick frozen (IQF), especially berries, and pear and apple slices. This method speeds the freezing process, reducing the formation of ice crystals. The grades for frozen fruits are U.S. Grade A (Fancy), U.S. Grade B (Choice or Extra

Figure 8-2

Standard Sizes of Canned Goods

no. 1/2		
Average Weight	*Cups per Can*	*Cans per Case*
8 oz. (225 g)	1	8
no. 1 tall (or 303)		
Average Weight	*Cups per Can*	*Cans per Case*
16 oz. (450 g)	2	2 or 4 doz.
no. 2		
Average Weight	*Cups per Can*	*Cans per Case*
20 oz. (565 g)	2-1/2	2 doz.
no. 2-1/2		
Average Weight	*Cups per Can*	*Cans per Case*
28 oz. (785 g)	3-1/2	2 doz.
no. 3		
Average Weight	*Cups per Can*	*Cans per Case*
33 oz. (930 g)	4	2 doz.
no. 3 cylinder		
Average Weight	*Cups per Can*	*Cans per Case*
46 oz. (1 k, 165 g)	5-2/3	1 doz.
no. 5		
Average Weight	*Cups per Can*	*Cans per Case*
3 lb. 8 oz. (1.258 k)	5-1/2	1 doz.
no. 10		
Average Weight	*Cups per Can*	*Cans per Case*
6 lb., 10 oz. (2.6 k)	13	6

Standard), or U.S. Grade C (Standard). When a government inspector has graded the fruit, the initials U.S. can be used. However, if the quality of the fruit meets the standards of the grade, packers may use the grade names, but not the initials, without outside inspection.

Processors often trim and slice fruits before freezing. Most frozen stone fruits have been peeled, pitted, and sliced. Other fruits, such as berries, are frozen whole. Packing fruit in sugar syrup before freezing not only adds flavor but also prevents browning. Frozen purées of some fruits are also available. Moisture-proof packaging is essential for frozen fruit, as is a constant temperature of 0°F–10°F (−18°C−12°C). Freezer burn can result from changes in temperature.

Dried Fruit

Drying substantially prolongs the shelf life of fruit. Almost any fruit can be dried, but the most common dried fruits are plums (prunes), grapes (raisins, sultanas, and currants), apricots, and figs. Fruit may be sun-dried or passed through a commercial dryer that quickly extracts moisture. The commercial procedure is more cost effective for producing large quantities of dried fruits.

Dried fruits remain moist and soft because they retain between 16% and 25% of their moisture. They are often treated with sulfur dioxide to retard the oxidation which causes browning, and to extend their shelf life. Dried fruits are very versatile. They are often used in baked goods such as breads, muffins, and pastries, added to salads or stewed for chutneys and compotes, or eaten raw. Before using dried fruits, reconstitute them by soaking them in hot water or another liquid such as wine, brandy, or rum. Some dried fruits benefit from being simmered for a short time before use.

When purchasing dried fruits, check the list of ingredients for sulfites, a preservative. People who are allergic to sulfites can suffer a life-threatening reaction to them. Store dried fruits in airtight containers, away from moisture and sunlight.

chapter 8　Fruit

CHAPTER 9

Herbs & Spices/Nuts & Seeds

Herbs and spices are used to enhance the flavor of baked goods and pastries and, as such, are considered flavorings. The addition of herbs and spices to bakery items can create new tastes and flavors without the addition of added fat or calories.

KEY TERMS

herbs

spices

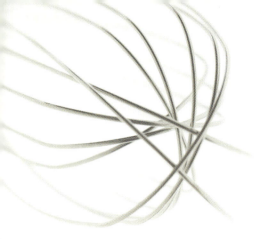

Herbs

Herbs come from a large group of annual or perennial aromatic plants. The leaves, stems, buds, or flowers are used whole, ground, or chopped to add flavor to many savory baked goods and desserts. They can be purchased fresh or dried. Herbs are available in a variety of forms and flavors to accentuate the flavor of baked goods and desserts.

Using Herbs

Although fresh herbs are preferable, they may not always be a viable choice. Dried herbs are commonly used because they are readily available and have an extended shelf life. If purchasing dried herbs, keep the quantities relevant to usage. After they have been opened, their shelf life decreases dramatically. Dried herbs have a more concentrated savory flavor than do fresh herbs. Cut the amount by either a third or a half when adding them to baked goods or desserts such as breads, pizza, and biscotti.

Storing Herbs

Proper storage and handling is essential in maintaining the full flavor of herbs. Fresh herbs should be checked for insects and bad spots prior to refrigeration. To maintain freshness, keep them loosely wrapped in a damp cloth or paper towel at a cool temperature (35°F–45°F/2°C–7°C). Dried herbs should be stored in airtight containers away from sunlight, moisture, and excessive heat.

Types of Herbs

CHERVIL
Chervil is parsley-like in appearance, slightly peppery in flavor, and available fresh, dried, crushed, or ground. It is use in baked goods such as breads, biscuits, quiche, cakes, and fruit preserves.

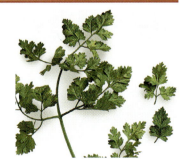

CHIVES
Chives are the long, green, stem-like leaves of the onion family that produce edible flowers and buds that are mildly onion-flavored. Chives are available fresh, frozen, or dried, and are used to flavor breads.

Dill

Dill is a feathery-leafed herb with a strong, distinct flavor that is available fresh, whole, dried, and chopped. Dill is used in specialty breads and butter.

Mint

Mint is a leafy, intrusive perennial herb with an array of flavors and intensity—spearmint, peppermint, apple mint, and chocolate mint are a few varieties. Suppliers often handle only the more common fresh varieties, peppermint and spearmint, and various dried versions. Use to flavor desserts.

Rosemary

Rosemary, an evergreen shrub of the mint family, has needlelike leaves and a strong, distinct flavor. The fresh form is preferred; however, it is available dried, chopped, or ground. Use in breads and infuse in olive oil and vinegar.

Thyme

Thyme is an aromatic, perennial shrub-like herb, that has many varieties (some with a hint of lemon). It is available fresh, dried, chopped, or ground. Pair thyme with other herbs, and use in savory baked goods.

Basil

Basil is a fragrant, leafy plant easily grown for consumption. It is a member of the mint family, and is available in many varieties. Some varieties contain a hint of cloves, mint, or licorice. Basil is available fresh, dried, and crushed and is often used on pizza and in preparations with tomato.

Oregano

Oregano is closely associated with wild marjoram. This aromatic green, bushy perennial has a distinct peppery flavor and is often associated with Mediterranean cuisine. It is available fresh, dried, chopped, or ground. Oregano is an herb commonly used on pizza and flat bread.

Spices

Spices are the berries, fruits, flowers, bark, seeds, and roots of plants or trees grown in tropical regions. They are available mostly in dried form and can be purchased whole or ground. Spices add flavor to various baked goods and desserts and can be added directly to foods during the cooking process or infused into hot foods in a sachet. Toasting certain spices can intensify the flavor of food.

Using Spices

The pairing of flavors is a learned skill. Some spice and dessert combinations are well-suited, and may be used together. These include nutmeg and custard, and allspice and puddings.

Storing Spices

Whole spices keep their flavors much longer than do their ground versions. Spices should be kept in airtight containers away from sunlight, moisture, and excessive heat. Stale spices can be quickly identified; they will often fade, lose their pungent aroma, and develop a bitter, almost musty flavor. They should be discarded and replaced.

Spice Blends and Mixes

Spice blends and mixes are a combination of aromatic ingredients—spices, herbs, and sometimes dried vegetables. Pie spice is a combination of sweet spices that is added to pie fillings prior to baking. The ingredients in pie spice usually include cinnamon, cloves, nutmeg, and ginger.

Types of Spices

Allspice

Allspice, the dried, unripe berry from the Jamaican pepper tree, has a clove-like flavor and is available whole or ground into powder. It can be used to flavor pickles, meats, consommés, casseroles, and sauces. Ground allspice is used mostly in baked goods and puddings.

Anise seeds

Anise seeds are small, dried seeds with a strong licorice-like aroma and flavor. Use sparingly to avoid the overwhelming licorice flavor. Anise seeds are available whole or finely ground and can be used in baked goods and confectioneries.

Cardamom

Cardamom is the aromatic pod from the perennial cardamom plant with a flavor similar to ginger and pine. It is available in pod, seed, or ground form. Use in yogurt and baked goods.

Cinnamon

Cinnamon is the sticklike quills or inner bark of the cinnamon tree. Use in baked goods, desserts, and preserves.

Fennel (seeds)

The fennel plant is used as a vegetable, and the seeds are employed as a spice. Both parts have a mild licorice-like flavor. Whole seeds are used in baked goods such as breads, rolls, biscuits, and crackers.

Ginger

Ginger is an underground stem or rhizome often associated with Asian cuisine. Fresh ginger is available peeled and sliced, pickled, or liberally coated in sugar crystals. Fresh ginger is added to fish and poultry and to Chinese or Japanese dishes. Ground ginger has a more global appeal, especially for baked goods.

Mace

Mace is the fibrous growth that forms around the shell of nutmeg. Mace is available dried and has a flavor similar to that of nutmeg, but not as strong. Use to flavor cakes and preserves.

Nutmeg

Nutmeg is a medium to large aromatic orb that is most fragrant and flavorful when freshly grated. It is also available whole or ground into powder. Pair with other sweet-pungent spices such as cinnamon, cloves, ginger, and cardamom. Use in custards, baked goods, and desserts.

Nuts and Seeds

A variety of nuts and seeds can be added to foods to enhance natural flavor, add color, and create texture. They also provide nutritional density to foods. Nuts are available shelled, unshelled, roasted, toasted, blanched, chopped (finely or coarsely), slivered, sliced, or processed into nut butters. The toasting and roasting of nuts brings out their natural flavors. Nuts are a bakery staple. When purchasing nut products, save on food costs by ordering broken pieces to use as chopped nuts in formulas.

Seeds can also be used to add flavor and textures to food or to garnish bakeshop items. The size and shape of the seeds create a visual interest. Poppy and sesame seeds, both white and black, are often used for baked goods but can be incorporated into savory dishes as well.

Storing Nuts and Seeds

Both nuts and seeds should be stored in airtight containers with limited exposure to light. Since they are prone to rancidity and pest infestation, purchase quantities appropriate to usage.

Nuts

There are many nuts to choose from. Keep in mind that any baked goods and desserts containing nuts or nut oils should be specified on the menu to avoid potential health risks. Nuts and nut oil can be highly toxic to those who are allergic to nuts. When purchasing nuts, make sure that they come from a current crop yield. Nuts that have not been shelled will keep longer than those removed from their shells or those that have been roasted.

Nut Butters

Nut butters are smooth spreads made from whole nuts that can be used in doughs, frostings, fillings, and ice cream. They have an extended shelf life and can improve flavor, richness, and taste. Nut butters are sometimes used as a substitute for oil and fat because they are plant based and do not contain cholesterol. Varieties of nut butters include roasted almond, peanut, pecan, filbert, and walnut. Nut butters are available in pails, cans, or jars.

Types of Nuts

Almonds

Almonds are tear-shaped single nuts with a medium brown covering. Their flavor ranges from sweet to bitter. Sweet almonds are added to foods while bitter almonds are used as a source for almond extracts and flavorings. Almonds can be purchased whole in shells or out of shells, skinned, sliced, chopped, or as a paste.

Brazils

Brazils are the seeds of enormous hardwood trees. The nuts are encased in a hard outer-shelled orb, containing a group of orange-like segments. They are harvested only after they fall to the ground. The whole nuts may be purchased in the shell or shelled and can range in size from medium to large. The larger nuts are most often used in confectioneries and baked goods.

Cashews

The cashew is the edible kidney-shaped seed of a tropical evergreen tree. Cashews, always hulled, are available either salted or unsalted. For baking, it is preferable to use the unsalted variety because salted cashews tend to contain too much salt.

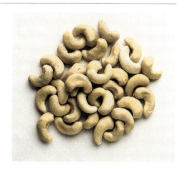

Chestnuts

Chestnuts contain more starch and less fat than most nuts. Large, rounded, and glistening brown, these nuts are often roasted whole. Chestnuts can be purchased fresh, dried, or canned. The naturally sweet, meaty nut pairs well with sweets and savory foods such as purées and butters. Chestnuts can be toasted, roasted, blanched, or steamed prior to peeling, but they are always cooked.

Hazelnuts/Filberts

These two names are used interchangeably to refer to nuts that are found in the wild (hazelnuts) or are cultivated from the European species (filberts). The largest commercial producer of hazelnuts is Turkey. They are most often used in baked goods, desserts, and confectioneries, but are also added to savory foods. Shelled filberts (hazelnuts) tend to be more shelf-stable than most other nuts.

Macadamias

Macadamias are creamy white nuts encased in a hard shell that must be removed by machine. They are sold shelled and usually vacuum packed in small glass jars. Macadamia nuts are high in fat and flavor. Combine them with chocolate, coconut, and fruits that are fresh or dried.

Peanuts

Peanuts, also called groundnuts, are members of the legume or pea family. High in protein, vitamins, and minerals, these nuts are considered highly nutritious. Unshelled in their fibrous coverings, peanuts keep well in cool, dark environments for several months. Shelled, they are not as shelf-stable and should be kept in cool areas or in the refrigerator. Well-known by-products of peanuts include peanut oil and peanut butter. Peanuts are sold in the shell, shelled, raw, roasted, salted, unsalted, or sugar-coated. Use as a flavor boost for subtle foods and to thicken sauces. Add to confectioneries, ice creams, puddings, and baked goods.

Pecans

These nuts can be used in sweet or savory dishes and are often added to cakes, cookies, and candies.

Pine nuts (pignoli)

Pine nuts are the tender kernels of pine cones that have a soft shell, no skin, and are high in oil. Pine nuts are fairly perishable and should be kept tightly covered in a cool, dark pantry, refrigerator, or freezer. Use in baked goods, desserts, confectioneries, and savory dishes.

Pistachios

Pistachios are usually green in color but can range from yellow to creamy beige. They are available in the shell or shelled, salted or unsalted. The naturally beige shell is sometimes dyed red with no benefit or purpose. Use in baked goods, desserts, and ice cream.

Walnuts

Walnuts are the edible fruits of the walnut tree. The most common species is the Persian walnut, also known as the English walnut. Walnuts are graded by size: small, medium, and large. They are available in the shell, shelled, halved, or chopped. Use in baked goods, desserts, and confectioneries. Add to savory foods and condiment products.

Types of Seeds

Like nuts, seeds can be added to many bakeshop or pastry shop products to enhance a food's natural flavor, nutritive value, or visual appeal.

Poppy seeds

Common poppy seeds are the small, round, gray to blue-black seeds of the poppy plant. The Indian version is a tiny off-white seed and is usually used as a thickener and in baked goods.

Pumpkin seeds

These are the seeds entangled in the fibrous strands of the interior of a pumpkin. They are available in their shells, shelled, salted, unsalted, roasted, or raw. Use as a garnish and in cookies, muffins, and breads.

Sesame seeds

These tiny flat seeds, either white, yellow, or black, come from an annual herb. The seeds have more flavor if toasted and can be used whole, pressed into oil, or pulverized into a flavorful paste called tahini. Many ethnic foods are enhanced by the tiny seeds. They can be sprinkled on baked goods or mixed with salt to form a seasoning mix.

Sunflower seeds

The seeds of the sunflower are grown primarily to be pressed into oil. Sunflower seeds are also a popular snack food when toasted and salted. Native to North America, this member of the daisy family is prized in Russia, where the seeds are used in sweets.

chapter 9 Herbs & Spices/Nuts & Seeds

CHAPTER 10

Liqueurs, Wines, Spirits, Coffee & Tea

Alcoholic flavorings are often used in the bakeshop or pastry shop. They may be incorporated as an ingredient (in a sabayon), or as a flavor medium (for poaching). Many liqueurs, wines, and spirits complement bakeshop or pastry shop items. Some are standard shelf products, while others are rare and reserved for signature items.

KEY TERMS

liqueurs
brandy
Champagne

spirit
arabica
robusta

green tea
oolong tea
black tea

withered

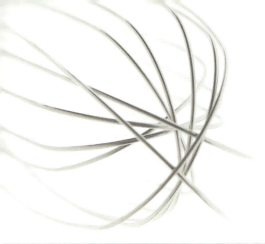

Liqueurs

Liqueurs are very popular as flavorings because their flavors blend well with chocolate, vanilla, and fruits. There is no international definition for liqueurs, although the producing countries usually regulate their production requirements through legislation. Liqueurs are characteristically much sweeter than other spirits and wines, and can vary widely in alcohol content.

Although liqueurs come in a great variety of flavors, they generally fall into these classifications:

Fruit-flavored These liqueurs are popular and bear a label that identifies the fruit that flavors the liqueur, such as Midori®, made with melon.

Seed-based Although one seed predominates in the liqueur, the beverage usually is made from several kinds of seeds.

Herbs Among the most complex, herbal liqueurs may contain a dozen or a hundred herbs; except for mint or anise seed, however, the herbal flavor does not predominate.

Crème Called crème liqueurs for their creamy texture and sweet taste, crèmes take their name from the dominant flavoring ingredient, such as a fruit (crème de banana).

Peels The most frequently used peels are those in the citrus family, such as orange or lemon.

Fruit brandies These fall into three general groups: brandies flavored with apples and pears; those using stone fruits like plums, peaches, or apricots; and those flavored with berries, such as blackberries or elderberries.

Nut liqueurs Among these are such favorites as Amaretto (almonds) and Frangelico (hazelnuts). *See Figure 10-1* for a list of some liqueurs often used in the bakeshop and pastry shop.

Brandy

Brandy is, by definition, a distillate of wine, although certain types of brandies are distilled from fruits. The French refer to brandy as *fine* to distinguish it from fruit brandies. Brandies are excellent flavorings for many desserts and baked goods and may serve as an accompaniment with desserts and after-dinner friandise. The making of brandy is strictly controlled in France, but each house or firm has its system of grading.

- VS—Very Special or Superior is aged no less than 3 years.
- VSOP—Very Special (Superior) Old Pale cannot be aged for less than 5 years.

Liqueurs

Figure 10-1

Liqueurs Used in the Bakeshop and Pastry Shop

AMARETTO (AMARETTO DI SARONNO)
- *fruit-based liqueur made in Saronno, Italy*
- *made from both sweet and bitter almonds*
- *has a sweet almond flavor*

CHAMBORD
- *raspberry-flavored liqueur*
- *made from black raspberries*
- *has a dark, rich color*
- *has a distinctively sweet taste*

COINTREAU (SOMETIMES CALLED CURAÇAO)
- *made from a blend of the peels of the bitter oranges of the island of Curaçao and other oranges*
- *chiefly from France and Holland, the house of Cointreau in Angers gave the liqueur its family name to distinguish it from the many versions, including Triple Sec*
- *clear and has a bitter-sweet orange flavor*

CRÈME DE BANANA
- *sweet, white- or yellow-colored liqueur*
- *banana flavor*

CRÈME DE CACAO
- *sweet chocolate-flavored liqueur*

CRÈME DE MENTHE
- *sweet mint-flavored liqueur*
- *can be clear or bright green*

CRÈME DE NOISETTE
- *made from hazelnuts*

FRANGELICO
- *hazelnut-flavored Italian liqueur*

GRAND MARNIER
- *orange-flavored liqueur*
- *made with cognac*
- *two types*
- *Cordon Rouge and Cordon Jaune (Cordon Jaune has a lower alcoholic strength)*

VANDERMINT
- *Dutch chocolate mint liqueur*

SAMBUCA
- *licorice-flavored liqueur from Italy*

- XO—Extra Old or Extra Ordinary are house blends that must be aged at least 6 years, but are usually aged 12 years or more.

Armagnac is one of the great brandies of the world, made in the Gers department in the southwest of France. Armagnac is slightly gentler in character than cognac, with a pronounced but delicate aroma.

Cognac is considered to be the finest of all brandies. It is made from wines in the Charente region in the west of France. The geographical area is divided into seven specific cognac producing regions that are strictly regulated by law. Most cognacs are blends of several brandies. The skills of each firm's blender marries different brandies to establish and preserve the brandy's qualities.

Fruit brandies, referred to as *alcohol blanc* and *eaux de vie* by the French, are actually not brandies, but more like liqueurs because they are distilled from fruit rather than from wine. Kirsch, or Kirschwasser, is distilled from cherries; Framboise from raspberries; Fraise from strawberries; and Poire Williams from pears.

Wines

No single volume can possibly contain the amount of information needed to accurately represent the world's wines; this section merely suggests the use of wines as an excellent flavoring ingredient. Wine is obtained from the fermentation of the

Port pairs well with a variety of foods including cheeses, nuts, and even chocolate

Wines can be paired with many desserts or enjoyed in place of a dessert

Calvados is distilled from apples, Navan is a cognac with the infusion of vanilla

juice of grapes. The fermentation takes place according to the laws controlling the district or region of the grape's origin and in accordance with that region's local traditions and practices. There are many growing regions throughout the world, but some of the chief regions include France, Italy, and the United States. Wines pair with many fruits and flavors, including chocolate, and add a distinct taste to many desserts.

Champagne is, technically, the wine that comes from a defined region in the north of France called Champagne. Most Champagne is a blend of Pinot Noir (black) and Chardonnay (white) grapes. Wines produced outside this region are sometimes referred to as Champagne, but should be more accurately termed sparkling wines. Champagnes and sparkling wines are bottled before second fermentation takes place so the fermentation occurs in the bottle. The expellation of carbon dioxide gives the wine its characteristic bubble. Champagnes and sparkling wines are sometimes used in making desserts and dessert sauces, and are especially good accompaniments to desserts.

Madeira is a fortified wine (fortified with brandy) from the island of Madeira off the coast of Portugal. It is characteristically sweet and often used to flavor cakes and custards. Marsala, another fortified wine, is made in the Sicilian town of Marsala. The sweet version of this wine is used in classic zabaglione.

Sherry, like Champagne, must be from a defined region in southern Spain around Jerez de la Frontera, although many sherries are also made outside this region and in many other countries. The term *sherry* has come to be associated with fortified wines that have characteristics similar to a true sherry. Sherries are excellent flavoring components and accompaniments to many desserts.

Spirits

Webster's defines a *spirit* as the "liquid containing ethyl alcohol and water that is distilled from an alcoholic liquid or mash." There are many so-called spirits, and many are useful in baked goods and dessert making.

Rum is perhaps the most commonly used spirit in the pastry shop because it complements ingredients such as vanilla, chocolate, fruits, and creams. Rum is distilled from sugarcane and was once, like molasses, a by-product of sugar making. Today, rum is produced outright for its own profitability. Puerto Rico, the Virgin Islands, Jamaica, Cuba, Barbados, and Guyana all produce quality rums, both light and dark.

Whiskey is distilled from grain, usually wheat, barley, or rye; although most whiskeys made in the United States are distilled from corn. Scotch Whiskey is perhaps the most well known and is made from either malted barley or unmalted barley with the addition of grains. Scotches are sometimes aged for up to 20 years.

Another popular whiskey is bourbon. Bourbon is made only in the United States, originating in Bourbon county. Kentucky Bourbon is distilled from at least 51% corn. Both scotch and bourbon are frequently used as flavorings in cakes, cookies, and dessert sauces.

Coffee

Coffee trees produce berries that resemble cherries. The berries, bright red when ripe, normally contain two seeds, or beans, that are processed, roasted, ground, and brewed into liquid coffee. Although more than 25 wild species of coffee trees have been identified, only two, *C. arabica* and *C. robusta* (now reclassified as *C. canephora*), are cultivated for commercial use.

Initially planted from nursery-grown seedlings, the coffee tree begins producing fruit in four to six years and continues to do so for about 40 years. A self-pollinating evergreen, the coffee tree blooms with fragrant white flowers and produces about 2,000 berries—enough to produce a pound and a half of roasted coffee beans. Most growing regions have a single yearly harvest, but regions with two distinct rainy seasons have two harvests.

Coffea Arabica

About 75% of the world's coffee production comes from *C. arabica* plants. **Arabica,** the highest-quality coffee, grows best in a narrow band on either side of the equator at altitudes ranging from 2,000 to 6,000 feet (610 to 1,829 m). In Central and South America and East Africa, where the majority of coffee grows, year-round temperatures of about 70°F (21°C) and abundant rainfall provide the plants with ideal conditions. Most *arabica* fruits ripen at different times on the same plant and grow in rugged terrain, so harvesting is labor-intensive. Only ripe berries yield superior coffee, and they are usually picked by hand. *Arabica* is also lower-yielding and more susceptible to disease than *robusta*. Still, consumers willingly pay a higher price for *arabica*'s sweet aroma and winey, fresh, and slightly acid taste.

Coffea Robusta

Coffea canephora, or **robusta,** is hardier than *arabica*, and grows well at lower altitudes in wet valleys and tropical forest climates. Like *arabica*, a *robusta* plant begins to mature four to six years after it is transplanted. It is easy to cultivate, and growers may use machines to harvest it. Large coffee roasters often use these less expensive *robusta* beans as a supplement in their blends and for making instant coffees. *Robusta*'s heavy, earthy aroma and flavor also make it ideal for making espresso with its characteristic thicker viscosity.

Varietal and Specialty Coffees

About 10% of the world's coffee bean crop is classified as gourmet or specialty quality coffee. A varietal coffee is a coffee from a single growing region. Antigua Guatemalan, Jamaican Blue Mountain, Hawaiian Kona, Costa Rican Tarrazu, Kenyan AA, Ethiopian Yergachev, or Sumatran Gayo Mountain are all examples of varietal coffees.

In the bakeshop and pastry shop, coffee is used to flavor ice creams, granitas, mousses, puddings, custards, sauces, soufflés, cakes, and candies.

Figure 10-2 Major tea-growing regions of the world

Tea

Tea is processed from the leaves of *Camellia sinensis*, a tree or shrub that grows best at higher altitudes under damp, tropical conditions. Tea leaves are hand-harvested or plucked from the plants' youngest shoots, called the first and second flush. About 4,200 pounds (1,905 kg) of fresh tea leaves produce 1,000 pounds (454 kg) of finished tea. Although all tea comes from basically the same plant species, there are approximately 1,500 grades and 2,000 different blends of teas. Three general classifications of tea define the level of enzymatic oxidation that leaves undergo during processing. Tea grows in many other regions, but India, China, Sri Lanka, Indonesia, and Kenya account for most of the world's commercial production and exports. Teas often take their names from the regions in which they are grown. *See Figure 10-2.*

Green Tea

Green tea is unfermented and yellowish-green with a slightly bitter flavor. Steaming or heating the leaves immediately after picking prevents enzymatic oxidation that turns the leaves into black tea. After heating, the leaves are rolled and dried. Green teas need to be stored properly and served fresh. Green teas are produced primarily in China, Taiwan, and Japan.

Examples of green tea include:

Gyokuro the finest grade of exported Japanese tea.
Tencha the powdered tea used in Japanese tea ceremonies.
Sencha the most common Japanese tea, popular in restaurants and sushi bars.
Gunpowder the highest Chinese grade; it is rolled into tiny balls.
Imperial a grade from Sri Lanka, China, or India.
Shou Mei Chinese green tea known as "old man's eyebrows."

Oolong Tea

Oolong tea combines the characteristics of black and green teas. The leaves are partially oxidized; then the process is interrupted, and the leaves are rolled and dried. Oolong is often flavored with scented agents such as jasmine flowers. It is also graded by leaf size and age.

Examples of oolong tea include:

Formosa Oolong a delicate, large-leafed oolong tea with a taste reminiscent of ripe peaches. Unique and expensive, it is suited for breakfast or afternoon tea.
Black Dragon a tea from Taiwan.
Pouchong grown both in China and Taiwan.

Black Tea

When steeped, *black tea* is strongly flavored and amber or coppery brown in color. Plucked leaves are dried, or *withered,* and then rolled in special machines to release the enzymes that give the tea its color, distinct taste, and aroma. The leaves are then "fired," or heated and compressed, to their final shape.

Teas are also sorted by leaf size and according to their brewing time. Larger leaves take longer to brew than smaller leaves. Size classifications include souchong (large-leaf tea), pekoe (medium-leaf tea), and orange pekoe (the smallest whole-leaf tea).

Broken tea leaves are categorized as either fannings or dust and are used for tea bags. Broken orange pekoe leaves produce a darker brew and are also commonly packaged in tea bags. Black teas are the best known and most popular in the West, but most teas sold in the United States are blends. Even the same tea blend or type of tea will vary in taste from one blender to another. Some of the most popular types of teas are:

Assam a rich black tea from Northern India, valued as a breakfast tea by connoisseurs.
Ceylon a full-flavored black tea with a delicate fragrance; ideal for iced tea because it does not turn cloudy when cold.
Darjeeling a full-bodied black tea with a Muscat flavor, grown in the foothills of the Himalayas.
Earl Grey a popular choice for afternoon tea, Earl Grey is flavored with oil of bergamot.
English Breakfast a full-bodied, robust blend of Indian and Sri Lankan teas.
Lapsang Souchong a tarry, smoky flavor and aroma, best suited to afternoon tea or as a dinner beverage.

Pastry chefs use tea to flavor ice cream and custards.

CHAPTER 11

Cooking Techniques

Many of the cooking techniques used in the kitchen are also common in the bakeshop or pastry shop, although the ingredients utilized vary. Bananas may be coated with sugar and caramelized for flavor/sweetness and then roasted and served with ginger ice cream. Fruit is often poached in a simple syrup with spices and served with panna cotta or a cookie. Cheese beignets that are deep-fried may be accompanied by an assortment of poached fruit and herbs for a sweet and savory dessert.

KEY TERMS

conduction
induction cooking
convection cooking
radiation
baking

carryover cooking
roasting
sautéing
deep-frying
recovery time

grilling
broiling
poaching
simmering
boiling

blanching
steaming

179

Heat Transfer

Cooking transfers energy from a heat source to food. Successful cooking in the bakeshop or pastry shop depends in part on an understanding of a variety of cooking methods. There are four ways to transfer heat to and through food: conduction, induction, convection, and radiation.

Conduction

Heat transfer by *conduction* takes place because of the direct physical contact of one item with another, followed by the inward movement of heat throughout a food item.

In a filled saucepan on a stove top, for instance, heat moves from the burner to the saucepan. The metal of the saucepan then conducts heat to the food. Because this type of conduction requires direct contact, it is a fairly slow means of heat transfer. Some materials conduct heat more rapidly than others. Aluminum and copper conduct heat faster than stainless steel. Glass and porcelain conduct heat more slowly. Liquids transfer heat more rapidly than gases.

Induction

Induction cooking uses magnetic energy to rapidly heat cookware. A glass or ceramic stove top houses electromagnetic energy coils under the unit's surface, that produce a high-frequency electromagnetic field that penetrates the metal of the cookware and creates a circulating electrical current to produce heat. The heat from cookware is transferred to the food by conduction. Benefits of induction include efficiency, speed, and safety. Aluminum and copper cookware cannot be used because they lack magnetic qualities.

Convection

In *convection cooking,* heat spreads through air or water. Convection can be natural or mechanical.

Natural Convection

Warm liquids and gases tend to rise, while cooler liquids and gases fall. This continuous, natural movement distributes heat. When a pan of water is placed on an electric burner or gas flame, for instance, the liquid at the bottom of the pan becomes warm and rises to the top. The cooler liquid at the top of the pan then falls to the bottom. This natural process creates a circular motion that spreads heat throughout the liquid. Natural circulation of heat is much slower in thick liquids because the hot liquid often does not rise quickly enough from the heat source to cause burning on the bottom of the pan. Stirring is a safeguard against burning when heating thick liquids. Stirring also helps foods heat faster and more evenly.

Mechanical Convection

This type of convection involves outside forces. Fans in convection ovens and convection steamers are used to circulate air or water faster. The increased circulation helps distribute heat quickly and evenly.

Radiation

Radiation transfers energy to food by waves. When the waves come into contact with the food, they convert into heat energy. There are various types of radiation, but only two, infrared and microwave, are used in the bakeshop and pastry shop.

Infrared

Infrared cooking occurs when a heated electric or ceramic element reaches a high temperature and gives off waves. The waves travel at the speed of light in all directions to cook the food as it absorbs the waves. Heat transfer through infrared radiation occurs in broilers, toasters, and special infrared ovens.

Microwave

Microwave cooking is the fastest type of heat transfer. A special oven generates microwaves. These invisible waves of energy enter the food and cause the water molecules in the food to move around. The molecules rub against one another, creating friction and producing heat that spreads through the food to cook it.

Types of Cooking Techniques

All cooking involves heating food. However, each technique uses a unique method to transfer heat to food to produce a variety of results. Cooking techniques can be classified as dry, moist, or a combination of both. *See Figure 11-1.*

Figure 11-1

Technique	Medium	Method
Baking	Air	Dry
Roasting	Air	Dry
Sautéing	Fat	Dry
Deep-frying	Fat	Dry
Grilling	Air	Dry
Broiling	Air	Dry
Poaching	Water or other liquid	Moist
Simmering	Water or other liquid	Moist
Boiling	Water or other liquid	Moist
Blanching	Water or other liquid, Steam	Moist
Steaming	Steam	Moist

This classification is based on the medium used for cooking.

Cooking Techniques

Dry Cooking

The dry technique transfers heat through hot metal, hot air, hot fat, or radiation. Any moisture that exudes from the food evaporates, or escapes into the air. Baking is an example of dry cooking.

Moist Cooking

The moist technique uses liquid other than oil to transfer heat. The liquid medium may be water or steam. Blanching is a good example of moist cooking.

Dry Cooking Techniques

In dry cooking techniques, food is cooked by a direct application of heat or by surrounding it with hot air or hot fat. Descriptions of specific dry cooking methods follow. The necessary mise en place (ingredients, equipment, tools, and serving pieces) and guidelines for testing for doneness are also indicated.

Baking

Baking is the method of using dry, hot air to cook food in a closed environment. Heat surrounds the food by radiation and convection and then spreads through it by conduction. The cooking medium contains no perceptible fat or liquid, and the moisture that the food releases in the form of steam evaporates in the heat of the oven or through vents. Because baking is a dry technique, steam is often used during the baking process to achieve desired results. For example, when baking bread, the injection of steam promotes proper starch gelanization and crust formation. Baking is complete when the food is cooked to a desired doneness and proper color is achieved.

Mise En Place

Scale all ingredients for cake batters and doughs, cookies, pastries, bread, and rolls. Oil all pans if necessary. Wash, peel, core, or pit fruit and then cut as required.

Method

The baking process is simple. Place the food in a pan or on a baking sheet that allows air to circulate, and then preheat the oven to the proper temperature and bake the food until the desired degree of doneness is achieved.

Testing for Doneness

Fruit should be baked until just tender and should reach 135°F (57°C) or higher for holding. Always follow suggested baking times. When done, baked foods such as bread and pastry goods will be lightly browned.

Carryover Cooking

Be sure to take carryover cooking into consideration during the baking process. *Carryover cooking* is the continued cooking that occurs even after a food item is removed from the oven. The cooking continues until the heat spreads through the food and the temperature throughout is stable.

STEP PROCESS

Roasting Fruit

Roasting, a technique associated more with savory cooking, can be used to add a new dimension of flavor and depth to many fruits.

To roast fruit, follow these steps:

1 Prepare fruit by sprinkling it with sugar, syrup, or honey. Toss gently to combine.

2 Arrange fruit on a sheet pan with parchment paper or a silicone mat. Place the sheet pan of fruit in a preheated oven and roast until sugar caramelizes and the fruit starts to brown.

3 Turn fruit over and return to the oven.

4 Roast until the desired caramelization has occurred.

Roasting

Roasting, like baking, also uses dry heat in a closed environment, usually an oven. Roasting is used for fruit and sometimes vegetables in the bakeshop or pastry shop. As in baking, convection transfers heat to the surface of the food, and then through the food.

Mise En Place

Prepare fruit and vegetables as required. Select a sheet pan and cover it with parchment paper or a silicon mat to prevent sticking.

Method

Roasting is a relatively simple process. Place the sugar coated fruit or oiled and seasoned vegetables on a sheet pan containing parchment paper or a silicon mat. Preheat the oven to the proper temperature and cook the product until the desired degree of doneness is achieved.

STEP PROCESS

Sautéing Apples

Sautéing is a quick technique so foods should be cut into even size pieces that cook quickly. A tasty glaze can be made in the sauté pan by adding liqueur, distilled spirits or wine to the pan and reducing to concentrate the flavors.

To sauté diced apples, follow these steps:

1 Preheat the sauté pan and add clarified butter. Heat the butter almost to the smoke point and add the fruit.

2 While the fruit is browning in the clarified butter, sprinkle it with sugar.

3 Remove the pan away from the flame and add liqueur, wine, or distilled spirits.

4 Return the pan to the flame and ignite. Allow the flame to subside before serving.

Although most fruit and vegetables can be roasted at temperatures between 325°F–425°F (163°C – 218°C), the specific temperatures may be dependent on the type and size of a fruit or vegetable and the desired result. When roasting fruit it is important to remember that the degree of ripeness also impacts the roasting time and temperature.

Testing for Doneness

Firmness and proper caramelization are the best ways to determine doneness.

Sautéing

Sautéing is a dry cooking method that cooks food quickly in a sauté pan with a small amount of fat. With the help of fat, conduction conveys heat from the pan to the food. Heat then spreads through the food by conduction. Sautéing is best suited to delicate foods that cook relatively quickly, such as fruit or vegetables. Clarified butter, sugar, liqueur, distilled spirits or wine, and spices can be added when sautéing fruits to create a glaze or syrup when the sugar caramelizes.

Mise En Place

Because sautéing is a quick cooking method, it is important to have all the necessary ingredients, equipment, and tools ready. Prepare food as required. Have clarified butter, sugar, liqueur, distilled spirits or wine, and spices on hand. Avoid whole butter, because its milk solids burn easily at the high heat used in sautéing. Choose a sauté pan that accommodates the food without crowding to allow the tossing or flipping of smaller pieces.

Preparing Food for Sautéing

Before sautéing, wash and peel fruit or vegetables (core or pit fruit if necessary) and then cut as required.

Method

To sauté, preheat the sauté pan before adding the clarified butter. Add just enough clarified butter to coat the bottom of the sauté pan. When the clarified butter nears the smoking point, add the fruit or vegetables. Do not overcrowd the pan. Too many fruit or vegetables in the pan will lower the temperature, inhibit browning, and cause the fruit or vegetables to emit natural juices. After the initial browning, lower the heat to allow for even cooking. Turning the fruit or vegetables occasionally or flipping the sauté pan to toss the fruit or vegetables also helps to cook contents evenly.

Testing for Doneness

An inspection of the firmness and texture of the fruit or vegetables is the best way to determine doneness.

Deep-Frying

Deep-frying cooks food by complete submersion in hot fat. Deep-frying is a relatively quick cooking method carried out at high temperatures. Deep-frying is most often used for cooking doughnuts and fritters.

Mise En Place

Scale all ingredients for doughnuts and fritters. For doughnuts, mix the ingredients and ferment. Then divide, cut, and portion the dough. For the fritters, combine, sift, stir, and fold all ingredients into the batter before deep-frying. A deep-fryer or deep pan is necessary, as is a good quality fat with a high smoke point.

Method

The proper fat temperature is critical in deep-frying. Preheat the deep-fryer from 360°F–385°F (182°C–196°C) for doughnuts and from 360°F–375°F (182°C–191°C) for fritters. Fat at these temperatures seals the surface of the dough or batter so that it does not become greasy but still allows the dough or batter to cook completely. When the fat is hot, add the dough or batter to the deep-fryer and allow it to float freely, or place the dough or batter in a basket and submerge. Use tongs to turn the doughnuts or fritters as needed. Remove when golden brown and fully cooked and drain on an absorbent surface. Deep-fried foods are best when served immediately.

STEP PROCESS

Deep-Frying Apple Fritters

Proper deep-fat frying results in food with a crisp exterior and a moist interior. *Follow these steps to deep-fry apple fritters:*

1 Dip the apple slices into the batter and allow the excess batter to run off.

2 Carefully drop the apple slices into the hot fat. It is best to hold the apple slices in the fat a few moments before releasing. This helps prevent the batter from sticking to the basket.

3 Cook the fritters approximately 4 to 6 minutes, turning halfway through.

4 Remove the fritters from the fat and drain.

5 Sprinkle fritters with powdered sugar if desired.

Although a deep pan is traditionally used for this method of frying, most deep-frying in the bakeshop or pastry shop is done in commercial fryers equipped with wire baskets for submerging food in hot fat. Commercial fryers have many advantages. As with all cooking methods, take into account *recovery time*, or the time it takes for the fat to return to the required cooking temperature after the food has been submerged. Commercial fryers generally have a much shorter recovery time and are designed to maximize the life of the frying fat.

Testing for Doneness

The color of doughnuts or fritters can be an indication of their doneness. Deep-fried doughnuts or fritters should be golden brown on all sides. You can also remove the doughnuts or fritters from the fat and cut them open to test for doneness.

Grilling

Grilling uses radiant heat from below to cook food on an open grid. In commercial bakeshops or pastry shops, gas or electric grills are common heat sources for grilling. Fruits are suitable for grilling.

STEP PROCESS

Grilling Pineapple

Grilling helps intensify the flavors of fruit. Lightly brushing the fruit with simple syrup improves sweetness and helps create distinctive grill marks.

Use these steps as a guide when grilling:

1 Lightly brush the fruit with simple syrup.

2 Place the fruit on a clean, preheated grill.

3 To create cross-hatch grill marks, turn the fruit 90 degrees after one set of grill marks has developed.

4 Turn the fruit over and finish cooking.

Mise En Place

Wash, peel, core, or pit fruit and then cut into desired shapes and sizes. Season, coat, or prepare as the formula indicates.

Method

Properly grilling foods involves preheating the grill. Lightly brush fruit with simple syrup, if required, before placing it on the grill. Once on the grill, leave the fruit in the same position until it displays distinct grill marks on its underside. When the grill marks are created, use tongs to turn the fruit. It is best to turn fruit over only once during cooking. Cook fruit to completion on the other side, adjusting the fruit's position on the grill or the grate's distance from the heat source as needed.

Creating Grill Marks

Place the fruit on the grill presentation side down. Cook until grill marks develop where the fruit makes contact with the grill. To make cross-hatch markings, lift and turn the fruit 90 degrees after one set of grill marks has developed, and cook until cross-hatch marks appear. Then turn the fruit over, and finish cooking.

Testing for Doneness

Fruits should be tender and lightly browned. It is important to remember that the degree of ripeness will also determine doneness.

Broiling

Although similar to grilling in some ways, *broiling* differs in that it uses radiant heat from above, rather than below, to cook food. Fruits lend themselves to this dry cooking technique. Commercial kitchens usually have either a grill or a broiler. Full heat is customary for broiling, and temperatures are higher than those used for grilling. Infrared broilers can reach temperatures that exceed 1,500°F (816°C). The procedure for broiling is much the same as that for grilling. Place the fruit on a preheated metal rack under the heat source. The hot rack will make grill marks on the fruit. As in grilling, turn broiled fruits over just once to cook on both sides.

Moist Cooking Techniques

In moist techniques, food is cooked by totally or partially submerging it in hot water, or another type of liquid, or by surrounding it with steam. As with the other cooking methods, mise en place (ingredients, equipment, tools, and serving pieces) and guidelines for testing for doneness will depend on the food and the formula. Procedures for specific moist cooking methods follow.

Poaching

Poaching cooks food gently in a flavorful liquid, such as wine, fruit juice, cider, or simple syrup. Convection transfers heat from the liquid to the food. It is an excellent technique for cooking fruit. Poaching is most often done on the stove top but may be done in the oven as well.

Mise En Place

Prepare fruits as required. Have cooking liquid available. Choose a pan suited to the quantity and type of fruit being cooked.

Method

Heat the poaching liquid to a simmer. Slowly submerge the fruit in the hot liquid. Reduce the heat so that the surface of the liquid moves just slightly during the cooking process. Slight bubbling may occur at the higher temperature ranges. Maintain a cooking temperature of 160°F–185°F (71°C–85°C). When the cooking process is complete, chill the fruit in the liquid to maximize flavor and deepen the color of the fruit.

Testing for Doneness

Firmness and texture are the best ways to determine doneness.

STEP PROCESS

Poaching Pears

Poaching cooks food gently in a flavorful liquid.
To poach pears, follow these steps:

1 Peel and core the pears. Slice pears in half if desired. Keep the pears submerged in cold water and lemon juice to prevent browning.

2 Place the poaching liquid in a saucepan deep enough to cover the fruit. Add the fruit and bring to a gentle simmer.

3 Cover the fruit with cheesecloth to weigh it down and prevent the fruit from rising to the surface. Continue to simmer gently. Do not let the liquid boil.

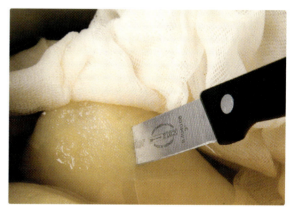

4 Test for tenderness by gently piercing fruit. Poached fruit should be tender but firm enough to hold its shape.

Simmering

Like poaching, *simmering* cooks food by mild convection in a liquid. Simmering, however, occurs at a slightly higher temperature than poaching, with more consistent bubbling on the surface of the cooking liquid. Although it produces more bubbles than poaching, simmering is still a less turbulent technique than boiling.

Simmering can also reduce a liquid, such as a sauce, to achieve a more concentrated flavor. It is used to make fruit sauces such as compotes or coulis, stirred custards, and some creams.

Mise En Place

Prepare fruit sauces, stirred custards, and creams according to the formula. When simmering fruit, select a cooking liquid, such as fruit juice, wine, or simple syrup that will augment the flavor of the fruit. Have an appropriate pan, slotted spoon, spider, or colander available.

STEP PROCESS

Making a Wine Reduction

Simmering is perfect for reducing a liquid such as a sauce. The resulting reduction will have a more concentrated flavor.

To simmer and reduce a sauce, follow these steps:

1 Put wine, sugar, and any desired seasonings into an appropriate pan and bring to a boil. Reduce heat to a simmer. Bubbles will pop rapidly.

2 Continue simmering until the mixture is reduced to the desired level and the proper flavor is achieved. As the mixture reduces, the bubbles will pop more slowly.

Method

Simmering is similar to poaching. First, bring the liquid to a boil. Add the fruit to the liquid by completely submerging it. Adjust the heat to maintain a cooking temperature of 185°F–200°F (85°C–93°C). A good simmer can be evidenced when steam bubbles rise slowly to the surface of the liquid. Remove the fruit when fully cooked. Chill the fruit in the cooking liquid to retain the flavor, and serve it as an accompaniment to a dessert. When preparing fruit sauces, stirred custards, or creams, it is important to constantly stir or whisk to avoid burning or scorching. In the preparation of sauces such as sabayon, a waterbath is used to simmer the sauce.

Testing for Doneness

For stirred custards and creams, check for proper consistency.

Boiling

Boiling also transfers heat from liquid to food by convection but with more liquid and greater agitation than in poaching or simmering. In boiling, liquid rises to the boiling point and remains at that temperature throughout cooking. Although the high temperature of this moist technique cooks foods quickly, the very brisk circular motion of the liquid can damage delicate foods. Fruit sauces, stirred custards, and some creams use the boiling method.

Mise En Place

Prepare food items as specified in the formula. When boiling fruit, have the appropriate cooking liquid ready. Select a saucepan that will accommodate both the food and the liquid. Have a slotted spoon, spider, or colander available as needed.

Method

Bring the cooking liquid to the boiling point—212°F (100°C) at sea level. Add fruit for sauce to the boiling liquid and return the liquid to a boil. The liquid should continue to move rapidly, and large bubbles should rise and break the surface. Stir the contents of the pan occasionally to prevent sticking and to ensure even cooking. When the fruit is done, strain if necessary. Chill and store for future use. When preparing stirred custards and creams, make sure that the product is continuously stirred or whisked to prevent scorching or burning.

Testing for Doneness

Cook fruit to desired tenderness. For stirred custards and creams, check for proper texture.

Blanching

Cooking foods briefly and partially as a first step to other cooking processes is called *blanching*. Blanching can be used to remove undesirable odors and flavors from some foods and it can also set texture and color. Blanching is a cooking technique used to loosen the skin of nuts. Certain nuts, such as almonds, can be purchased blanched.

Mise En Place

Prepare nuts as required. Have the water and a saucepan on hand. A slotted spoon or spider should be available for removing the nuts from the boiling water.

Method

Blanching is usually a two-step process:

1. Completely submerge the nuts into boiling water and lightly cook them.
2. Remove the nuts from the stove top and pour the boiling water and nuts into a bowl to cool. Carefully remove the skin with your thumb and index finger and then toast the nuts slightly in the oven or on the stove top until lightly browned.

Testing for Doneness

The skin should come off of the product easily.

Steaming

Steaming is a method for cooking desserts, such as puddings, in a closed environment, such as a covered mold or basin surrounded by a waterbath. As liquid heats in the covered pan, it turns into vapor. The dessert does not touch the liquid but rests above it in a rack or a basket. Convection transfers heat from the steam to the dessert. Steam conveys much more heat than boiling and cooks desserts faster than other moist cooking methods.

Mise En Place

Prepare batter for pudding. Have water or other cooking liquid ready as appropriate. Also have a wire rack or basket at hand if needed.

Method

Steaming involves placing a wire rack or stainless steel steamer basket in a pan containing a small amount of water. The rack or basket should just clear the surface of the liquid so that the steam can circulate around the pudding. Cover the pan, and bring the liquid to a boil. When the liquid boils and begins to vaporize, place the pudding in a single layer on the rack or in the basket. Cook covered to desired doneness.

Testing for Doneness

The pudding should have a firm texture.

CHAPTER 12
Dessert Presentations

Restaurant patrons frequently order dessert as the culmination of a satisfying dining experience. Delicious and elegant desserts provide dining establishments a final chance to impress customers. Their preparation is a science and their presentation an art. In recent years, emphasis has been placed on visual appeal and the creative presentation of desserts on the plate.

Chocolate is a popular dessert item that can be plated in a creative manner. Fruit offers variety and versatility in presentation and appeals to the more health conscious. Sorbets and ice creams are popular as single dessert items or as accompaniments to selections, such as pies and tarts.

KEY TERMS

crunch
theme

zones

pièce montée

blueprint

Plated Desserts

Attractive plate presentation is the result of careful attention to texture, temperature, shape, flavor, and color. Creative plating enhances a dessert's visual appeal and increases its perceived value.

Components of a Plated Dessert

A plated dessert has four basic components: the main item, the sauce, the garnish, and the crunch. Plated desserts that include a frozen item such as a small scoop or quenelle of ice cream or sherbet have that additional fifth component. *See Figure 12-1*.

Main Item

The main item, or entrée item, is the focal point of the dish. Plated, it is usually 3–5 ounces (85–140 g) in weight and is the item listed on the menu.

Sauce

Sauces complement or contrast the flavor and color of the main item. One or more sauces may be used, but their compatibility for flavor should always be evaluated. The total weight of all sauces should be between 1 and 2 ounces.

Figure 12-1 A small scoop of mango sorbet adds a fifth component to this plated dessert of rich chocolate cake with poached pears, caramel and chocolate sauces, a chocolate curl garnish, and a powdered tuile cookie for crunch.

Techniques for Saucing a Plated Dessert

In addition to the important role that dessert sauces play in enhancing the flavor of a plated dessert, sauces may be used to create visual appeal. Painting a plate with sauce introduces color, creates flow, adds visual balance, and highlights the main item. Dessert sauces can also have a pragmatic function, such as anchoring an item on the plate.

A sauce's consistency and appearance are often critical to the success of the design of a plated dessert. Most sauces should be smooth in texture and thick enough to coat the back of a spoon. Sauces should also be thick enough to hold a line when dispensed through a sauce bottle.

Multiple sauces that are "married" for feathering into a design, must be of the same consistency. The combination of incompatible sauces will cause heavy sauces to sink into a thin sauce and ruin the design. Similar consistencies will also prevent one sauce from bleeding into another.

Sauce can be portioned onto a dessert plate by ladling or by applying it with a plastic sauce bottle or sauce gun. A sauce gun, also known as a portion gun or fondant funnel, is a funnel-shaped container with a spring-loaded cover across the bottom opening that controls the flow. The cover can be opened or closed with one hand while holding the funnel upright. This tool is useful for dispensing a specified portion of sauce onto the dessert plate.

Chocolate, food gels, and thickened sauces are often used as barriers for sauce—a way of holding the sauce within a certain design. Piped items should be piped thinly and neatly.

STEP PROCESS

Making a Web Pattern with Sauces

For this presentation, crème anglaise forms the base sauce and a contrasting raspberry sauce is used to create the web pattern.

To make a web pattern with two sauces, follow these steps:

1 Ladle a thin layer of crème anglaise onto the plate. Tilt the plate to even the sauce. Use a squeeze bottle to pipe the raspberry sauce.

2 Starting from the center, pull a toothpick through the raspberry sauce ending near the edge of the crème anglaise. Repeat to form evenly spaced spokes.

3 For a dramatic effect, start at the bottom circle of the raspberry sauce, equidistant between the spokes, and pull the toothpick toward the center.

STEP PROCESS

Using Chocolate Sauce as a Border for Another Sauce

A small parchment pastry bag is ideal for applying a border of chocolate sauce to help shape the pattern of the main sauce. Warm the chocolate sauce slightly and apply even pressure.

To make a decorative border with chocolate sauce before adding the main sauce, follow these steps:

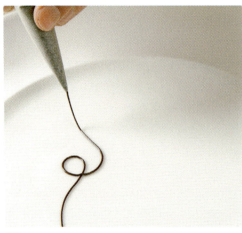

1 Pipe chocolate sauce into a thin pattern that best highlights the main item on the plate. For best results, do not allow the tip of the bag to touch the plate when piping.

2 Begin to apply the main sauce using a plastic squeeze bottle.

3 Fill in with the remaining sauce.

4 Complete the sauce design.

Garnish

Garnishes vary widely in sophistication from the basic powdered sugar dusted on a plate to an intricately stenciled and shaped tuile cookie. Simple or complex garnishes add visual appeal and perceived value to a plated dessert. Garnishes used on a plated dessert must be edible and complement the main dessert item with flavor and a contrast of shapes. *See Figure 12-2.*

STEP PROCESS

Using Two Sauces as a Plate Decoration

Sauces add delicious components to plated desserts either by complementing or contrasting the flavors of the main item. Sauces also add decorative components enhancing the overall appearance of the entire dessert.

The following techniques can be used when decorating with two sauces:

1A Pipe dots of one sauce onto the plate. Then add dots of the other sauce.

1B Use a toothpick or skewer to pull one sauce through the other.

2A Pipe rows of one sauce onto the plate leaving an even space between rows. Pipe the second sauce between the rows.

2B Use a toothpick or skewer to make a swirl pattern.

3A Use a squeeze bottle to pipe concentric circles of sauce onto the plate with a space between circles. Pipe the other sauce into the space.

3B Feather swirls dragging the toothpick or skewer through sauces in a circular pattern. The back of the skewer can be used to make broader swirls.

Figure 12-2 A fruit tartlet becomes an attractive plated dessert resting on a circle of confectionary sugar with the addition of a mixed berry sauce and a spiral garnish made of caramelized sugar.

Crunch

The *crunch* component offers textural contrast to the main item. Crunch may be in the form of a decorative cookie, or as a container for the main item. The crunch component can also be found in a garnish such as a tuile or almond lace garnish. This component should never overwhelm or detract from the main item on the plate. *See Figure 12-3.*

Types of Plating

There are generally two ways to plate desserts: banquet style and à la minute. The type of plating also depends on the kind of dessert served.

Banquet Style

Banquet plating is used to accommodate a large number of diners at a function. At times as many as several hundred plates are needed. Banquet plating often requires special planning and organization because desserts are usually prepared ahead of time and a change in texture, color bleeding, and shrinkage may occur. Large hotels and banquet halls usually use this method.

À la Minute

This type of plating, literally translated as "by the minute," is standard for white tablecloth restaurant service where the pastry chef prepares items as they are ordered. "À la minute" requires a great deal of prep work or mise en place to reduce the time needed to deliver the item to the customer. Hot desserts such as soufflés, beignets, and fritters are prepared and served by chefs on the hot line. Sorbets, ice creams, and delicate cookie and sugar garnishes can be offered "à la minute."

Figure 12-3 Crisp fruit slices make a flavorful and crunchy garnish.

Slice fruit into 1/16-inch (2-mm)-thick slices using a mandoline. Dip into simple syrup.

Arrange fruit slices on a nonstick baking sheet and bake at 180°F–200°F (80°C–94°C) for about 1 to 2 hours until desired degree of crispness is achieved.

Plating Contrasts

Contrasts are an integral part of plate design that provide appeal for the customer. Most plates feature contrasts in at least one or two of the following areas:

Texture

Texture contrasts are very effective in producing visual and mouthfeel appeal. Smooth, velvety-textured components, crunchy components, and chewy-textured items may be used together to make a dessert more interesting.

Temperature

Temperature contrasts are another important component. The combination of hot and cold items on the same plate adds to the overall appeal of a dessert. Always remember to serve hot items hot, and cold items cold.

Shape

The presence of visual contrasts are also important. Use components of different shapes and sizes on a plate to draw attention to the main item.

Flavor

Although most desserts are usually sweet, everything on the plate need not be sweet. Savory components may be added to balance the sweet flavors. Only use contrasting flavors to complement the flavor of the main item.

This advanced plated dessert is created using a variety of baking and culinary applications. Israeli couscous is cooked in pear nectar sweetened with honey. The couscous is placed in a mold and soft goat cheese mixed with crème anglaise is added in the center. Once unmolded, the couscous is topped with pear mousse and a slice of dried pear. It is then finished with a mirror glaze. A scoop of pear sorbet and Florentine cookie pieces are added to provide complimentary flavor and crunch. A spiced wine reduction is used as a sauce and then small cubes of the sauce solidified with gelatin are added. Basil infused syrup and a basil leaf garnish enhance the Mediterranean theme. Flaked sea salt adds an exciting dimension to this savory and sweet dessert presentation.

Color

Color is one of the easiest and most visually exciting ways to add contrast to a plate. The addition of each new color should promise the introduction of increased flavor. Avoid adding color just for the sake of appearance.

Tips for Plating

Dessert plating requires a great deal of mise en place as well as time management and organizational skills. The following general guidelines may be useful to remember:

- Consider the kind of plate to be used before designing the dessert. Its size, color, and shape will greatly influence the final plate design.
- Make designs and contrasts interesting but simple. Complex plate design confuses the eye and the palate.
- Balance the various aspects of the plate—textures, shapes, flavors, and colors. Asymmetric or symmetric balance may be used, as long as the main item is the focal point of the plate.
- Do not allow stronger, contrasting flavors to overpower subtle flavors.
- Make sure that the components of the plate work together as a single offering. They should come together harmoniously, rather than "just fill space."
- Use the placement and shape of the components to create a flow or movement that leads the eye toward the main item.
- Take care that all the items on the plate are edible.
- Avoid unnatural food colors such as blue.

Planning a Buffet

The presentation, appearance, and flavor of the desserts and pastries served are paramount to the success of planning a buffet. As in all menu planning, gastronomics, economics, and practicality must also be considered. Buffet planning begins with a ***theme***, or central concept or motif, and a menu. Design, table placement, available space, the number of ***zones***, or buffet areas, and décor are then assessed.

Theme

A buffet is a carefully planned presentation of a multitude of dessert and pastry items from which customers make their selections. Before determining the dessert and pastry items that will be served and presented, the theme must be selected. The theme sets the mood of the event or affair, and may span a spectrum from black-and-white formal to casual dining. A theme may be event focused, as in the case of a wedding, charity luncheon, or holiday party. The theme dictates the selection of dessert and pastry items, the menu design, décor, music, lighting, linens, and flatware. In deciding the theme and the buffet offerings, cost must be considered.

The Menu

The buffet menu is generally à la carte with a representative variety of dessert and pastry choices. All selections should coincide with the theme of the buffet and both whole dessert items and pastry platters should be offered. When planning a menu consider gastronomic, economic, and pragmatic aspects.

Gastronomic Plan a menu that integrates a variety of ingredients, colors, and textures. Give guests choices that incorporate ingredients such as chocolate, fruit, and nuts. Also present a variety of colors and food textures. Guests should have choices such as custards, creams, mousses, puff-pastry items, and cake-based desserts.

Economic Pay close attention to cost when planning a menu. Creating and presenting visually appealing and satisfying offerings within budget is a challenge that requires a careful assessment of both ingredients and time costs.

Pragmatic When planning a menu, consider the equipment available, the staff's skill level and the type of service needed. Also assess the religious or ethnic dietary needs of guests.

Buffet Design

Buffet design involves determining the arrangement of a buffet. The number of zones; table placement, size, and shape; space for service; centerpieces and other decorations are all considerations.

Zones A zone is a buffet area designed for the smooth and speedy flow of patron traffic. A single buffet zone may serve 75 to 125 people. For larger groups, additional zones may be needed to ensure rapid service and shorter lines. Each zone should contain the same menu items. A zone may consist of a single table that patrons approach along one side. A single table may also serve as two zones for larger groups, offering identical desserts and pastries on both sides of the table with patrons forming a line on each side. Another arrangement to accommodate large groups is several zones on a single table that guests approach on one side. *See Figure 12-4.*

Table Placement, Size, and Shape The configuration of the dining room and seating arrangements often dictate the size and shape of the buffet table or tables. Tables of various shapes strategically placed in a dining room can influence the buffet's initial visual impact on guests and contribute to an impression of abundance and beauty.

Space for Service The table arrangement should also accommodate servers and staff. Servers or runners need to replenish the buffet and keep it neat and appealing. They should have enough space to do so without disturbing dessert or pastry displays or guests. Convenient access to the kitchen is essential for quick replacement of desserts and pastry items. Additional space may also be needed when servers prepare desserts at the buffet table. Staff on site can also describe items to guests.

Presentation

Presentation of dessert and pastry items is key to the success of a buffet. Dessert and pastry selections should be plentiful and visually appealing to satisfy the eye and stimulate the palate. Dessert presentation may be simple to allow the natural colors and textures to take

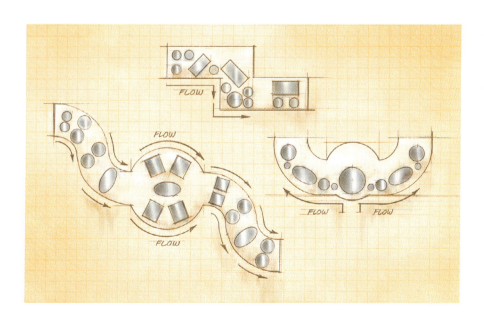

Figure 12-4 The size and shape of the dining room, along with the theme, food selection, and number of guests influence the size and shape of the buffet table.

STEP PROCESS

Plate Dusting

In this example, cocoa is used to decorate two plates using both a commercially available stencil and one made from an aluminum pie tin. When dusting the rim of the plate, be sure to leave a significant amount of the rim undecorated to allow servers to handle the plate.

To dust plates using stencils, follow these steps:

1A Lay the stencil over the area you wish to decorate. Put cocoa into a sifter and tap the sifter to distribute the cocoa evenly into the desired area of the stencil.

1B Carefully lift off the stencil and use a pastry brush to remove any unwanted cocoa from the plate.

2A Make a decorating stencil by tracing a design onto an appropriate medium and cut out the design. Place the stencil over the plate and tap the cocoa from a small sifter into the opening in the stencil.

2B Continue tapping the cocoa into the stencil to fill in the cut out.

2C Carefully remove the stencil.

center stage, or more elaborate or intricate. To prepare, consider how the food will be presented and how the desserts and pastries will look on the buffets. Decide on the focus of the presentation, and how the guest's visual, textural, and flavor interest will be engaged throughout the buffet. Present contrasts in food color and appearance, shapes, height, texture, flavor, and temperatures.

HOT FOOD ITEMS When serving hot food items, such as cobblers, crisps, fondues, and sauces, follow the proper holding and serving procedures as outlined by HACCP. Serve hot foods hot, possibly from chafing dishes placed near the end of the buffet table.

COLD FOOD ITEMS Cold foods should also be held and served following HACCP guidelines. Serve cold foods cold, and present them artistically.

STEP PROCESS

Making Caramelized Cages

Cages can be made using the inner or outer bowl of a ladle. Cool the hot caramel just enough so that it can be picked up and drizzled with a fork or spoon. The lines of caramel can be randomly swirled or applied in a pattern. A very thin coating of vegetable oil applied to the ladle before adding the caramel facilitates easy removal when the caramel has fully cooled.

To make caramel cages, follow these steps:

1. Holding the ladle with one hand, drizzle and swirl hot caramel into the bowl of the ladle.

2. Use a hot knife to trim the edges of the caramel.

3. When the caramel has cooled, remove the cage from the ladle.

Centerpieces

A centerpiece (*pièce montée*) enhances the appearance of the buffet table. It provides a focal point when guests first enter the dining room and reinforces the theme of the buffet. A dessert or pastry buffet centerpiece often incorporates vases of flowers, ice carvings, chocolate sculptures, and sugar showpieces. Color, contrast, shape, size, medium, and cost are all factors to consider when planning a centerpiece. The availability, ease of storage, facility of transport, and the risk of breakage should also be evaluated.

Ice Carving

Ice carvings are some of the most impressive and sought-after centerpieces for dessert or pastry buffet tables. The cost of producing ice sculptures varies, depending on the availability of ice, the skill of the artisans, and cost of storage. When not produced in-house, ice sculptures can be expensive.

Platter Design

Platters are most often used to display individual pastries on a buffet. Platter shapes and sizes, and their material composition vary greatly. The selection of platters and the arrangement of pastries on them is a critical part of the aesthetics of a buffet table.

STEP PROCESS

Making Caramelized Sugar Decorations

Caramelized sugar decorations can be used as garnishing components or bases on plated desserts. They can be free-formed patterns as some of the ones below or they can be shaped using tools and Flexipans.

The following are some techniques used to make caramelized sugar decorations.

Drizzle caramelized sugar onto parchment paper for attractive garnishes.

A nonstick baking sheet can also be used.

Inverted Flexipans of all shapes can be used.

A clean and sanitized sharpening steel or a wooden dowel can be used to make spirals of caramelized sugar.

Blueprints

After choosing the appropriate platters for a particular buffet menu, develop platter *blueprints* or drawings of how the platters will look with pastries displayed on them. Each platter should have a blueprint that details the pastries that will be presented, the garnishes for the platter, and the space requirements based on platter size. A platter blueprint provides a visual of the finished pastries and their locations on the platter, and ensures that each platter offers color and texture contrasts, as well as diverse shapes and heights.

STEP PROCESS

Working with Tulip Paste

Tulip paste is used to make thin cookies that can be used as bases or cups for main items such as ice creams and sorbets, or to provide an attractive crunch on the plate.

Follow these steps to work with tulip paste:

Prepare tulip paste according to the desired formula.

1A Lay a template over a nonstick baking sheet and spread the tulip batter into the template.

1B Use a spatula to smooth the tulip paste.

1C Sauce can be piped onto the tulip paste to make an attractive design.

1D Bake in a 350°F (177°C) oven for 5-6 minutes.

1E Remove from the pan immediately: mold into shape while hot, as desired.

2A Use the same technique to make spoons from tulip paste.

2B To form a three-dimensional spoon, place the hot cookie onto an actual spoon.

2C Put another spoon on top and press lightly.

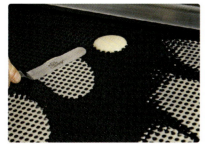

3A To make an attractive wafer, tulip paste can be spread onto a bubble frieze mat.

3B Remove from the mat when baked and shape immediately.

Confectionary sugar can be used to finish tulip cookies if desired.

Figure 12-5 Linear placement is the most effective way of arranging pastry items on a platter.

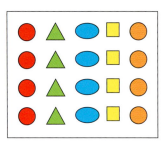
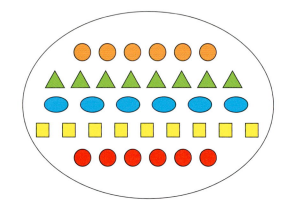

Composition of a Pastry Platter

A balanced platter has a variety of individual pastry items with varying flavors, colors, textures, heights, and shapes. When composing a pastry platter care should be taken to ensure that the individual pastry items are arranged and presented in a visually pleasing manner.

Guidelines for Presentation

Careful planning of the layout of pastry elements is vital to the success of the presentation. For eye appeal and ease of service, follow these guidelines:

- Maintain a natural frame on the platter.
- On bordered platters, position the pastries at least 1 inch (2.5 cm) from the inner edge. On borderless platters, position the pastries at least 2 inches (5 cm) from the edge.
- Use the one-quarter concept. Visualize an imaginary grid over the platter blueprint and later over the finished presentation.
- Confirm the alignment of portioned pastries. Shingle or line up the pastry neatly in a row toward the guest.
- Do not crowd the platter or leave it looking empty. The amount of pastry presented on the platter should look appetizing.
- Make portions consistent. For example, if there are eight portions of one pastry item, there should also be eight portions of every other pastry item.
- Confirm that the portion sizes are correct. *See Figure 12-5.*

chapter 12

Dessert Presentations

CHAPTER 13

Breads

Flour, water, salt, and yeast are the fundamental ingredients of bread. With an understanding of the relationship of these four simple ingredients, a baker can produce a wide variety of breads. Breads have been the foundation of Western diets for six thousand years.

KEY TERMS

whole-grain flour
refined white flour
hard wheat
soft wheat
red wheat
white wheat
winter wheat
spring wheat
extensibility
roller milling
stone grinding
extraction rate
ash content

falling number
yeast
wild yeast
naturally leavened or sourdough
manufactured yeast or baker's yeast
amylase
fermentation
preferments
lactic acid
acetic acid
hydration

sponge
pâte fermenté
poolish
biga
gluten window
carotenoid pigments
folding
retarding
couche
underproofed
overproofed
baisure de pain/kissing crust

scoring
oven spring
grigne
lean dough
levain or sourdough starter
pentosans
spelt

An Introduction to Bread

A simple analysis of breadmaking reveals that the fundamental process has not changed for six thousand years: flour, yeast, water, and salt are combined into dough; the dough is kneaded and left to rise as gluten strands trap carbon dioxide produced by the fermenting yeast. The dough is divided and shaped into loaves; the loaves are left to rise again and then baked.

Many variables affect the finished loaf including: flour selection, temperature and humidity in the bakeshop, hydration of the dough, mixing method, and fermentation time. Fermentation is the key to great flavor, a hearty crust, and a long shelf life. Bakers must understand and control these and other variables to produce good bread.

During the late 1980s and early 1990s an artisan bread renaissance occurred around the world, including in the United States. European baking techniques that had not been widely used resurfaced. Bakers began to enlist the old ways to make bread—very wet dough, long and cool fermentation, and use of the hands rather than assembly-line production. American consumers were eager to buy the European-style hearth breads that were becoming available in the United States and these artisan breads competed significantly with traditional American sliced white bread.

The characteristics of American pan bread are familiar: soft white bread with a golden crust domed over the edge of a loaf pan. American pan bread can be made with artisan techniques that include careful mixing, controlled fermentation, and proper shaping. Many quality bakeries include pan breads in their selection.

Hearth breads are hand-shaped loaves with a dark crust that offers wonderful flavors, and a cut that reveals stretched gluten strands (evidence of a carefully mixed and developed dough). The aromatic interior of the loaf, referred to as the crumb, is cream-colored and scattered with random holes (the sign of a wet dough).

The Four Basic Ingredients of Bread

Flour, yeast, water, and salt are the fundamental ingredients of bread. Using these ingredients alone, a wide variety of leavened breads can be made. Of the four ingredients, flour and water are present in all breads. Yeast is left out of unleavened or flat breads, and salt is sometimes omitted in regional traditional breads.

Flour

The most important ingredient in bread is flour. The most commonly used is wheat flour.

Wheat Flour

Of all the cereal grains, only wheat contains the abundant high-quality gluten necessary to produce light, easily digestible loaves of leavened bread. Other cereal grains are sometimes used alone or blended with wheat to make bread, but wheat flour is the foundation of leavened breads. An individual grain of wheat, called a berry or kernel, is about ¼ inch long and consists of three main parts: the bran, the endosperm, and the germ.

The bran—roughly 14.5% of the total wheat berry—is a protective coating for the wheat berry and a source of insoluble fiber, minerals, and B vitamins. Inside the bran, the endosperm, a starchy interior, makes up about 83% of the total wheat berry. The germ, only 2.5% of the wheat kernel, is the embryo of the wheat berry where a new wheat plant germinates, or sprouts. The germ contains important vitamin E and oils.

Whole-grain flour contains all three parts of the wheat berry, while *refined white flour* contains only the endosperm. Because of the oils contained in the germ, whole-grain flour tends to go rancid more quickly than flour that does not contain the germ. Refrigeration is recommended for whole-grain flours when possible.

There are many varieties of wheat. Bakers and farmers classify the varieties of wheat by three factors: the hardness (hard or soft), the color (red or white) of the wheat kernel, and the growing season (winter or spring).

- Hardness of the wheat kernel

  Hard wheat is higher in protein (approximately 11%–14%) than soft wheat and is more appropriate for breads and other yeast items that rely on a gluten matrix to trap carbon dioxide. Hard wheat is coarse to the feel, does not pack when squeezed, and absorbs more water than soft wheat.

  Soft wheat contains about 6%–10% protein and is more frequently used in cakes, cookies, and pastry items. Soft wheat is soft and silky, packs when squeezed, and absorbs less water than hard wheat.

- Color of the wheat kernel

  Red wheat has a bran pigment that makes it darker than white wheat. Some people find that red wheat imparts a bitter or astringent taste, especially in unbaked dough. The majority of wheat grown in the United States is red wheat.

  White wheat is often considered to have a more mellow or nutty taste.

- Growing Season

  Winter wheat is planted in autumn. It germinates soon after planting then lies dormant in the field during the cold winter months. In spring the sprouts resume growing, and by early summer mature wheat is ready for harvest. Winter wheat is used in artisan hearth breads because of its high-quality gluten and superior fermentation tolerance.

  Spring wheat does not overwinter in the field as winter wheat does. It is planted in spring, grows and matures over the summer, and is harvested in autumn. Spring wheat generally contains more protein than winter wheat (higher quantity), but is usually of a lower quality.

Gluten

Gluten strands begin to form when flour is hydrated, and they continue to develop as dough is kneaded or folded, as well as during fermentation. Two main gluten-forming proteins are present in the endosperm: glutenin and gliadin. Each of these proteins is tangled until the flour is hydrated and kneaded. The long molecules then untangle and align to form gluten strands. Glutenin builds strength and structure that helps the dough retain gas, while gliadin increases the extensibility of the dough. *Extensibility* is the ability of the dough to stretch and not pull back. Extensible dough is easier to shape and is necessary for a long, thin loaf of bread such as a baguette. Extensibility also contributes to larger volume in the baked loaf.

To achieve the proper amount and development of gluten for a particular baked good, the baker first selects flour with the appropriate protein content. Both the quality and the quantity of protein (i.e., gluten) must be considered during flour selection. Flour containing 11.5%–11.7% protein is well-suited for the production of artisan hearth breads.

The Milling Process

Milling is the process of transforming grain kernels into flour. *Roller milling*, the dominant method used in North America, shears grain open with corrugated rollers that allow the germ and bran to separate from the endosperm. *Stone grinding*, on the other hand, retains some of the germ and bran. Stone-ground flour is more nutritious and flavorful than roller-milled. Because stone-ground flour retains some of the oil-rich germ, shelf life is reduced. *See Figure 13-1.*

The percentage of flour produced from a given weight of wheat kernels is known as the *extraction rate*. For example, if 60 lbs. (27 kg) of flour are obtained from milling 60 lbs. (27 kg) of wheat kernels, the extraction rate is 100%—this is whole-wheat flour, because the whole-wheat berry is used. Most white flours have an extraction rate of about 72%. All the wheat in white flour comes from the endosperm with all the germ and bran removed. Higher extraction flours ferment more quickly than low extraction flours because the yeast has more food with which to metabolize.

The *ash content* of flour refers to the mineral content of milled flour and is a good indicator of the extraction rate—a higher ash content means a higher extraction rate. The ash content is determined by incinerating a flour sample and measuring the amount of uncombusted minerals in the ash left behind. Minerals tend to give a grey or flecked appearance to flour and increase the rate of fermentation. Good ash content for white flour is about 0.4% to 0.6%.

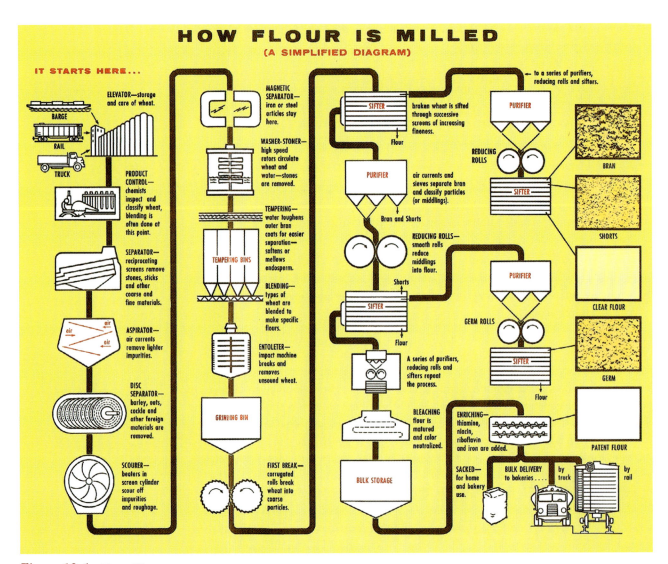

Figure 13-1 The milling process.

Enrichments and Other Additives

Vitamins and minerals are lost when the bran and germ are removed. In addition, heat generated during the milling process reduces some of the nutritional value of the wheat. Federal law requires that wheat flour be enriched to replace or augment some of these losses. Because the mandates only address a fraction of the nutrients lost during milling, so-called enriched flour is still less nutritionally complete than whole-grain flour.

Common additives in wheat flour include: B vitamins, niacin, thiamin, and riboflavin. Folic acid, a B vitamin required by the human body for cell growth and blood production, and iron are also added. Malted barley is included to aid fermentation, improve flavor, and to ensure good crust color.

Maturing

Ideally, flour is allowed to age naturally after milling. Freshly milled flour is "green" and not suitable for use. Dough made with "green" flour has sluggish fermentation, is hard to shape, and produces loaves with low volume. Natural aging improves the quality of milled flour, and produces bread with a creamy crumb. It is advisable to use flour that has aged a minimum of three to four weeks.

To avoid the high cost of storing flour while it ages and to get the milled flour to market sooner, some flour producers artificially age their flour. Flour can be artificially aged by treating the flour with bleaching agents such as chlorine gas and oxidizing agents such as benzoyl peroxide and potassium bromate. Bleaching and oxidizing agents whiten the crumb of baked goods, reduce mixing time, and increase elasticity.

Unfortunately, there are known harmful side effects from aging agents, such as the carcinogenic nature of bromate. (California does not allow products to be bromated.) In response to the artisan bread renaissance and a greater interest in healthy foods among American consumers, most mills that produce flours for hearth breads offer unbleached and unbromated flour.

Millers produce flours with the specified extraction rates and protein levels requested by the baking industry. Millers can usually supply a "spec sheet" for each mill run that lists many performance values such as ash content—an indicator of the extraction rate—and the "*falling number*"—an indicator of enzymatic activity. Bread bakers knowledgeable about wheat and flour are able to make appropriate flour selections.

Yeast

Yeast is a living, single-celled fungus that multiplies when fed simple sugars in a warm, moist environment. The metabolism of yeast is directly proportionate to temperature.

Thinking in Terms of the Baker's Ratio

In a bread formula, the quantity of each ingredient is listed both by weight and by a system called the baker's ratio or baker's percentage system. The baker's ratio expresses the weight of each ingredient as a percentage of the total flour weight (a weight equal to 100%).

The baker's ratio is the industry standard for communicating bread formulas. By looking at the relative amount of all ingredients in a formula, a trained baker can anticipate the consistency of the dough and choose appropriate production techniques. The baker's ratio also makes it easy to create and modify formulas. It provides a template for product development, and is useful when resizing formulas to accommodate production needs.

Any unit can be used to calculate the baker's ratio as long as the units are consistent for all ingredients in the formula. Metric units are easiest to use for calculating the baker's ratio, and serve as the industry standard for artisan bread formulas.

Although there are many types of yeast, two species are used to leaven bread: wild Saccharomyces exiguus and manufactured Saccharomyces cerevisiae. *Wild yeast* occurs naturally in the environment and is captured by the baker and maintained or "cultured" by regular feeding and care. Breads leavened with wild yeast are called *naturally leavened* or *sourdough* breads. *Manufactured yeast*—also called *"baker's yeast"*—is produced in a lab specifically for use in breadbaking. Louis Pasteur discovered this yeast in the mid-1800s. This discovery streamlined bread production and had a huge impact on the baking industry.

Since the beginning of leavened breads thousands of years ago, until the discovery and production of manufactured yeast in the 1800s, all yeasted breads were made with wild yeast. Today, due to a revived interest in naturally leavened breads, bakers have added wild yeast breads to their repertoire. A baker's ability to produce fine leavened loaves using only wild yeast is considered a measure of skill.

Fermentation

In yeasted doughs, the complex carbohydrates in flour are broken down into simple sugars by enzymes, in particular the enzyme *amylase*. The yeast metabolizes the simple sugars to generate energy, as well as the by-products carbon dioxide gas, organic acids, and alcohol. This biological process is known as *fermentation*. Yeast ferments dough slowly in a warm (75°F [24°C]), moist environment with an abundant food source in the form of simple sugars. The high heat of the oven causes fermentation rates to spike and to stimulate a yeast feeding frenzy. Yeast activity continues as the dough temperature climbs, until the yeast is killed at 138°F (59°C), when fermentation ceases. Fermentation is largely responsible for the aromas, flavors, and shelf life of good bread.

The By-products of Fermentation

Yeast generates its energy by metabolizing simple sugars in the dough. *Preferments* (made by combining a portion of flour, liquid, yeast, and sometimes salt) and doughs are usually slightly warmer than the surrounding environment by 1°F to 2°F (–17.2°C to –17°C). The carbon dioxide, alcohol, and organic acids produced by fermentation are responsible for the attributes that characterize leavened bread.

CARBON DIOXIDE The build up of CO_2 gas produced during fermentation is trapped in the gluten matrix of the dough, causing the dough to rise, or leaven.

ALCOHOL Alcohol contributes flavor and aroma to fermented breads. Alcohol expands during the initial baking stage to increase the volume in a baked loaf. The alcohol created during fermentation dissipates in the baking process.

ORGANIC ACIDS The organic acids produced by fermentation accumulate in the dough to create a tightening effect on gluten that increases the strength of the dough. Organic acids strongly influence the flavor and aroma of a baked loaf and extend its shelf life. A variety of organic acids can be produced by manipulating the fermentation environments. Bakers can create breads with particular characteristics and develop efficient and feasible production schedules.

Lactic acid and *acetic acid* are the two organic acids most commonly produced during fermentation. Each develops at different rates and times to impart unique characteristics to the final loaf. Lactic acid imparts a mild but complex flavor. It is produced in the early stages of fermentation in wet preferments and doughs. Its production rate increases with temperature and it mildly retards microbial growth. Acetic acid, on the other hand, imparts a more sour flavor. It develops in later stages of fermentation in drier preferments and doughs. Its production rate remains flat as the temperature rises, and it strongly retards microbial growth.

The Effects of Improper Fermentation

Bakers must control the rate of fermentation and determine when it is at its peak. Dough that is fermented too long, or overfermented, yields a pale loaf with a thick, dry crust, and an unpleasant flavor. Underfermented dough, on the other hand, results in dense bread with undeveloped flavor, and a reduced shelf life.

Aiding Fermentation

Malted barley flour is sometimes added to the final dough in a baker's ratio of approximately 0.5%. Malted barley flour produced by sprouting barley contains diastatic enzymes that aid in fermentation.

Manufactured or Baker's Yeast

Four types of manufactured yeasts, known as baker's yeasts, are commonly used in bakeshops.

Fresh, compressed, or cake yeast is a solid block comprised of 70% water, and is easier to scale than dry yeast. High-quality fresh yeast should be tan in color, firm, and break cleanly. It should also smell fresh and slightly fruity. Fresh yeast must be refrigerated. Its shelf life is one to three weeks. The baker's ratio of fresh yeast is approximately 1% for lean dough, but may increase depending on the type of bread.

Active dry yeast was developed during World War II to extend the yeast's shelf life and to provide soldiers with fresh bread. Before it is added to dough for mixing, active dry yeast should be "activated" through rehydration at 100°F (38°C) for approximately 10 minutes in a 5 to 1 ratio of water to yeast. This activation period allows the water to permeate the layer of dead yeast to properly hydrate, and to reach the viable yeast so it can begin fermentation. A formula that calls for fresh yeast uses 40%–50% as much active dry yeast.

Instant active yeast is produced by a sophisticated and gentle drying process that does not damage the outer layer of yeasts cells. An activation period is not required, so it may be added directly to the dough when mixing. The convenience of not having to activate instant active yeast makes its use very popular in bakeshops. A formula that calls for fresh yeast would use 30%–40% as much instant active yeast.

Osmotolerant yeast is used in dough that contains a high percentage of sugar to resist the hygroscopic pull of sugar. If osmotolerant yeast is not used in dough with a high presence of sugar, fermentation is impaired because the yeast does not adequately hydrate. If a formula calls for fresh yeast, 30%–40% as much osmotolerant yeast should be substituted.

When dry yeast is substituted for fresh yeast in a formula, the amount of water needed may increase. The baker's percentage of dry yeasts is usually about 0.3%–0.5% for lean dough. Active dry, instant active, and osmotolerant yeast have a longer shelf life than fresh yeast. They should be stored at room temperature until opened, then stored at 38°F to 40°F (3°C to 4°C).

Individual cells of dry yeast are invisible to the naked eye; there are approximately 10 billion cells per gram of dry yeast. *See Figure 13-2.*

Wild Yeast

"Wild" yeast *(Saccharomyces exiguus)* occurs naturally in the environment, but collects and is visible as the white film on grapes, plums, and other fruits. It has fewer natural yeast cells per unit of weight than manufactured yeast cells and metabolizes sugars more slowly. It is the yeast used in sourdough production.

Physical Leavening

Although yeasted breads are characterized primarily by the biological leavening process of fermentation, like all baked goods containing water, they are also physically leavened.

Figure 13-2 Ripening poolish, osmotolerant yeast, instant active yeast, and fresh yeast.

Water

Water is necessary for gluten development, and the hydration and temperature control of dough. Warmer or wetter dough ferments more quickly than colder or drier dough. Water permits starches to swell, and provide elasticity to the dough to make it more malleable. Water also helps to physically leaven bread.

The percentage of water needed in a formula is known as ***hydration***, and is an extremely important variable to understand and control. Fermentation rate, volume, and crumb structure are all affected by hydration. Hearth bread formulas often strive for the highest manageable hydration. *See Figure 13-3 for more information on dough hydration.*

Depending on its hydration, bread can lose 20% or more of its weight during baking as the water turns to steam. The steam created during this dehydration process contributes to a steamy oven environment that benefits crust formation.

Water Temperature Formula for Achieving Desired Dough Temperature

Dough temperature control is one of the most important factors in achieving desired formula results. When dough temperature is too low, fermentation is inhibited, and when it is too high, the dough overferments. Optimal fermentation temperature is 75°F (24°C) for most yeasted dough and 78°F (26°C) for most sourdoughs. The water temperature formula should be used prior to mixing so that the dough is at the desired temperature when the mixing is complete.

Dough temperature results from the physical temperature of the room, that of the ingredients (flour, water, and preferment), and the "friction factor" (the amount of heat added to the dough by mixing). Bakers must manipulate these temperatures to end up with the desired dough temperature. The easiest way to accomplish this is by adjusting the temperature of the water in the formula.

The following Water Temperature Formula is used to determine the water temperature necessary to achieve the desired final dough temperature in a dough with a preferment. (If a dough does not have a preferment, the desired dough temperature is multiplied by 3.)

$$T_{water} = [(4 \times T_{final}) - (T_{flour} + T_{room}) + T_{preferment} + \text{friction factor}]$$

Figure 13-3

Hydration Range for Various Types of Bread

HYDRATION RANGE	DOUGH CONSISTENCY	EXAMPLE
60% or lower	Firm	American Style Pan Bread
66%–69%	Soft	Poolish Baguettes
70% and higher	Wet and sticky	Ciabatta

The friction factor is equal to the difference in the temperature of a dough before and after mixing. This factor is constant whenever a particular dough is mixed for the same length of time by the same mixer or hand mixing technique. In general, mixers increase the dough temperature from roughly 22°F to 30°F (–6°C to –1°C), while hand mixing increases the dough temperature about 5°F to 15°F (–15°C to –9°C).

EXAMPLE:

You are making a sourdough with a desired dough temperature (T_{final}) of 78°F (26°C).

The flour and ambient room temperature are both 72°F (22°C), and the preferment is 74°F (23°C). Having previously tested, you know that the friction factor of your mixer is 26°F (–3°C). Using the Water Temperature Formula, you determine that the water temperature is 68°F (20°C).

$$T_{water} = (4 \times T_{final}) - T_{flour} + T_{room} + T_{preferment} + \text{friction factor})$$
$$= (4 \times 78°F - (72°F + 72°F + 74°F + 26°F)$$
$$= 312°F - 244°F$$
$$= 68°F$$

The minimum temperature of water used in bread dough is 32°F (0°C, the freezing point of water), and a safe maximum is 115°F (46°C). Yeast can be killed when water temperature exceeds 138°F (59°C). If the water temperature is outside these limits, one or more of the other contributing temperatures must be modified. Flour temperature is the easiest to modify by chilling or warming the flour as needed. The water temperature formula does not need to be used for preferments.

Salt

Salt has a tremendous impact on the quality of bread. Although the ratio of salt to flour is only about 1.8%–2% by weight, its importance should not be underestimated. Salt regulates enzyme and yeast activity to help ensure the retention of residual sugars that create a dark, well-flavored crust. Dough without salt easily overferments, becoming gassy and sticky. Salt also enhances flavor and tightens gluten by forming strong ionic bonds with flour proteins to make them less flexible. It increases shelf life as well due to its hygroscopic nature.

Always measure salt by weight because its volume changes depending on the size of the crystals. Breads with salty ingredients such as olives may need a slight reduction of salt in the formula.

Preferments

The extended time necessary for fermentation presents both production and timing problems in the bakeshop. To decrease bulk fermentation times, artisan bread production often incorporates the use of preferments. Preferments are allowed to ferment for 20 minutes to 18 hours before the final dough is mixed. Preferments contribute a more complex flavor, increase dough strength, and increase shelf life to leavened breads.

A preferment can be leavened with manufactured yeast (called a "yeasted preferment") or it can be leavened with wild yeast (called a "naturally leavened preferment," levain or sourdough starter). One key distinction between these two types of preferments is that yeasted preferments are used in their entirety, while a portion of the naturally leavened preferment is always reserved to propagate more starter. Dough that is prepared without a preferment is called "straight dough." In a straight dough, all ingredients are mixed at once and the breadmaking process is executed "straight through."

Four Yeasted Preferments

The names used for the same preferments vary from country to country, and from region to region. Wet preferments (i.e., poolish) promote extensibility and produce lactic acid. Dry or firm preferments (i.e., biga or sponge) produce more acetic acid (vinegar).

Sponge

A *sponge* preferment contains flour, water, and yeast. It is usually 50% to 60% hydrated (i.e., the ratio of flour to water, by weight, is roughly 2:1), with a consistency that is firm rather than liquid. Sponge preferments can be used in any dough, but are typically employed in enriched or sweet dough products for the strength they provide. Sponge preferments compensate for the inhibited gluten development caused by fats and sugars. The flavor they provide depends on the amount of fermentation.

Pâte Fermenté or Old Dough

Pâte fermenté is composed of flour, water, salt, and yeast in the same percentages as in the final dough. The *pâte fermenté* method is an easy way to use a preferment without having to mix a separate preparation. Simply reserve a portion of dough from a daily production and use it as a preferment in future dough. Remove the portion after mixing, or following primary fermentation, and refrigerate it to slow down fermentation. To incorporate "old dough," cut the portion into small pieces and add it toward the end of the final dough mix cycle. Pâte fermenté can also be mixed as a separate preferment.

Poolish

A *poolish* contains flour, water, and yeast and is always 100% hydrated (i.e., the ratio of flour to water, by weight, is 1:1). A poolish provides the sweet, nutty, and wheaty flavors captured in a lactic acid profile. Due to the liquid nature and lack of salt in a poolish, the activity of the protease enzyme is rampant—mellowing the gluten for good extensibility. This extensibility is especially important in the shaping of breads. For this reason, poolishes are often used in baguettes. A poolish typically contains one-third to one-half of the dough's total water. It is ripe and ready to use when the surface has reached its maximum domed shape and is just beginning to recede. A ripe poolish exhibits small crevices in the surface and good bubble activity from the carbon dioxide. *See Figure 13-4.*

Figure 13-4 Just mixed, ripe, and overripe poolishes.

Biga

Biga is an Italian preferment, developed for flour made from the soft wheat grown in that region. The abundant acetic acid produced in a biga strengthens the gluten matrix of the dough, compensating for the low protein content of soft wheat. Biga, like a sponge, is 50% to 60% hydrated with a firm consistency. It consists of flour, water, and yeast, and produces a flavor profile of full fermentation. Bigas are good to use in doughs with higher hydration, but can be manipulated to benefit any dough.

Levain or Sourdough Starter— Naturally Leavened Preferment

A naturally leavened preferment relies on wild yeast for fermentation in contrast to preferments using manufactured yeast. See Naturally Leavened Breads, page 229 for more information on naturally leavened preferments.

The Steps of Breadmaking

Bread production is made up of a series of interrelated steps that may include scaling, mixing, primary fermentation, dividing, rounding, bench resting, shaping, final proofing, baking, and cooling and storing. Some breads, for example, ciabatta and other free-form loaves, are divided directly from the mass of dough into a final shape where bench rest or preshape are not necessary.

Scaling

It is important to accurately measure all ingredients. Weighing ingredients is more accurate and faster than measuring ingredients volumetrically and also allows the use of the baker's ratio system. Preferment and final dough ingredients may or may not be scaled at the same time. Scaling is also an important part of "mise en place."

Mixing

For thousands of years bread was mixed by hand in large troughs. Today, commercial bakers use planetary or spiral mixers. In planetary mixers, a dough hook kneads bread in a fixed bowl. In a spiral mixer both the hook and the bowl rotate. Spiral mixers are designed specifically for bread dough. They produce a more developed dough and heat the dough less than planetary mixers.

The Autolyse

An autolyse is an optional step used prior to mixing that reduces mix time, improves extensibility of the dough, and promotes a cream-colored crumb. Developed by the influential French baker Raymond Calvel in the mid-1900s, *autolyse* roughly translates from the French into "self-destruction." Flour, water, and sometimes a preferment are mixed together until just combined, and then allowed to rest for 20 to 30 minutes. During this resting period, the naturally occurring enzymes destroy weak gluten bonds, and allow stronger gluten bonds to form more quickly once the mixing process begins. Salt and yeast are not included in the autolyse because they tighten rather than relax gluten. Salt and yeast are added just before the final mix begins.

Bread containing whole-grain flour benefits from the use of an autolyse because the absorbent bran has an opportunity to thoroughly hydrate and ensure proper hydration of the final dough. The autolyse is best suited for lean dough and is generally not used for enriched dough used in items such as brioche.

STEP PROCESS

The Steps of Breadmaking

The steps to making bread are similar to a journey the baker embarks upon taking simple materials such as water, flour, salt, and yeast, and ending with a complex, beautiful, and tasty loaf of bread.

To make yeast-leavened bread, follow these basic steps:

1 Scale the ingredients, and measure all liquids.

2 Mix the liquid and the dry ingredients.
In the development stage of mixing, the gluten is developed and oxygen is absorbed into the dough.

3A Floor time, or bulk fermentation.

3B During bulk fermentation, much of the dough's flavor is being developed.

3C As an added step, the dough may be turned out onto a bench and folded.

3D Folding the dough before dividing helps expel accumulated carbon dioxide and strengthens the dough.

4 Divide and scale the dough.

5 Rounding or preshaping dough.

6 Bench rest makes final shaping easier.

7 Shaping.

STEP PROCESS

8 Panning and final proof: Assorted pans, baskets, and couches are some of the ways to hold shaped loaves during the final proof.

9 Most bread doughs are scored before baking.

10 Baking.

Relative humidity is an important consideration that affects the moisture content of flour. Dough hydration must be adjusted to accommodate seasonal changes in humidity. Before mixing, calculate the necessary water temperature based on the desired dough temperature listed in the formula.

Proper mixing also requires use of the senses—one should smell the ripeness of the preferment, taste it to ensure that salt has been added, listen for the different sounds the dough makes as it develops, watch the gluten develop as dough comes together, and feel the dough to check its hydration. To avoid errors or omissions during mixing, it is prudent to develop and follow a routine and to make necessary adjustments as quickly as possible.

Ingredients should be mixed on the first speed until all ingredients are thoroughly incorporated. At this stage, dry ingredients will become properly hydrated and gluten molecules begin to align. The mixer is then set on second speed. On second speed gluten begins to develop and organize, and the dough becomes smoother, stronger, and more cohesive. As the dough becomes cohesive, it pulls away from the sides of the mixing bowl. This is called the "clean-up" stage.

Mixing on second speed is continued until proper development is achieved. Proper development is usually determined by pulling a small sample into a thin membrane called a *gluten window*. Depending on the type of bread, the gluten window may not be achieved until the end of mixing, or further along in the fermentation process.

See Figure 13-5 showing three stages of dough development: incorporation, clean-up stage, gluten window.

Artisan breads tend to limit mixing in favor of dough development because extensive mixing incorporates too much oxygen into the dough. This oxidation destroys *carotenoid pigments* responsible for desirable flavor development and creamy coloration, and yields bread with a white crumb and inferior flavor.

Incorporation of Additional Ingredients

The objective of proper mixing is to develop the gluten adequately so that the dough can retain carbon dioxide and create the desired crumb structure. Large or sharp ingredients such

Figure 13-5 Three stages of gluten development.

Incorporation Clean-up stage The gluten window

as nuts, whole-grain soakers, olives, or dried fruit can damage gluten by severing gluten strands. To avoid this damage, add these ingredients at the end of the mix after gluten has developed and mix on first speed until just incorporated. Moisture in the additional ingredients will affect overall hydration, so it is important to adjust accordingly. Olives, for example, are moist, so the dough should be mixed slightly dryer than usual to allow for the moisture from the olives. Olives should be thoroughly drained, pitted, and mixed gently on first speed. Avoid overmixing, or the olives will be mashed into paste.

Common additional ingredients are numerous and may include:

- ***Dried Fruit***   Raisins, currants, and cranberries can be soaked in advance of mixing or used in their dried state. They should be used at room temperature so that the dough temperature is not altered.
- ***Nuts and Seeds***   Toast nuts lightly to release their full flavor, allowing ample time for the nuts or seeds to cool completely before mixing so that the dough temperature is not inadvertently increased.
- ***Whole-Grain Soakers***   Hydrate whole or cracked grains for up to 24 hours prior to mixing. This soaking period allows the grains to become fully hydrated before they are incorporated in the dough, so that the dry grains don't absorb the water necessary for proper dough hydration. Soaking also softens the sharp edges of the grains to minimize gluten damage. Salt from the final dough may be added to the soaker to prevent undesirable fermentation, especially when the soaker is stored in warm conditions.

Three Common Mixing Techniques

Mixing methods greatly affect dough strength and required fermentation time. Three mixing techniques commonly used in bakeshops today include the short mix, the intensive mix, and the improved mix.

SHORT MIX   Before industrialization and the advent of mechanical mixers after World War II, bread dough was mixed by hand. This gentle way of mixing produces dough that tends to be slightly underdeveloped. By extending the length of the first fermentation, the strength of the dough can be increased.

The short mix is intended to replicate hand mixing, so mix time is limited and primary fermentation is extended. With the short mix there is little chance of oxidation, so this method generally yields a creamy crumb and superior flavor.

INTENSIVE MIX   When mechanized mixers became widely used, bakers soon understood that more mixing created dough that was strong enough to compensate for greatly reduced fermentation times. Mechanized mixers also allowed for greater production and less labor to bake the same amount of bread. But, while productivity increased, bread quality declined sharply. The satisfying tastes, aromas, and texture of well-fermented bread were replaced by a bland flavor and a white crumb indicative of oxidized dough. Intensive mix is used primarily for enriched and sweet doughs that contain fats and sugar (discussed later in this chapter).

IMPROVED MIX The improved mix combines elements of the short and intensive mixing methods to produce dough that is not overdeveloped yet strong. The improved mix is used for most artisan bread production. Mix on first speed for three to four minutes, then on second speed for three to four minutes, until the dough is smooth, slightly tacky, and slightly shiny.

Primary Fermentation

After mixing, the dough is put into tubs, covered, and allowed to ferment. This primary fermentation is also referred to as "floor time" or "bulk fermentation." The majority of the dough's flavor is developed during this stage. The baker may fold the dough one or more times during primary fermentation. *Folding* strengthens the dough, expels some of the accumulated carbon dioxide, stimulates fermentation by incorporating oxygen, and gives the baker an opportunity to monitor the dough's progress. Primary fermentation can be as short as 45 minutes but is often as long as several hours.

Dividing

Dividing is the step of cutting a large piece of dough into individual sized pieces. Divide the dough gently, quickly, and accurately with a dough cutter or bench scraper, and adjust the weight as necessary by adding or removing dough.

Hydraulic and volumetric dough dividers are used in high-production bakeshops. They reduce labor but may handle dough roughly, degassing the dough, and compromising the interior structure of the final loaf.

The Retarding Process

Making good bread is a long process that often requires many hours. In planning a workable production schedule that yields finished products early in the day, bakers often manipulate the timing of a dough's progress by chilling the dough to slow down, or retard fermentation. An extended production timeline makes production more convenient for the baker because bread can be mixed one day and baked the next. (Recall that yeast activity is regulated by temperature.) Extending fermentation in this way is called *retarding* the dough.

The long, cool fermentation imparts certain characteristics to the finished bread. Retarded breads tend to have a stronger sour flavor because the conditions are ideal for the production of acetic acid, and loaves often have small blisters on the crust from gas collected just beneath the outer crust as it migrates to the surface.

To retard during primary fermentation, reduce fermentation time at room temperature and place the dough in a cooler or retarder for 12–18 hours. After bulk retarding, divide, shape, proof, and bake as usual. To retard during final proof, divide and shape as usual, and chill the shaped loaves overnight. The following day, loaves can be immediately baked to provide fresh bread the first thing in the morning.

Commercial retarders keep dough at the ideal retarding temperature (approximately 45°F–50°F [7°C–10°C]) and humidify the air so that the surface of the dough does not get dry and crusty. This equipment is not always found in a bakeshop, but a regular walk-in or other standard refrigeration unit can be substituted. The temperature for standard refrigeration (39°F [4°C]) is a bit too low for retarding, so nearly all yeast activity stops. In this case, additional fermentation and/or proofing time may be necessary before or after the retarding period.

Rounding

Rounding is also called preshaping. Immediately after dividing, gently work each weighed-out piece of dough into a somewhat smooth and uniform shape in anticipation of final shaping. Place preshaped dough on lightly floured proofing boards and load the boards onto a covered rack.

Bench Rest

Bench rest is the time when the preshaped loaves simply rest on the bench or on a covered rack. This rest period allows the gluten to relax, facilitating final shaping. Cover loaves resting on the bench with plastic so that a skin does not develop. Rest time is sufficient when loaves have noticeably relaxed.

Shaping

After an adequate bench rest, preshaped loaves are shaped into tight, well-shaped loaves as gently and efficiently as possible. Shaping techniques vary widely depending on the bread and the baker. After each loaf is shaped it is placed in or on parchment paper, bread pans, proofing baskets like German brotform or the French banneton, or in linen material called a *couche*.

To prevent sticking, couches and proofing baskets may need to be lightly floured with bread flour. The flour layer should be just heavy enough to prevent sticking. Excessive flour inhibits crust formation and imparts an unpleasant dry sensation in the mouth. Breads and rolls may also be slightly dampened and rolled in seeds for added flavor and a decorative appearance.

Final Proof

The final proof immediately precedes baking. During this stage, carbon dioxide continues to accumulate in the gluten matrix, and the loaves increase in volume by about 66%–75%. A warm (75°F–78°F [24°C–26°C]), slightly humid environment is ideal for the final proof, but a proofer/retarder may also be used. It is important to cover shaped loaves during final proof to prevent the dough from drying out. Proof times range from 45 minutes to several hours (or overnight if the dough is retarded).

Determining when loaves are properly proofed requires intuition and careful attention to cues learned through experience. All the variables that are part of breadbaking—dough temperature, hydration, fermentation rate, etc.—affect final proof time. Generally, bread is ready to bake if it springs back lightly when touched and has increased in volume by about 66%–75%. Dough that springs back quickly when touched has not achieved full proofing volume and is considered *underproofed.* At the other extreme, *overproofed* dough does not spring back when touched. Overproofed loaves may collapse when transferred or scored. If optimum proofing cannot be achieved due to timing conflicts in the bakeshop, it is preferable to bake dough that is slightly underproofed rather than overproofed.

Characteristics of underproofed dough:
- Loaves will burst and become misshapen in the heat of the oven because the rapid production of carbon dioxide cannot be contained within the underproofed dough.
- Baked loaves will feel heavy, have low volume and a dense crumb that seems somewhat raw, as compared to the dense crumb of overproofed bread.
- Crust will be a dark reddish brown due to an excess of residual sugars in the dough.

Characteristics of overproofed dough:
- Baked loaves will be flat, with low volume and a dense, crumbly dry crumb due to excess acid buildup, which deteriorates gluten.

- Crust color will be pale because of a lack of residual sugars.
- Loaves may be overbaked in an attempt to achieve a darker color resulting in a thick, dry crust.
- Unpleasant flavors and aromas of overfermentation may occur.

Baking

Baking is the culmination of the breadmaking process. This step includes setting up mise en place at the oven, scoring, loading and steaming the oven, and the actual baking.

Mise En Place for Baking

To facilitate swift loading and scoring, the baker should have all necessary tools on hand ahead of time.

When loading the oven, use the hearth space completely to maximize production in each load. Avoid wasting space, but allow enough space for air to circulate between the loaves. Loaves should also be spaced sufficiently apart to keep them from touching each other as they expand in the oven. The undesirable soft, pale sidewalls that result from loaves touching is called *baisure de pain*; the translation from French means *kissing crust*.

Scoring

Immediately before baking, most loaves are scored with a cutting instrument called a lame (pronounced *lahm*). **Scoring** provides a predictable spot for the loaf to burst. If the dough is not cut, a loaf may not split at all, yielding a dense crumb. It may also burst randomly at the weakest spot in the loaf and create a distorted loaf. The cuts allow the loaf to expand fully during the oven spring period of baking to create an open crumb.

In addition to its functional aspect, scoring also has aesthetic merit. While some types of bread have characteristic score patterns that are traditionally used, bakers can also develop signature score patterns to identify particular loaves in their production. With practice, graceful italic letters or other artistic score marks can greatly enhance the beauty of a loaf of bread.

When scoring a loaf, only the corner of the lame blade should come into contact with the dough. The baker should stand back slightly from the loader or peel, and make cuts with swiftness and precision. To avoid snagging, the blade should be moving when it comes in contact with the dough.

Steaming the Oven

Before dough is placed in a deck oven, steam must be introduced into the bake chamber. Convection ovens should be steamed immediately after they are loaded to avoid losing steam out of the open door. Most convection and deck ovens have built-in steam injection systems, but steam can also be added by spraying water directly into

---

Oven Spring

Oven spring is the dramatic increase in volume that occurs during the first minutes of baking due to the rapid production of carbon dioxide as the fermentation rate spikes in the heat of the oven, and the expansion of water into steam. The loaf swells as both gases are caught in the gluten matrix of the dough. Maximum oven spring is achieved when the dough is extensible, when the oven is hot enough, and when the oven chamber is adequately steamed.

a brick or pizza oven. Vent the steam once the bread begins to color, and finish baking in a dry oven.

In addition to the steam injected into the bake chamber, steam is also produced by the loaves as water in the dough evaporates in the heat of the oven. Steam affects yeast-raised doughs in several ways:

- Provides a humid environment that allows the crust to stay supple so it can expand to its greatest volume during oven spring. Dough that is placed in a dry oven quickly develops a skin that prevents the loaf from full expansion.
- Promotes a shiny crust by aiding the Maillard Reaction.
- Allows cuts to open dramatically and form distinct "ears." Too much steam, however, may cause cuts to stick together and fail to open.

If steam is absent during the first minutes of baking, the loaves will be dull, pale, lacking in flavor, and the cuts will fail to open.

Baking

When loaves are placed in a properly heated and steamed bread oven, a number of baking phases take place. These phases are not strictly successive, that is, they may overlap each other.

- The loaf swells rapidly due to oven spring.
- Steam condenses on the surface of the loaf, delaying the formation of the crust. This supple outer layer allows for greater oven spring.
- Proteins begin to coagulate at 140°F–160°F (60°C–71°C) giving structure to the loaf. By this time oven spring has ended.
- Starches gelatinize when starch granules absorb water and swell. This process begins at about 105°F (41°C) and is mostly complete by 200°F (93°C). Gelatinization also contributes to structure.
- Alcohol and water evaporate. Dough can lose approximately 20% of its scaled weight during the baking process due to evaporation.
- When surface temperature exceeds 212°F (100°C)—the boiling point of water—crust color and development begins. At this temperature, residual sugars (that remain in the crust after fermentation) begin to change due to caramelization and the Maillard Reaction. Caramelization and Maillard browning are separate processes. Caramelization occurs when sugars break down, while Maillard browning occurs when sugars break down in the presence of proteins. Caramelization and the Maillard Reaction contribute significantly to the satisfying taste of well baked bread. If sugar, steam or high heat are absent, these processes will not occur and the baked loaf will be dull, pale, and lacking in flavor. *See Figure 13-6 for baguette troubleshooting chart.*

Bread Ovens

Ovens are usually the most expensive piece of equipment in the bakeshop. It is important to understand the type of oven needed for different kinds of products.

BRICK OVENS These are single-chamber ovens that are usually heated with wood. A fire is burned in the firebox and heats up the thermal mass. After a firing that can last 8 hours, coals and embers are raked out, the hearth is swept, and the oven cools to a proper baking temperature. Dough is peeled onto the hearth and is baked with residual heat stored in the thermal mass of the oven. It is important to learn the characteristics of an individual brick oven to consistently bake good bread. Steam is injected manually.

DECK OVENS Deck ovens are ideal for baking hearth loaves. Because there may be several decks, each with separate controls, this oven fueled by gas or electricity offers ver-

Figure 13-6

a. underbaked
b. overbaked
c. baked too close together
d. no steam
e. no scoring
f. poorly scored
g. underproofed
h. poorly shaped

Baguette Troubleshooting Chart

satility and quick temperature recovery. Deck ovens are steam injected. Loading is facilitated with a device called a loader.

CONVECTION OVENS Fans circulating the hot air and a rotating rack in the convection oven ensure even baking. This type of steam-injected oven is ideal for baking pastries and often has the capability to lift and rotate a rack full of product. *(See Chapter 5: Equipment, Hand Tools, Smallwares, and Knives for more information on ovens.)*

Unloading the Oven

Hearth breads are properly baked when they have achieved a rich, dark crust color. The *grigne*, the area on the surface of the crust that was exposed by the score marks, should also be hard and well colored. Fully baked bread will sound hollow when thumped on the bottom, offer resistance when squeezed, and feel lighter than it looks. Smaller loaves should be baked at a higher temperature for a shorter time, and larger items require a lower temperature and longer bake time.

Enriched breads such as brioche or American style pan breads darken in the oven, but do not achieve a color as dark as hearth breads. Their crust will also stay soft due to the sugar and fat in the dough. It is important to bake loaves completely to the center because water in underbaked dough will migrate throughout the baked bread, soften the crust, and decrease the shelf life.

Cooling and Storage

Loaves should be placed in a single layer on a cooling rack to allow the circulation of air. *See Figure 13-7*. Products such as rolls, sweet dough, and enriched breads may cool on a sheet pan. Anything baked in a pan or tin should be removed from the baking container after 10 minutes to prevent a soggy crust caused by condensation.

The crust of well-baked hearth breads will crackle as it cools and constricts. Sweet dough items should be glazed while still hot.

Although warm bread fresh from the oven is appealing, bread should cool fully before it is sliced to allow time for the crumb to set. The flavor profile is more complex and nuanced in room temperature bread.

Hearth breads should always be bagged in paper rather than plastic bags so that the crust stays crisp. Storing hearth breads in plastic will cause the crust to get tough and gummy. Enriched dough, such as brioche, should be bagged in plastic bags. Sweet-dough pastry items, like sticky buns, can be wrapped in plastic or stored in pastry boxes.

Figure 13-7 Baked loaves of bread should be cooled on a rack before storing or slicing.

Bread wrapped in plastic freezes well. When defrosting bread, allow it to thaw completely before removing the plastic so that water lost in the form of ice crystals inside the plastic can be reabsorbed into the loaf. *See Figure 13-8.*

Types of Breads

The varieties of bread are numerous and interconnected, although most can be categorized as lean-dough breads, enriched breads, or decorative breads.

Lean-Dough Breads

Lean-dough breads are made with the four basic bread ingredients and contain little or no added sugars or fats. They have a chewy crumb and a well-formed crust, and may contain whole grains, herbs, dried fruit, nuts, or cheese. Lean-dough breads include yeasted artisan hearth breads, naturally leavened sourdoughs, whole-grain breads, hard rolls, and flat breads. High hydration, open crumb structure, thorough fermentation, and a hardy crust are characteristics of lean-dough breads.

Yeasted Artisan Hearth Breads

There are many artisan hearth breads that illustrate the characteristics described above. *See Figure 13-9.* When bakers refer to bread as "yeasted," they are referring to bread leavened with manufactured yeast rather than wild yeast that is used in naturally leavened bread production. Two classic hearth breads and their traditional preferments are baguettes and ciabatta.

BAGUETTES Baguettes, the iconic bread of France, originated in Paris in the 1920s. In France a baguette is defined as containing 100% refined white flour, baker's yeast, salt, and water. Additives, other than ascorbic acid (vitamin C), are illegal. In the United States, the term baguette is sometimes used more loosely and refers to any loaf with the characteristic long, thin baguette shape. It is not uncommon to see sourdough or whole-grain "baguettes" in American bakeries. In recent years, the artisan bread renaissance has helped to reinstate much of the baguette's former romance and glory.

Figure 13-8 Hearth breads should be bagged in paper to maintain a crisp crust.

Proficiency in the production of baguettes is a standard measure of skill in the artisan bread industry worldwide. While there are various baguette formulas, the "Baguette with Poolish" method is a benchmark of the Johnson & Wales bread program.

CIABATTA Ciabatta, from the Lake Como area of Italy, is made from a very wet dough. The high hydration of the dough yields a very open crumb. The name means slipper in Italian, which the shape is said to suggest. Due to the extremely wet dough, a longer mixing cycle and/or more folds are necessary to develop sufficient strength. One production advantage of ciabatta (and other freeform doughs such as *pane pugliese* and *pane francese*) is that the loaves do not require the shaping step. Pieces of dough are simply cut from the bulk dough, adjusted to achieve the correct weight, and placed on a well-floured couche.

Ciabatta is perfect for sandwiches when sliced horizontally because the holes trap the dressings of a sandwich and the durable crust keeps them from dripping through. It is often used for panini and open-faced sandwiches, and is a versatile table bread.

Figure 13-9 Artisan breads: couronne, pain rustique, and boule.

Naturally Leavened Breads

Naturally leavened breads, often referred to as sourdoughs, are the oldest style of leavened breads. They are leavened solely with a *levain* or *sourdough starter*—a flour and water culture populated with natural or wild yeast. Wild yeast (*Saccharomyces exiguus*) is present in the air, collects on growing grain, and is visible as the white film on grapes. The levain is a naturally leavened preferment.

One key distinction between naturally leavened and yeasted preferments is that yeasted preferments are used in their entirety, while a portion of the naturally leavened preferment (sourdough starter) is always reserved to propagate more starter. The levain is "built" for production and kept healthy with regular feedings of flour and water. Naturally leavened breads are an important and respected part of breadbaking history and enjoy great popularity in today's industry.

Sourdough Is Local Food   Manufactured yeast is grown under controlled conditions. The resulting fermentation rate and flavor profile are consistent around the world if the baking conditions are constant. The characteristics of wild yeast and bacteria, however, vary from locale to locale to create bread that is indicative of a certain geographic location. For instance, San Francisco sourdough can only be authentically made in that city.

The taste and acidity of naturally leavened bread (and yeasted breads to some extent) can be controlled by manipulating the lactic and acetic acid balance. Acid balance is manipulated by flour selection, proofing, and the amount of starter used. Whole-grain flours increase fermentation and acetic acid production. The extended, cool fermentation in bulk or final proof retarding provides conditions that favor acetic acid production. The greater the amount of starter used in the dough the more sour the bread is. Because of the slower fermentation rate, naturally leavened bread is better suited to retarding than yeasted bread.

The Starter   Sourdough starters are preferments that influence breads the way yeasted preferments do. All preferments yield a more complex flavor, increased dough strength due to the tightening effect of proper acid levels on gluten, and an increased shelf life.

Naturally leavened preferments can have a stiff or liquid consistency and are made from white or whole-grain flour. Liquid white flour starter is hydrated from 100% to 125% and is very popular. Its high hydration makes it easy to feed. Stiff wheat starters are usually hydrated at a dough consistency of about 69%. Rye starters are composed entirely of rye flour

Figure 13-10 Sourdough starters.

and are usually stiff. Acidifying rye flour using a rye sour will be discussed later in this chapter. *See Figure 13-10.*

CARE AND FEEDING OF AN ESTABLISHED SOURDOUGH STARTER Sourdough starters are fed water and flour on a regular schedule. The frequency of the feedings depends on the hydration of the starter and the type of flour. In general, starters and sours are fed 6–12 hours before mixing, but feedings and fermentation times can be manipulated to coincide with a bakeshop's production schedule. Cool water feedings, low hydrations, and retarding can make a starter ripening schedule flexible. Underripe starters will appear inactive, lengthen the fermentation and proofing times of the dough, and may compromise the volume of the bread.

Overripe starters will begin to break down due to excess acidity. Overripe liquid starters "fall," and have a pungent smell. In addition, alcohol in the form of a gray liquid collects on the surface. Overripe stiff starter also deteriorates, takes on a gray cast, and seems lifeless and uncharismatic. *See Figure 13-11 for a comparison of naturally leavened versus yeasted breads.*

LEVAIN DE PÂTE Sometimes bread is made using both manufactured yeast (dry or in a preferment) and a sourdough starter. Bread made this way is called *levain de pâte*. This hybrid leavening method imparts characteristics of both methods, depending on the quantity of wild or manufactured yeast.

Whole-Grain Breads

Whole grains can be incorporated in bread dough in the form of flour—very fine to very coarse—or by adding crushed, broken, or intact grain berries to the dough. Finer flour produces lighter bread with more volume than coarse whole-grain flour. Broken or whole grains are incorporated in the form of a soaker. *See "Incorporation of Additional Ingredients" for more information on grain soakers.*

When using whole grains in any form, the ratio of water to flour (the hydration) should be increased to accommodate the high absorbency of the wheat bran. Whole-grain breads may be made entirely of whole grain or may contain a percentage of white flour. Check with the current FDA requirements for the proper labeling of whole-grain bread. *See Figure 13-12.*

Figure 13-11

		Naturally leavened breads	**Yeasted breads**
How Naturally Leavened Breads Differ from Yeasted Breads	**Crust**	thicker crust	thinner, crisper crust
	Shelf life	longer shelf life	shorter shelf life
	Crumb texture	chewier crumb	more tender crumb
	Length of production time	longer production time	shorter production time
	Consistency of flavor profile	flavor profile varies regionally, depending on local strains of yeast and bacteria	consistent flavor profile
	Nutrient Bioavailability	more nutritious	less nutritious
		natural fermentation neutralizes phytic acid, a substance in bran that resists mineral absorption by the human body	yeasted fermentation retains 90% of bran's phytic acid, so mineral absorption is blocked

Beyond Wheat: Other Bread-Making Grains

The advent of leavened bread dramatically increased the popularity of wheat across the globe because wheat, above all other grains, contains the gluten necessary to produce a light, digestible loaf. Cultivation, consumption, and knowledge of other grains decreased with time as wheat became the dominant grain used in breadmaking.

Figure 13-12 Whole grains can be ground very fine to very coarse when added to bread dough.

RYE   Rye is still used extensively in the breads of eastern and northern Europe, and Russia. Breads with high rye content from these parts of the world are often referred to as "black breads." Rye is the only grain besides wheat with qualities suitable for baking leavened bread. The gluten content of rye is much lower than wheat. There is a modest amount of glutenin and a very small amount of gliadin present.

In addition to gluten, rye contains another gas-trapping substance called *pentosans*. Pentosans are long chains of sugars (polysaccharides) that have high absorption. Dough with a high percentage of rye flour must be well-hydrated. Whole rye flour has a relatively short shelf life because of the high oil content of the rye berries. The oils oxidize at room temperature and impart a rancid flavor to the flour.

SPELT   *Spelt* is an ancient, nutritious, flavorful subspecies of wheat. Because it is low in gluten, spelt does not support extended fermentation. Spelt declined in availability with the mechanization of agriculture because its kernels have a husk that cannot be easily removed. Recent advances allow more efficient husk-removal, making spelt flour generally available to today's bakers.

CORN   Corn *(maize)* is indigenous to the Americas, and was a primary grain in the traditional diet of many North, Central, and South American cultures. Corn is flavorful and easy to grow, but is nutritionally deficient in comparison to most other cereal grains. Corn is milled to various degrees of fineness to make grits, hominy, polenta, cornmeal, and corn flour. Corn contains no gluten.

OATS   Oats are thought to have originated around the Mediterranean Sea. Romans introduced them to the rest of Europe, and by the 6th century oats had become a staple in and around Scotland. Oats contain no gluten but have more fat and protein than other grains. The insoluble fiber may help to reduce cholesterol. In baked goods, oats help retain moisture. The inside of the oat kernel is called the "groat" (the outer husk is removed). Oats are widely eaten as a warm cereal or porridge, while rolled oats are used in many baking applications.

Hard Rolls

Most of the dough that is used to shape loaves of bread can also be divided into smaller pieces for rolls. Ninety (90) grams of dough (3 ounces) is a good size for a roll. Dividing may be done by hand but it is more common to use a divider. Rolls are familiar to American

Acidifying Rye Flour

Rye enzymes, unlike those in wheat, are not killed by the heat of the oven. If not deactivated, these enzymes continue to turn the bread's starch into sugar even after baking. This effect is minimal in breads with only a small percentage of rye flour, but when loaves contain more than 25% whole rye flour, a failure to deactivate the enzymes results in a gummy loaf. The acidic environment of a naturally leavened preferment deactivates the enzymes. A rye sour should be used in breads with a high percentage of rye flour.

Figure 13-13 Sticky buns are a popular sweet-dough item.

consumers. Because portion control is easy, and slicing is not necessary, rolls are popular in food service establishments. Lean-dough hard rolls include Kaiser rolls, bagels, and pretzels.

Flat Breads

Flat breads may be leavened or not. Either way, they are as their name implies—flat. They are descendants of simple breads that were baked long ago.

Unleavened Flat Breads   Unleavened flat breads include chapatti, crackers, and tortillas.

Leavened Flat Breads   Pizza and focaccia are examples of popular leavened flat breads. Other leavened flat breads include Indian naan and Middle Eastern pita and lavash.

Enriched Breads

Yeasted enriched doughs differ from lean dough because of the addition of fat, sugar, and sometimes eggs. Enriched breads like brioche are high in fats, but contain less sugar relative to the high percentage of fat. Enriched dough that contains a high percentage of both fat and sugar is generally referred to as "sweet dough." Familiar sweet-dough items include cinnamon rolls, fruit-filled coffee cakes, and sticky buns. Sweet-dough items should be glazed or iced while warm. *See Figure 13-13*.

Some Examples of Enriched Breads

Because of the rich ingredients used, many traditional enriched breads are baked for special occasions rather than as everyday breads. Kugelhopf and panettone are two enriched breads that are baked especially during the holiday season.

Pan Breads   American pan breads that are sliced for sandwiches often include some sugar or other sweetener, and some fat. Soft dinner rolls are also made from dough similar to that used for American pan breads. Pan de mie is an example of a European-style sandwich bread that is baked in a lidded mold called a Pullman pan.

Techniques for Sweet and Enriched Doughs

The addition of fats, sugars, and eggs in enriched and sweet dough makes it necessary to utilize a variety of techniques while still adhering to the basic steps of breadmaking.

Sweet and enriched doughs can be made with either a preferment or a retarded straight dough. When using a retarded straight dough, fermentation is slowed and extended, allowing for more flavor development. The cool temperature also firms the fats and eggs to make the dough easier to sheet or shape.

Sweet and enriched doughs require an intensive mix due to the effects of sugars and fats on gluten. Sugar weakens the gluten making the product tender (a characteristic that is desirable in this type of bread) while the fats coat and shorten gluten strands to tenderize the dough. Proper mixing ensures that fat is adequately incorporated into the gluten matrix.

Osmotolerant yeast is used in enriched dough formulas to counter the hygroscopic nature of sugar. For a shiny, smooth crust on brioche, challah, croissant, or sweet-dough products, proofed loaves should be egg-washed just before baking. Bake in a convection oven at approximately 350°F (177°C), and glaze or finish, if desired, while still hot.

BRIOCHE Brioche is a classic French yeasted pastry item that utilizes an intensive mixing method to incorporate a high ratio of butter into the dough (up to 50% to 60% of total flour weight). The name comes from the French word *brier* meaning "to pound," referring to the pounding required to make cold butter pliable. (This step is *not* unique to brioche, and is necessary in the enriched dough products listed.) Brioche has numerous traditional shapes and is versatile in use. It can be employed for both sweet and savory applications, and can also be laminated. *See Figure 13-14.*

STOLLEN Stollen is a traditional Christmas bread made in Dresden, Germany. Stollen always contains nuts and candied fruit. An optional ingredient in stollen is almond paste, which is rolled into a log and placed in the center of the loaf. While still hot, loaves of stollen are brushed or dunked in melted butter and then dredged in granulated or powdered sugar.

CHALLAH Challah is the braided egg bread traditionally served on the Jewish Sabbath and other Jewish holidays. Challah is often braided into a straight four-strand braid, but round loaves and other braiding styles are also common.

Figure 13-14 Traditionally shaped brioche.

Decorative Breads

There are two distinct categories of decorative breads: one group is made from regular bread dough shaped into special decorative shapes, and intended to be served and eaten. Many of the traditional shapes used for these decorative breads were developed for European artisan breads and are symbolic or meaningful to certain regions or events.

Dough used to make purely decorative breads, on the other hand, is called *decorative* or *"deco" dough*, and, depending on the style of the decorative piece, may or may not contain yeast. Decorative dough is called "live" dough if it contains yeast and "dead" dough when no yeast is used. Even though deco dough is not intended for consumption, the convention is to use only food-grade ingredients when formulating deco dough. Depending on the desired colors and textures, a wide range of ingredients may be used. Turmeric produces a deep golden tone, paprika is for red, and powdered dried spinach produces a rich green color. Decorative bread pieces are often built by assembling many pieces of deco dough that have been previously baked. This type of decorative bread may be used to create handmade marketing tools for a bakery such as plaques or signage, and for making elaborate competition showpieces.

CHAPTER 14

Quick Breads

Quick breads are familiar staples in the American diet. Biscuits and muffins are popular breakfast quick breads, corn bread is served with lunch and dinner, and scones and banana bread are served anytime with a cup of coffee or tea.

KEY TERMS

quick breads
chemical leaveners

bench tolerance
tunneling

rubbing method or biscuit method

creaming method
blending method

Quick Breads

Quick breads are chemically leavened by carbon dioxide (CO_2) gas that is produced by a chemical reaction. Products containing baking powder or baking soda such as cakes, cookies, pancakes, and waffles are all quick breads. *See Figure 14-1.* Other breads are leavened by CO_2 gas produced through the biological process of living yeast feeding on sugars in the dough. Quick breads are also distinguished by their mixing methods, which minimize gluten development. These methods create a tender, rather than gluten-strengthened, crumb.

Figure 14-1 Collectible trading cards were used to market many baking products in the 19th century.

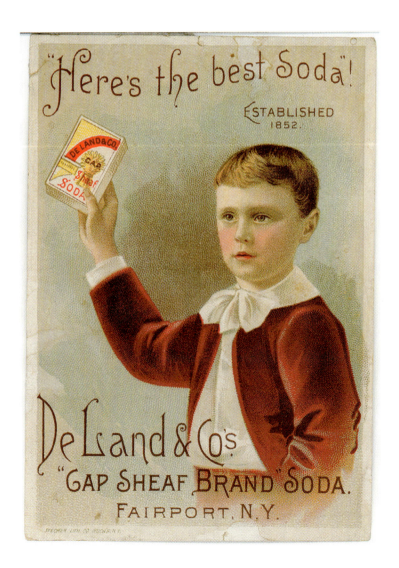

Quick breads serve the same function and fill the same carbohydrate niche in a meal as do yeasted breads. Their production time is much shorter because there is no fermentation. Production methods are also relatively simple as no special bakeshop or pastry shop equipment is required.

Chemical Leaveners

Chemical leaveners produce carbon dioxide (CO_2) in the presence of moisture and/or heat. When carbon dioxide collects in the air cells of the minimal gluten matrix, the cell walls stretch and become thinner to create a tender product. Chemical leaveners also help to tenderize quick breads by weakening gluten.

Leavener provides volume and tenderness in quick breads. Insufficient quantities of leavener in a formula result in low rise and less tender crumb in the finished product. Too much chemical leavener causes the product to rise too high and to then fall before or during the bake. Excess leavener also produces odd colors, a chemical flavor, and a dry, crumbly texture.

In a humid kitchen, prescaled chemical leaveners are partially activated by moisture in the air, which causes the release of some of the carbon dioxide before the batter is mixed. This release produces a heavy, dense final product. For this reason, prescaling chemical leaveners is not recommended. Chemical leaveners work best when fresh. Store in airtight containers in a cool dry place and be mindful of expiration dates.

Bench tolerance indicates the amount of time dough or batter can retain reliable leavening action between the mixing and baking steps. Bench tolerance increases with the use of chemical leaveners and allows large batches of quick bread items to be mixed well in advance of baking. For instance, muffin batter can be mixed, refrigerated, and used over several days without a reduction in leavening or volume.

Baking soda (sodium bicarbonate) must be combined with liquid and an acid to effectively create carbon dioxide. Some acid sources common to quick bread production are buttermilk, honey, molasses, cream of tartar, and fruits or fruit juices. It is necessary to use an acid to efficiently break down baking soda into carbon dioxide, water, and a salt residue. Failure to use an acid with baking soda (or the use of too much baking soda) creates a greenish color and a chemical flavor in baked goods. The ratio of baking soda to flour, by weight (i.e., the baker's percentage), ranges from 0.5% to 1.5% for most quick breads.

Baking powder contains baking soda and an acid/salt such as cream of tartar, or monocalcium phosphate. Baking powder also contains a filler such as cornstarch that helps to absorb moisture, and to prevent the baking soda and other acidic ingredients from reacting during storage. When double-acting baking powder is sifted with dry ingredients and a liquid is added, the baking soda and the acidic ingredients combine to create a rapid chemical reaction that produces carbon dioxide. This gas is then trapped in the dough or batter. When the product is baked, a slower chemical reaction takes place and produces more gases. The remaining gases expand during the baking process, causing the product to leaven. This double action makes baking powder a reliable leavener and greatly increases bench tolerance. The amount of double-acting baking powder in a quick bread formula generally ranges between 1% and 7% of the amount of flour, by weight.

Physical Leaveners

Quick breads, like all baked goods, are also leavened by physical leaveners. The physical leavening occurs in the heat of the oven as molecules of air and water trapped in a dough or batter expand and turn to steam. The rapid expansion of steam swells air cells in a dough or batter to produce a volume of steam that is over 1600 times that of liquid water.

Air is incorporated by sifting dry ingredients, creaming fats and sugars, and whipping eggs and sugar. The final mixing process also incorporates air. The proper incorporation of air is vital in achieving a light final product. Heavy, dense quick breads are often created by

Doughnuts and Fritters

Doughnuts and fritters are made from dough or batter that is fried rather than baked. Many doughnuts are made from yeasted doughs, but cake-style doughnuts and fritters are quick breads that are mixed using either the creaming or blending method. Yeasted doughnuts are mixed according to the intensive mixing method for yeasted doughs discussed in Chapter 13: Breads. They are proofed to proper volume before frying.

Hydration of doughnut batter is adjusted according to the production method. A batter-like consistency is used for an automatic depositor that cuts doughnuts directly into frying fat while a firmer dough is appropriate for hand-cut cake doughnuts. Quick bread doughnut dough should rest for approximately 20 minutes before frying to allow the gluten to relax.

Frying fat must be clean and between 360°F to 385°F (182°C to 196°C) for doughnuts and from 360°F–375°F (182°C–191°C) for fritters. Doughnuts and fritters will absorb frying fat that is too cool, producing a greasy and heavy doughnut or fritter. Fat that is too hot will darken and break down prematurely. Replace fat when it becomes dark to avoid rancid flavors and excessive browning.

Hold the frying screen at the surface of the hot fat until the doughnuts or fritters float free. Fry approximately 1 minute, flip, submerge, and fry for another minute. Lift the screen out of the fryer, allow excess fat to drain into the fryer, and deposit doughnuts or fritters on absorbent paper. Doughnuts are finished in a variety of ways. They may be glazed, dredged in powdered sugar, or iced. The shape and type of doughnut usually determines the finish.

Cracked surfaces on cake doughnuts may be due to

- overmixed or stiff consistency batter.
- an insufficient rest period between cutting and frying.
- dough that is too cold or fat that is too hot.
- too short a fry period on the first side.

Flat cake doughnuts may be caused by

- overhydrated, warm, or undermixed dough.
- a low fat level in the fryer.

mixing and handling methods that fail to incorporate enough air cells. No new air cells form during baking, so proper mixing is vital to successfully leaven quick breads.

Types of Quick Breads

Quick breads can be made from dough or batter. Quick bread dough has a consistency that makes it ideal for gently rolling and cutting into various shapes before baking. The rubbing mixing method, also called the biscuit method, produces soft dough. Dough-consistency quick breads include biscuits, scones, and Irish Soda Bread.

Muffins and loaf quick breads are batter products that are produced by either the creaming or blending mixing methods described below. Some quick bread products such as pancakes, waffles, and corn bread have thin batters referred to as pourable batters. Other quick bread products such as blueberry muffins or banana bread can be deposited with a spoon or dropped with a portion scoop.

Mixing Methods

Mixing should accomplish the following objectives:

- Achieve a uniform and complete mixture of all ingredients.
- Form and incorporate air cells.
- Develop the desired grain and texture in the finished product.

Mixing is necessary to create minimal gluten development. Excessive gluten development due to overmixing quick breads produces dense, tough, and distorted shapes in the

STEP PROCESS

The Rubbing, or Biscuit, Method

The rubbing, or biscuit, method produces not only biscuits but also scones, shortcakes, and soda breads. It is very similar to the method used to make pie dough.

To use this method, follow these steps:

1 Sift all dry ingredients.

2 Rub or cut the fat into the sifted dry ingredients.

3 Add liquid ingredients (often milk and eggs blended together), and mix lightly, until just slightly developed and incorporated.

4 Turn the dough onto a lightly floured work bench, and knead several times to add volume and flakiness.

5 Allow the dough to rest about 15 minutes. Use a rolling pin to roll the dough about 1 inch (2.54 cm) thick.

6 Cut out biscuits close together to avoid excess scraps. Place about 1 inch (2.54 cm) apart on parchment lined sheet pans.

---

final product. Overmixing may also cause *tunneling*—the formation of holes within the baked good. Care should be taken to mix ingredients until incorporated.

There are three basic mixing methods for producing quick breads: rubbing, creaming, and blending.

The *rubbing method,* also called the *biscuit method,* involves rubbing or cutting fats into flour to create a mixture of flour and large fat flakes.

Fats should be solid and cold to prevent blending with the flour. Large, flat shards of fat the size of a dime create flaky products, although many formulas specify pea-sized pieces of fat. The water in the fat turns to steam in the oven and creates a light and flaky product through physical leavening. Cutting the fat into the flour with the fingers or with a pastry blender is best to prevent overmixing. A mixer fitted with a paddle and carefully operated on low speed can also yield good results. Fat coats the gluten strands to limit water absorption and minimize gluten development. After combining the fat and flour, cold eggs and liquid ingredients are added and mixed until just combined. Knead the dough several times on a lightly floured bench to add volume and flakiness. Pieces of fat should be visible in the dough. Roll the dough to a uniform thickness, rotating the dough several times so that the gluten develops in all directions to avoid distorted shapes.

Biscuits and scones should be cut with a clean, sharp, floured cutter or knife. *See Figure 14-2.* Twisting a cutter will smear the layers of fat and dough and create products that are less flaky. Placement of the cuts should be as close as possible to each other to reduce scrap dough that needs to be rerolled. Allow dough to rest and relax for 20–30 minutes before baking. Scones and biscuits may be eggwashed before baking or left plain for a more rustic appearance. Sanding or coarse sugar can also be applied to eggwashed scones. *See Figure 14-3.*

The *creaming method* produces a cake-like texture by mixing solid fats with crystalline sugars and then adding dry ingredients and any liquid ingredients. **(See Chapter 24: Cookies and Petits Fours, for more information on the creaming method.)** Fats and sugar should be creamed on second speed of a 20-qt. mixer or third speed on an 80-qt. mixer until the mixture lightens in color and increases in volume. The sharp edges of the sugar crystals help to cut air cells into the fat. Overcreaming or undercreaming will have a dramatic effect on the finished product, resulting in less volume and poor texture.

Numerous factors contribute to proper creaming, including the speed of the mixer, and the temperature of the fat, the eggs, and the bakeshop or pastry shop. To properly incorporate air cells, the fats should be 65°F to 75°F (18°C to 24°C). Cold fats require a longer mixing time for maximum creaming; fats above 75°F (24°C) are too soft and cannot hold the proper amount of air cells, resulting in poor volume. Bakeshop or pastry shop temperature also affects creaming. Creaming takes place faster in cooler weather than in extreme heat. Mixing at higher speeds creates friction and reduces or destroys the proper amount of air cells. Temperature and mixture speed adjustments and allowances for mixing times should be evaluated.

The addition of eggs must be gradual to allow them to emulsify into the sugar and fat mixture. Egg yolks contain lecithin, a natural emulsifier that coats the surface of the air cells and holds liquid without curdling. Curdling occurs when the amount of liquid exceeds the retention capacity of the air cells. It is important to

Figure 14-2

After rolling the dough, scones can be cut into pie-like slices using a French knife. If using a standard biscuit cutter, cut out biscuits close together to avoid excess scraps.

Figure 14-3

Before baking, scones may be brushed with egg wash. Sanding or coarse sugar can also be used.

STEP PROCESS

The Blending Method

Use the blending method to make muffins and loaf breads such as blueberry muffins and banana bread, as well as pancakes and waffles. Muffins made by this method have a breadlike texture similar to that of loaf breads.

To apply the blending method, follow these steps:

1 Blend liquid ingredients with sugar.

2 Add remaining dry ingredients to the liquid mixture.

3 Add water, and blend together.

4 Add additional ingredients such as berries or nuts. The batter is ready to be stored (depending on bench tolerance) or panned.

scrape the bowl and paddle frequently at this stage to ensure thorough incorporation of the ingredients.

In the final stage, add all the sifted dry ingredients and mix on low speed until just incorporated. If more liquids are added to the mixture, alternate dry and wet ingredients. Begin and end with dry, until all ingredients are incorporated. Ingredients such as nuts, chocolate chips or blueberries can then be added and gently mixed or folded.

The *blending method* is used in quick bread formulas that call for oil. When using this method, the dry ingredients—with the exception of the sugar—are sifted together and set aside. The oil and eggs are blended together with the sugar. The dry ingredients are then added to the egg mixture and mixed on low speed until just incorporated. Water or other liquids are added according to the formula. Ingredients such as nuts, chocolate chips, or blueberries can then be added in and gently mixed or folded.

STEP PROCESS

Panning Quick Breads

When panning quick breads, it is best to scoop the batter from the edge of the mixing bowl to prevent overmixing.

After mixing the batter, follow these methods for panning quick breads:

1 A portion scoop is used to deposit batter into prepared muffin tins or paper baking molds.

2 For loaf breads, scale an appropriate amount of batter into loaf pans.

3 Streusel topping can be applied before baking.

4 To make an attractive split in the top of a loaf bread, dip a dough scraper in vegetable oil and swipe the dough scraper down the length of the batter.

Panning, Baking, and Cooling Quick Breads

Once quick breads are mixed they can be stored (depending on their bench tolerance), panned, and baked. Loaf pans and muffin tins should be sprayed with pan release or lined with parchment or paper liners. To prevent overmixing the batter when transferring it to the loaf pans or muffin tins, scoop the batter from the outside edge of the mixing bowl. Fill loaf pans or muffin tins two-thirds full.

A predictable split on top of the loaf is sometimes desired to achieve a uniform appearance or to differentiate one product from another. To create a split in the baked loaf, dip a dough scraper in vegetable oil and make a shallow swipe down the length of the batter in the loaf pan. Streusel or crumb topping may be added at this point.

Muffins and loaves can be made with the same type of batter, but it is important to adjust baking times and temperatures for the size of the product. Muffins are baked at a high temperature for a short time, while larger loaves are baked at a lower temperature for a longer time.

Biscuits and scones should be cooled directly on the sheet pans, while muffins and loaf breads are cooled for roughly 10 minutes before turning them out onto a cooling rack.

Freshly baked quick breads, still warm from the oven, have a wonderful aroma and delicate tender crumb, and are a customer favorite. They may be made daily or throughout the day and should be served immediately when possible. Although best when freshly baked, quick breads can be wrapped in plastic and stored frozen for up to two weeks or at room temperature for one to two days.

CHAPTER 15

Laminated Doughs

Paper-thin layers of shiny golden-brown crust, a light, airy crumb, and rich flavors characterize laminated dough products. Laminated doughs are used to make items such as croissants, Danish, and puff pastries.

KEY TERMS

laminated dough
détrempe

block or French method
threefold

fourfold
blitz or Scottish method

turn
bookfolds

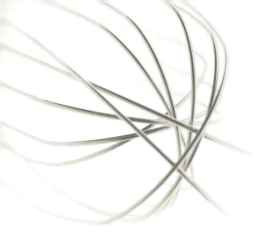

An Introduction to Laminated Dough

Although the finished product is often described as "pastry," success with laminated dough requires sound breadmaking principles. To make croissants and other *laminated dough* products, the baker encases a layer of fat (preferably good quality butter) in dough, rolls it out, and folds it over. The rolling and folding process is carefully repeated to create a finished dough that has many very thin layers of fat and dough. Once placed in the oven, water in the fat layers expands and turns to steam, lifting and separating the dough to create the flaky layers that characterize laminated dough products.

Unyeasted Laminated Dough

Unyeasted laminated dough, such as puff pastry, is leavened solely through physical means. Products made from unyeasted laminated dough are often softer in texture and have a short shelf life. *See Figure 15-1.*

Puff-Pastry Ingredients

Puff-pastry dough is made with water, flour, butter, and salt, and is referred to as *détrempe.* Without yeast or a chemical leavener the détrempe alone does not rise much when baked, other than a little oven spring as the water turns to steam. But when many layers of fat are rolled in, the physical leavening of this added moisture, mostly from the butter, turns to steam and puffs up the dough into a light, buttery pastry. This finished product has many applications, and is a standard preparation in most bakeshops and pastry shops.

Figure 15-1 Puff pastry using the blitz method.

Water is necessary to hydrate the flour and for gluten development. Water in the dough also contributes to physical leavening. (*See Chapter 13: Breads, for more information about the functions of water in dough.*)

The flour used in puff pastry is often a combination of hard-and soft-wheat flours. Hard-wheat flour is used to create a dough that is strong enough to support the high percentage of rolled-in fat. Hard-wheat flour is also able to withstand the repeated rolling and folding necessary to create the many flaky layers. Soft-wheat flours are sometimes added for lightness in the dough.

As with all laminated dough, butter is the preferred fat used for puff pastry. Butter is superior in flavor and mouthfeel (no waxy sensation), and contains a great deal of water, which creates the steam necessary to "puff" the product. The quality of the finished product is dictated by the quality of the fat, so it is important to use the best quality roll-in fat available.

Salt, in addition to providing flavor, strengthens the strands of gluten in the flour. A very small amount of acid, such as lemon juice, may also be added to relax the gliadin in the gluten to make the dough more extensible. When acid is added, the dough is usually easier to roll and there is less shrinkage when the pastry is baked.

Making Puff Pastry

Correct techniques and procedures are essential for consistent success in making puff pastry. The alternating layers of fat and dough should be evenly spaced, and the dough should be developed and relaxed enough to accommodate a generous rise. Avoid excess flour in the détrempe, and work the dough gently. Too much flour and overworked dough creates a rubbery mixture, difficult to manipulate, and likely to shrink when baked. The détrempe and the fat should be of the same consistency before beginning the lamination process to achieve an even structure of alternating layers. After mixing, and again after each fold, give the dough a minimum 30-minute rest period to allow the gluten to relax.

There are many methods of rolling the fat into the dough. The most common is the **block or French method.** In this technique the first fold is made after spreading the block of rolled-in fat over two-thirds of the dough. In another version of the block method, the dough is rolled into a square slightly larger than the block of fat. The fat is then placed diagonally on the dough, the flaps of dough are folded toward the center and sealed, and the first fold is made. After the initial fold, fold the dough again, using either a single turn (*threefold*) or a double turn (*fourfold*). The dough should be refrigerated for at least 30 minutes after each fold. A combination of these types of folds is also possible. Either two fourfolds or a threefold/fourfold/threefold/fourfold alternation is customary for puff pastry. Too many folds will result in a compact product, while too few folds will produce an inadequate number of layers needed for a flaky texture. *See Figure 15-2.*

The **blitz or Scottish method** is easier and faster than either version of the block method. Instead of using a block of fat, the fat is added to the dough and mixed only enough to incorporate. Large pieces of fat are visible throughout the dough. The dough is then folded using the same procedures as those used in the block method.

The blitz method is popular in high-volume bakeshops or pastry shops because it reduces procedure time and quickly creates dough. It produces a slightly more compact product than the block method and is used primarily when the dough does not need to puff much as when making fleurons or Napoleons.

Rolling puff pastry too thin or too hard can damage the structure of the layers. Be careful not to roll over the edges of the dough because this will prevent the sides of the dough from rising. Use equal pressure when rolling out the dough so that the butter is

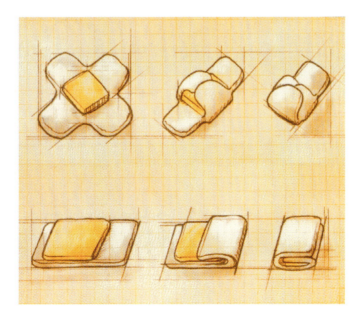

Figure 15-2 Block Method of Roll-in, Threefold, and Fourfold

(a) Block Method—In this version of the block method, place the block of fat on a square of dough, and fold the flaps toward the center.

(b) Threefold and Fourfold—After rolling the dough in a rectangle and placing the fat on it, fold using either a single turn (threefold) or a double turn (fourfold).

evenly dispersed. If the dough is too thick relative to the butter layer, the butter tends to melt during the baking before it can create the steam needed to produce flakiness.

Baking

Puff pastry has no yeast so it does not need to be proofed, but it does require a rest period of 30 minutes or more in a cool place before baking. Bake small puff-pastry items at 415°F–425°F (213°C–218°C), and use slightly lower temperatures for larger items. Keep the temperature above 400°F (204°C). Low oven temperatures allow butter to melt before it can turn to steam, resulting in puff pastry that does not puff up. The oven temperature should be uniform throughout the oven so that the pastries rise uniformly. To avoid fluctuations in temperature, do not open the oven door during the bake.

Storing Puff Pastry

Puff-pastry dough can be kept for up to a week. If kept too long, the flour and water in puff-pastry dough, as in pie dough, begins to ferment and oxidize, turning the dough gray and giving an off-taste. Dough should be covered at all times. Puff-pastry dough freezes well. For best results, cut all the pieces at one time, wrap well, and freeze. Take them out of the freezer as needed, defrost in the refrigerator, and bake.

Items Made from Puff Pastry

Puff pastry can be used to create a variety of items, both savory and sweet. *Vol-au-vents* and *bouchées* are puff-pastry containers with savory applications. *Fleurons* are small, crescent-shaped pieces of puff pastry used as a garnish for many savory dishes.

Dessert items made from puff pastry include the following:

- **Napoleons** Also known as mille-feuilles à la crème ("thousand leaves with cream"), napoleons are among the best-known puff-pastry items. Napoleons are rectangular with a pastry cream filling. They are finished with white fondant icing and chocolate feathering.
- **Palmiers** These puff pastries may be large or the size of petit fours. Palmiers are gently rolled out with granulated sugar and cinnamon and folded in a fourfold to look like a palm leaf, and are also often referred to as elephant ears.
- **Turnovers** Triangular or semicircular fruit-filled pasties that are usually baked.
- **Sacristans** These twisted puff creations are coated with almond or cinnamon sugar and resemble corkscrews. Their ecclesiastical name is derived from the use of a corkscrew to open wine for the sacrament of Communion.
- **Pithiviers** Named for the town of Pithiviers about 50 miles south of Paris, this large, dome-shaped gateau is a puff-pastry structure filled with frangipane. The top of the dome is scored with curved lines before baking. The finished product is meant to resemble the sun.

Yeasted Laminated Dough

Yeasted laminated dough is leavened both physically and biologically. In addition to extra leavening power, yeasted laminated dough enjoys other attributes of fermentation, which include a more complex flavor, stronger dough, and a longer shelf life. Croissants are popular examples of yeasted laminated dough products.

Historians believe croissants may have been created in Budapest, Hungary, in 1686 to celebrate the defeat of the Turkish Army; the shape represents the crescent moon on the Turkish flag. The original dough was an unlaminated sweet dough, but French pastry chefs began to laminate it in the 1920s. Croissant dough tends to be leaner than Danish dough (i.e., it usually contains less fat, sugar, milk, and eggs). Both croissants and Danish have a modest amount of sugar. White flour is most common in yeasted laminated products but whole-grain flour may be incorporated.

Croissants traditionally have no filling, but can be filled with almond paste, savory fillings, or chocolate batons. When filled with chocolate batons, croissants are traditionally rectangular rather than crescent-shaped, and are called *pain au chocolat.*

Danish dough is another well-known yeasted laminated dough product that is usually a bit sweeter than croissant dough and may include spices such as cardamom and/or nutmeg. Shapes of Danish vary, but most have a fruit or savory filling located top and center. *See Figure 15-3.*

Yeasted laminated dough is dry enough, with a mix time that is short enough, to require that dry yeast be dissolved in the liquids before the flour is added to prevent unhydrated yeast particles from impairing fermentation. If the baker's ratio of sugar is greater than 10%, osmotolerant yeast is used to ensure an active fermentation. Croissant and Danish dough can be altered by increasing or decreasing the ratio of butter to flour. A standard range for the baker's ratio of roll-in fat is 25% to over 50% of the flour weight. Other types of yeasted dough, such as brioche and sweet dough, can be laminated, although it is less common.

Figure 15-3 Streusel-topped Danish.

Types of Roll-In Fats

Laminated dough can be made with a variety of fats and fat blends. The type of fat used in laminated dough affects the quality, shelf life, and cost of the laminated item.

BUTTER Unsalted European butter contains a higher percentage of butterfat and a lower percentage of water than domestic butter, and is the preferred butter for laminating high-quality baked goods. Butter literally melts in the mouth because its melting temperature is approximately 80°F (27°C) and lower than human body temperature. The use of butter in baked goods ensures a pleasant mouthfeel and eliminates the greasy sensation associated with higher melting point roll-in fats. The relatively low melting point of butter makes its use difficult in warm bakeshop and pastry shop conditions. If the temperature in a facility is warmer than the melting point of butter, the baker or pastry chef needs to work swiftly and must keep the dough properly chilled. Unsalted butter is the best choice for flavor, aroma, and mouthfeel. Baking formulas are balanced for salt and the use of salted butter will upset that balance.

BAKER'S MARGARINE Baker's margarine has a melting point of about 90°F–100°F (32°C–38°C), and can be used as a substitute for unsalted butter in most laminated doughs when the quality and flavor of unsalted butter is not necessary. In addition to hydrogenated fat or oil, margarine contains water, air, flavoring agents, and coloring. The melting point of margarine is similar to body temperature allowing it to impart a characteristic mouthfeel. The unpleasant waxy sensation caused by the shortening residue is a disadvantage of baker's margarine, while its primary advantage is reduced cost. Margarine and unsalted butter are sometimes blended to lower production costs, and to still achieve some buttery flavor and aroma.

SHORTENING Shortening is 100% hydrogenated oil or fat that has a very high melting point. It may be used in the production of laminated items in warm temperatures but leaves a waxy feeling in the mouth. A benefit of using shortening is reduced cost but at the expense of a quality baked good.

PUFF-PASTRY SHORTENING Puff-pastry shortening is composed of 80%–100% fat, water, colorings, and flavorings. This shortening is designed to have a texture that is soft enough to roll out evenly between the layers of laminated dough. Regular shortenings have no flavor, are harder, and break up in colder dough to give products an uneven lift. Puff-pastry shortening has a high melting point for better steam leavening. It too, like other roll-in fats, leaves a waxy residue in the mouth.

STEP PROCESS

Making Yeasted Laminated Dough

Yeast laminated doughs are used to make items such as croissants, and Danish.
To make yeast laminated dough, follow these steps:

1 Mix dough according to formula.

2 Place the dough into a lightly oiled container.

3 Soften the butter until malleable. Roll butter into an even layer about ½ inch (13 mm) thick. Rolling in parchment paper makes the process easier. Chill overnight on a parchment-lined sheet pan.

4 Allow the butter block to sit at room temperature about 30 minutes before using. Put the dough on the sheeter and place the butter block on top of one half of the dough. Fold the other half of the dough over the butter block.

As an alternative, the butter block can be added to the center of the dough and edges of the dough can be folder over to meet in the center.

5 Gently pound the dough and butter so they stick together.

Making Yeasted Laminated Dough

The process of making yeasted laminated dough is very similar to that used in making hearth bread. Yeasted laminated dough is actually bread, although it seems more like pastry in its finished form. One of the keys to tasty yeasted laminated products is adequate fermentation. Principles learned during the process of breadmaking in Chapter 13: Breads (The Steps of Breadmaking) should be applied to yeasted laminated dough production as well. Care must be taken when laminating because as the laminating progresses the layers become increasingly delicate and can be easily torn.

6 Send dough into the sheeter with the folded edge, "the belly," first.

7 Sheet dough to 13 inches (33 cm) long.

8 Laminated dough after the first trifold.

9 Repeat the steps creating a second trifold then chill the dough in the freezer. By the second trifold, the dough will have distinctive stratified layers.

10 Complete the third trifold and then chill the dough approximately 2 hours. Place dough in freezer for 30 minutes then complete the final sheeting.

MIXING Laminated yeasted dough gains considerable strength during the laminating process. Dough that is too strong and elastic will create low-volume products that are tough and dense. A short mix must be used to avoid excessive dough development. A preferment can be employed or a straight dough can be mixed and retarded overnight. The retarding step allows time for fermentation and for chilling the dough, which is necessary to maintain the consistency of the fat and to prevent it from melting during the laminating process. Chilling the dough also gives the gluten time to relax before the dough warms.

PRIMARY FERMENTATION After the dough is mixed it should be given 2 hours of primary fermentation at room temperature. Following primary fermentation, the dough is rolled into an even layer approximately ½ inch to 1 inch (12 mm to 25 mm) thick, or slightly smaller than a sheet pan and covered. It is helpful to shape the roll-in fat at this point to save time on the day of production. The fat should be softened and malleable, with a consistency often referred to as "plastic." The fat should not be softened to the point of melting. Form the roll-in fat into an even layer the same width as the dough and exactly half as long as the dough. The dough and shaped fat should then be covered and placed in the cooler.

Preparing for Laminating Before the laminating process, the dough and fat should be of the same consistency. To achieve this, place the dough in the freezer for approximately 30 minutes and the butter at room temperature for the same amount of time.

The fat should be plastic enough to be rolled into a thin layer. Cold fat will shatter when rolled, and create damaged dough and uneven lamination. Plasticity can be tested by running the shaped fat over the edge of a work bench. If it bends without breaking, it is pliable enough with which to work. Fat that is too warm will melt and run out between the layers, soaking the dough with melted butter and preventing the product from rising evenly.

Folding in the Butter Place butter on one half of the dough and fold the other half over, squaring the corners of the dough. The butter should be visible on three sides. Gently pound the dough and butter together with a rolling pin to ensure that the dough and butter stick together.

Laminating Laminating yeasted dough can be accomplished by hand rolling, but commercially, it is more feasible if a reversible sheeter is used to execute a *turn*—the step of rolling out and folding the doughs to create the alternating layers of dough and butter. A sheeter moves dough back and forth between adjustable rollers that are gradually reduced to make the dough thinner with each pass. The folded-over edge of the dough (the belly) is sent first into the rollers, so that the dough and butter are not pulled apart.

Reducing the thickness of the dough too quickly increases the risk of tearing it between the rollers. The clearance between the rollers is decreased in increments of ¼″ (5 mm) for the first passes of the dough. Once the dough is approximately 1″ thick (25 mm), the thickness of the dough is reduced by 1/8″ (2 mm) with each pass.

As the thickness of the dough decreases, the surface should be checked for tackiness. If the surface of the dough gets tacky, a light dusting with flour will prevent the dough from catching in the rollers. The dough becomes increasingly supple and strong as laminating progresses and should be handled like a piece of fine fabric.

Sheeting dimensions vary depending on the size of the dough and the size of the final product. It is important to maintain a routine and to complete the correct amount of turns. Work must be done carefully, confidently and quickly. Generally, the first and second turns can be executed in immediate succession, before the dough is given a 30-minute period in the freezer to chill and relax. When the third trifold is completed, the dough is chilled in the cooler for 1½ hours and then in the freezer for an additional 30 minutes before the final sheeting.

Different methods or folding protocol may be used to achieve the desired number of layers. *Bookfolds*—similar to the fourfold described in the puff-pastry section—may be used in conjunction with singlefolds to complete the laminating process.

Final Sheeting Before the final sheeting, the work bench must be clear and ready to receive the large piece of dough. The dough should be sheeted to 20 inch (508 mm) long, rotated 90 degrees, and sheeted down to .11 inch (2.75 mm) in .12 inch (3 mm) increments. The surface of the dough must again be checked for tackiness throughout this last sheeting step. Dust lightly, roll the dough onto a large rolling pin, and finally carry to the bench and gently unroll.

Shaping Dough will shrink if cut immediately because of the elasticity created by the sheeting process. It is important to cover the dough and allow it to rest for several minutes before cutting it into pieces. Cut with a sharp knife to prevent crushing or smearing the layers of dough and butter. If the integrity of the layers is not maintained, the dough may not gain full volume during the proof and oven spring. (Some sheeters also have drop-down cutters that cut pieces from the dough as it passes through the sheeter on its last pass.) Stack pieces one on top of the other as they are cut to prevent them from drying.

STEP PROCESS

Shaping and Finishing Croissants

A perfect lean-dough baguette and a great croissant can be the standards used to identify a great baker or pastry chef.

To shape and finish croissants, follow these steps:

1 Cut sheeted dough in half lengthwise and place one half over the other.

2 Trim the dough edges to form a perfect rectangle. Use a straightedge to cut the dough into smaller rectangles and then into triangles. Slit the center of the wide end of the triangle to allow the dough to roll out more and keep it in a crescent shape when formed.

3 Take one triangle, point facing you, and gently stretch it out. Begin rolling from the back toward you (toward the point).

4 Stretch the point a bit before completing the roll.

5 Shape each croissant into a crescent with the point tucked under, facing the inside of the crescent. Place the croissants on a parchment-lined sheet pan and proof.

6 Brush the proofed croissants with egg wash before baking.

7 Bake according to formula and oven until golden.

Retarding and Freezing Laminated dough retards and freezes well both in bulk and when shaped. Bulk dough can be retarded before or at any point during the laminating process. Shaped products can be retarded, proofed, and baked on the next day. Danish dough for savory items can be shaped, retarded, and then filled and baked to better cater to customers looking for a lunch or dinner option.

Bulk dough can be frozen before or at any point during the laminating process. The laminating process continues once the dough is thawed. Individual, shaped items may also be frozen, "removed from the freezer" the day before baking, placed on a sheet pan, and allowed to thaw in a cooler or proofer/retarder. Final proof times on all frozen and retarded dough are extended unless the quantity of yeast is increased to compensate for the yeast damaged by freezing or slowed due to low temperature.

Proofing Sheet pans of shaped croissants or Danish should be placed in a proof box set between 75°F (24°C) and 78°F (26°C) to avoid melting the butter. Proof time is approximately 1½ hours for Danish and at least 2 hours for croissants. For Danish, the filling should be applied before the Danish are fully proofed. Well-proofed laminated items will jiggle slightly when gently shaken. Individual layers of dough will also be visible in these products.

Baking Yeasted laminated dough should be baked in a convection oven that has been heated to approximately 350°F to 360°F (177°C to 182°C).

Egg washing croissants by gently brushing from front to back rather than from side to side avoids tearing or smearing the delicate layers of risen dough. Danish not covered by filling can also be highlighted with egg wash if the Danish is not glazed or iced after baking.

Bake laminated dough products for 15–20 minutes or until golden brown. Underbaked products are bland and doughy, and lack flakiness. Freshly baked laminated items are delicate and may easily be damaged, so it is best to allow them to cool completely before removing them from the sheet pan. Laminated dough products should be packaged in pastry boxes or waxed paper bags for individual sale.

Filling and Finishing Danish Danish should be filled approximately 1/2 hour before baking. If filled immediately after shaping, the dough will be weighed down by the filling and the baked Danish will be dense and doughy beneath the filling.

The variety of fillings possible is limited only by the creativity of the baker or pastry chef. Simple jam is a fine filling, but a combination of cream-based filling topped with a fruit or fruit filling creates a full-flavor profile. Canned or fresh peaches on pastry cream, and raspberry marmalade on a cream cheese filling are two good choices. When using canned fruit, always drain the liquid to avoid saturating the dough. Savory fillings contrast well with the slight sweetness of Danish dough. Roasted and caramelized vegetables on a goat-cheese filling make tasty Danish dough tartlets. Choose brightly colored vegetables such as red or orange bell peppers to contrast with the golden crust.

Avoid prepared fillings with artificial flavors and colors. Scratch-made Danish deserve a scratch-made filling that customers will appreciate. If streusel or crumb topping is desired, add it immediately after filling the Danish. Return the Danish to the proofer/retarder until fully proofed and ready to bake. After baking, glaze or ice Danish while still warm.

STEP PROCESS

Shaping and Finishing Danish

These are just two of the many shapes that can be formed with laminated dough and filled to make Danish.

To shape and finish Danish pinwheels and baskets, follow these steps:

1 To make pinwheels, cut dough into 4-by-4-inch (10-cm) squares. Make cuts into each corner and brush the middle with egg wash.

2 Fold one edge of each corner into the center of the square pressing lightly. The egg wash will help keep the points in the center.

3 Proof danish and brush with egg wash. Add the filling in the center.

4 Add streusel topping.

5 To make baskets, start with the same 4-by-4-inch (10-cm) squares but do not make cuts at the corners. Brush the centers with egg wash and fold the pointed corners into the center.

6 Proof danish and brush with egg wash. Pipe filling into the center.

7 Top with fruit slices.

CHAPTER 16

Pies & Fruit Tarts

Brought to America by European settlers, sweet and savory pies offered food that traveled well, was fairly easy to prepare, and did not require high-quality ingredients. Pies offered a way of preserving fruit, because of the high sugar content of their fillings, and also allowed the use of fruit that was past its prime.

Tarts are the European version of the American pie. The two products have much in common, such as a bottom crust and a variety of fillings, but also have some differences. Pie shells are deeper than tart shells and can thus accommodate more filling. They are generally made with an unsweetened crust and can be either open-faced or covered with a top crust. Slices of pie often do not hold up without the support of the pie pan and are usually cut directly before service.

The doughs for the tart shells are richer and sweeter than pie shells and are most often referred to as short dough. Tart shells require less filling than pie shells because the traditional tart form is usually only an inch deep. Most tarts do not have a top crust and have sides that are usually fluted. Tart forms have a removable bottom, and unlike pies, tarts can be removed from their form and displayed in their entirety.

KEY TERMS

long-flake pie dough
short-flake pie dough
mealy pie dough
fluting
streusel topping
crumb crusts
lattice crust
weeping
bloomed
short dough
fraisage
blind baking

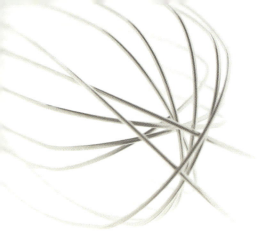

Types of Pie Dough

Dough is the most important element in a pie. The flakiness and tenderness of the baked dough are crucial to a pie's success, regardless of the filling. Flour, fat, cold liquid, and salt are the basic dough ingredients, with sugar or dry milk solids (DMS) as optional additions to either sweeten or enrich the dough. The way in which the fat is distributed into the flour determines the flakiness of the crust. To achieve a tender crust, great care must be taken when preparing the dough.

There are two types of pie dough: flaky and mealy. In both doughs, the cold fat is cut into the flour with ice-cold water used to bind the mixture. A flaky pie dough can consist of either a long or a short flake. To produce **long-flake pie dough**, cut the fat into nickel-sized pieces. Long-flake pie dough is used only as a top crust. **Short-flake pie dough** requires hazelnut-sized pieces of fat that can be used as either a top crust or a bottom crust. To make **mealy pie dough**, cut the fat into small, pea-sized pieces and incorporate them completely into the flour, leaving no lumps. Mealy pie dough is used only for bottom crusts. Mealy pie dough does not produce a flaky pie crust, and is not as tender as a long- or short-flake crust. Its denser nature makes it an ideal watertight lining for moist fillings such as custards, which are not cooked prior to putting them into the pie shell.

Dough Ingredients

Also known as 3-2-1 dough, pie dough usually contains 3 parts flour to 2 parts fat to 1 part liquid by weight. Salt, sugar, and dry milk solids (DMS) can also be added to the pie dough. It is important to understand the contribution of each ingredient to the quality of the dough.

Flour

Pie dough calls for pastry flour. Pastry flour contains a relatively small amount of gluten, just enough to hold the dough together and give it elasticity for rolling. Bread flour has an excess of gluten and produces an overly tough dough. Cake flour has too little gluten that causes the dough to crack and crumble when rolled out.

Fat

Butter, lard, and all-purpose shortening are the fats most commonly used for pie dough. Each has its pros and cons. Butter provides excellent flavor, but a low melting point and high water content that produces a dough that is difficult to handle. When baked, the crust will be mealy and crumbly. Lard's higher melting point results in a lighter, flakier crust; however, because the quality of lard is not consistent, it can give crust an undesirable flavor. Lard is also unacceptable for use in kosher, halal, and vegetarian diets. All-purpose shortening is the best choice for pie dough. Its high melting point, close to that of lard, produces a light, flaky crust. Shortening also has a consistently good quality and imparts no flavor to pie dough. The combination of both shortening and butter produces a flaky pie dough with good flavor.

Cold Liquid

Cold liquid helps the gluten in the flour to develop properly. Water is the most frequently used liquid because it provides needed moisture without adding extra components that will affect the quality and tenderness of the dough. Milk is an alternative liquid that enriches dough and adds to crust color. The additional fat and sugar in milk, however, may produce a crust that is more tender than flaky. (It is important to note that the sugar referred to here is contained in the milk's lactose.) Whether water or milk is used, the liquid should be ice-cold to preserve the solid state of the fat. It is the melting of the fat and the release of steam that give the dough structure and contribute to the flakiness of the baked crust.

Salt

The main function of salt is to enhance the flavor of the dough. It also acts as a gluten conditioner that slows the development and tenderizes the crust. For even distribution, salt should be sifted with the flour or dissolved in the liquid.

Sugar

When added to pie dough, sugar provides flavor and adds color. The golden color is the result of the caramelization of the sugar during the baking process.

Making Pie Dough

Although large amounts of dough may require machine mixing, hand mixing is best for pie dough. Mixing by hand allows for greater control over gluten development, fat distribution, the size of fat pieces, and ultimately the flakiness of the crust. Avoid overmixing or overworking the dough, as this will result in a tough crust.

To mix pie dough, use the rubbing or biscuit method, and follow these steps:

1. Sift the flour and salt together and place in a bowl, or place the sifted flour in a bowl and dissolve the salt in the water.
2. Using the hands or a pastry cutter, cut the fat into the flour. (If mixing by machine, use the dough hook attachment.) The fat should be well distributed and still in recognizable pieces relative to the type of dough being made.
3. Sprinkle the cold liquid over the mixture a little at a time, and mix until the dough holds together in a ball.

Wrap mixed dough tightly in plastic wrap, label and date, and refrigerate. Before using the dough, let it rest for at least 1–2 hours so that the gluten can relax. Chilling firms the flour and fat, and makes the dough easier to handle and roll to produce a flakier crust. Do not refrigerate dough longer than a week, or it will begin to oxidize, and the flour and water will start to ferment and turn the dough gray. For storage longer than a week, freeze the dough in 8- to 10-ounce (225- to 285-gram) portions. Thaw overnight in the refrigerator before using.

Shaping Pie Dough

After the pie dough is chilled, it is ready to be shaped. To shape the dough, roll it out to a size appropriate to fit a pie pan, tart pan, or as a top crust.

STEP PROCESS

Making Pie Dough

The most common error to be made when making a pie dough is overmixing.
To make pie dough, follow these steps:

1. Sift the dry ingredients into a bowl.

2. Cut or rub the fat into the flour.

3. For a flaky dough (see text on long-flake and short-flake pie dough), continue cutting or rubbing the fat into the flour until nickel- or hazelnut-sized pieces are formed.

4. For a mealy dough, cut or rub the fat into smaller pieces.

5. Sprinkle cold liquid over the mixture or add water into a well formed in the center of the mixture.

6. Mix until the dry and wet ingredients are combined and holds together in a ball.

7. Shape the dough for portioning.

8. Portion pie dough appropriately.

Rolling and Panning Pie Dough

Roll out the dough on a clean, flat surface lightly dusted with flour. Also dust the rolling pin and the dough with just enough flour to prevent the dough from sticking. The addition of too much flour will result in a tough product.

STEP PROCESS

Rolling and Panning Pie Dough

Roll out the dough on a clean, flat surface lightly dusted with flour. Also dust the rolling pin and the dough with just enough flour to prevent the dough from sticking. The addition of too much flour will result in a tough product.

To roll and pan the dough, follow these steps:

1 Lightly flatten the dough with a rolling pin.

2 Place the rolling pin at a diagonal, and roll the dough from the center to the outer edge.

3 Roll the dough to the outer edge in the opposite direction, also at a diagonal, to form a round about 1/8 inch (3 mm) thick and 1 to 2 inches (2.54 cm–5.08 cm) larger in diameter than the pan.

4 Roll the dough onto the rolling pin. Then place the pin over the pan, and unroll the dough. Gently place the dough in the pan, making sure that it fully contours to the sloping sides of the pan with no air pockets.

Single-Crust Pies

For a single-crust pie, fluting is the next step after panning the dough. *Fluting* is a decorative method of finishing a pie crust by making folds or pleats in the dough at regular intervals around the edge of the pie. After filling, certain single-crust pies are topped with a *streusel topping* (a crumbled mixture of flour, fat, sugar, and often spices and nuts) before baking. Other single-crust pies are finished with a meringue or a whipped cream topping.

STEP PROCESS

Making a Single-Crust Pie with a Streusel Topping

In this preparation, the streusel topping was prepared, shaped, and rolled out into blocks. It was then refrigerated and shredded using a box grater.

To make a single-crust pie with streusel topping, follow these steps:

1 Roll out pie dough and shape into a circle.

2 Press the dough into the pie pan. An empty pie pan can be used to press on top of the dough to help eliminate air bubles, if desired.

3 Roll the overhanging dough to make edges for fluting.

4 Flute the edges.

5 Shredding the streusel topping.

6 Add the filling and top with streusel.

Some single-crust pies do not use pie dough at all. Instead, they feature **crumb crusts,** usually made from finely ground cookies or graham crackers. Sweeten the crumbs with sugar as needed, and add spices and nuts as desired. Then moisten the crumb mixture with melted butter, and press into a pie pan. The crust is chilled and then baked before filling. Crumb crusts are used primarily for ice cream pies and other specialty pie fillings. *See Figure 16-1.*

Two-Crust Pies

Two-crust pies feature both a bottom crust and either a full top crust or a lattice crust. To create a full top crust, roll out the dough as for a bottom crust, but make the round large enough to hang over the edge of the pan by 1 to 2 inches (25–50 mm). Roll the crust onto a rolling pin, or fold it in half. After lifting and placing the crust on top of the filled pie, use fluting to seal the top crust to the bottom crust. Vents should then be cut into the top crust to allow the steam to escape from the fruit as it cooks. This will prevent the sealed fluted edge from bursting and the top crust from splitting. A *lattice crust* is a crust of interwoven strips of dough evenly placed across the top of a pie. The top crust, or lat-

Figure 16-1

To form a cracker or cookie crumb crust, prepare crumb mixture according to formula and press into pie tin. Flatten the bottom and push crumbs to the sides.

Press the crumb mixture to the sides of the pie tin.

Use an empty pie tin to press the crumbs into a smooth even crust.

tice, can then be brushed with an egg wash and sprinkled with sugar before baking to give the final product a wonderful sheen. See *Figures 16-2 and 16-3*.

Preparing Pie Fillings

Pies contain a variety of fillings. Some of these fillings are baked with the crust, while others are added after the crust has been baked.

Fruit Fillings

Fruit fillings consist of fresh, canned, frozen, or dried fruit mixed with sweeteners, spices, and starches (flour, cornstarch, tapioca, modified food starch, or pregelatinized starch). The type of fruit filling used often depends on the fruit itself. There are two types of cooked-fruit fillings; the cooked-juice method and the cooked-fruit-and-juice method. An uncooked method (homestyle method) of preparing fruit filling also exists but is not frequently used in the bakeshop or pastry shop because the final product often has an inferior top crust that is uneven after baking.

Figure 16-2

A lattice roller can be used to make quick and uniform lattice-topped pies.

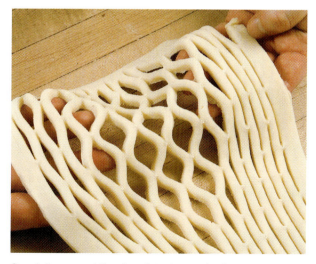
Carefully spread the dough to form a lattice topping.

Figure 16-3 Traditional lattice crust pie.

Cooked-Juice Method

This method is used for soft-textured fresh fruits, canned, and frozen fruits that require no cooking before baking. The fruit will retain its shape, flavor, and color because it is exposed to less heat and limited handling.

The general procedure for making cooked-juice fillings is:

1. Drain all juice from the canned fruit. If fresh fruits are used and there is not enough juice, use water in the same amount. If frozen fruits are used, thaw them first to extract their juices.
2. Reserve a small amount of the juice or water to dissolve the starch and create a slurry.
3. Bring the remaining juice or water and sugar to a rolling boil.
4. While the mixture is boiling, vigorously whisk in the dissolved starch.
5. While continuously whisking, bring the mixture to a second boil, allowing it to fully thicken to remove its cloudiness and turn translucent. Remove from heat.
6. Gently fold in the fruit.
7. Let the filling cool in a covered shallow pan.
8. Fill an unbaked pie shell with the specified amount of filling.
9. Assemble with a double crust, lattice crust, or streusel topping.
10. Bake at 425°F (218°C) until golden brown.

Cooked-Fruit-and-Juice Method

This method is used for firm, fresh fruits and dried fruits that require cooking before baking to soften the fruit. The fruit must be able to withstand the heat of the cooking process.

The general procedure for making cooked-fruit fillings is:

1. Dissolve cornstarch in a small portion of the cold liquid, either juice or water.
2. Bring the liquid, fruit, and granulated sugar to a rolling boil.
3. While the mixture is boiling, vigorously stir in the dissolved starch.
4. Stirring with a spoon, bring the mixture to a second boil, allowing it to fully thicken to remove its cloudiness and turn translucent.
5. Remove from heat and cool immediately in a covered shallow pan.
6. Fill the unbaked shell and follow the procedure as described for the cooked-juice pie.

Making a Two-Crust Pie with the Cooked-Juice Method

The cooked-juice method is particularly good when using soft fruit that will lose its shape and structure if overcooked.

To make a two-crust pie with the cooked-juice method, follow these steps:

1 Drain juice from the fruit, reserving a small amount to use with the thickener. Bring juice and sugar to a boil and thicken using a mixture of cornstarch and the reserved juice.

2 Remove the pan from the heat and gently fold in the fruit.

3 Cover fruit mixture, label, date, and refrigerate.

4 Scale cooled pie filling into unbaked shell.

5 Brush dough with egg wash.

6 Apply the top crust. Press to seal and form a decorative edge.

7 Brush with egg wash and sprinkle with sanding sugar if desired.

Homestyle Method

The homestyle method usually uses a variety of fresh seasonal fruits such as apples, peaches, and blackberries. Unlike cooked-fruit filling that has a characteristic gel, this method produces a filling that is virtually all fruit. The fruit is dredged in sugar, spices, and cake flour. The amount of coating is not specified, but rather feel, taste, sight, and experience determine the amount used. The consistency of the fillings is not very accurate because there is no precise measurement of the flour thickener or the natural liquid content of the fruit.

The general procedure for making homestyle pies is:

1. Peel, core, and prepare fruit accordingly. Berries and similar sized fruit are left whole.
2. Dredge the fruit in sugar, according to the sweetness of the fruit. The sugar and juices of the fruit will create a syrup.
3. Add spices.
4. Dredge the fruit mixture with flour, according to the juiciness of the fruit. During baking, the starch of the flour will gelatinize and cause the juices to thicken.
5. Fill an unbaked pie shell and top with a double crust, lattice, or streusel topping.
6. Bake at 375°F–425°F (191°C–218°C) until the fruit is tender and the crust is golden brown.

Cream Fillings

Cream fillings are essentially flavored pastry cream—a smooth, thick, stirred custard made from eggs, milk, sugar, thickening agents, and flavoring. These fillings are precooked and thickened with cornstarch and/or flour. Add the cooked cream filling to the prebaked crust, cool, and top with a meringue or whipped cream. If a meringue is used as a topping, bake it in a 400°F (204°C) oven until lightly browned. As an alternative, caramelize meringue beneath a salamander or by using a propane blowtorch.

Custard and Soft Pie Fillings

Custard fillings are dairy based (pumpkin pie), whereas soft fillings are syrup based (pecan pie). Eggs are added to these bases to thicken the fillings as the egg proteins coagulate during the baking process. Cornstarch is sometimes used by bakers and pastry chefs in custard and soft pie fillings to prevent the separating, or *weeping* of the delicate liquid. The preparation of these fillings has no formal baking procedure, but simply requires that the ingredients are combined and then placed in the unbaked pie shell. The most difficult part of preparing custard and soft pie fillings is the baking procedure. The delicate filling requires a low oven temperature for proper cooking and the crust requires a higher oven temperature for proper baking.

To properly bake the crust without overcooking the filling, follow this baking procedure:

1. Start with an oven temperature of 425°F (218°C) for 15 minutes; this will begin the proper baking of the crust.
2. Lower the oven temperature to 350°F (177°C) to cook the filling until it is set.
3. Shake the pie lightly by tapping the side of the pie pan to determine the doneness of the custard or soft pie. If it is firm and has no moving liquid, the filling is set.
4. Insert a knife into the center of the pie to see if it comes out clean indicating that the filling is set.

Chiffon Fillings

Chiffon filling is made with a cooked-fruit or cream filling base that is stabilized by the addition of gelatin. After the base mixture has cooled, a meringue is folded in to give the filling a light, airy texture. The filling is then added to the prebaked pie shell.

> ### Quiche
> A quiche is an open tart filled with a baked egg custard to which other fillings have been added. The filling may contain one or more types of cheese, along with complementary meat, seafood, and/or vegetables. Puff pastry or pie dough may be used for a quiche shell. As a quiche bakes, egg proteins in the custard coagulate and cause the filling to thicken. Overbaking causes the egg protein to curdle, resulting in a watery quiche.

Using Gelatin for Chiffon Fillings

Granulated gelatin must be *bloomed* or hydrated in a proper ratio of cold liquid for approximately 4 minutes before it is dissolved in a hot liquid or heated separately. Sheet gelatin must be bloomed in cold liquid for approximately 4 minutes to hydrate or soften it. When the gelatin is softened, the excess water is squeezed out. The gelatin is dissolved either by adding it to a hot liquid or heating it separately. When tropical fruits such as pineapple and kiwi are added, these fruits must be cooked to break down the enzymes they contain, which weakens the protein structure of the gelatin and prevents it from setting properly.

A base that contains dissolved gelatin is cooled or chilled until thickening begins and not allowed to set completely as it is impossible to work with a set base. Folding meringue into a set base causes lumps and an uneven filling. The filling base must also be stirred frequently while it is cooling or chilling for even thickening. If the gelatin begins to set before the meringue is folded into the filling base it can be melted again by heating it over a double boiler. The base must then be cooled or chilled again. Work quickly when folding cold meringue into gelatin stabilized bases to prevent the gelatin from setting too quickly. The pie shell must be filled immediately before the filling begins to set.

The general procedure for producing chiffon pies includes:

1. Preparing the fruit or cream filling base.
2. Blooming the gelatin in cold liquid.
3. Adding the bloomed gelatin to the hot filling.
4. Cooling the filling in an ice-water bath or chilling it in a refrigerator until the mixture is thickened but not set, taking care to stir frequently so that the mixture thickens evenly.
5. Preparing the meringue and folding into the cooled filling mixture.
6. Immediately pouring the mixture into a prebaked pie shell.
7. Chilling until set.

Specialty Fillings

This category includes fillings that do not fit into any other classification of pies. Specialty pies are usually made of ice cream, mousses, or Bavarian creams. Their crust can be either prebaked pie shells or crumb crusts. Another popular example is Boston cream pie, which has cake layers instead of prebaked pie shells or crumb crusts.

Fruit Tarts

Although tarts may have a variety of fillings, their base is made of a short-dough shell. **Short dough** is a mixture of sugar, eggs, butter, and pastry flour. The term *short* refers to the action of the butter, which shortens and tenderizes the gluten strands in the flour. Classic examples of tarts include fruit tarts with pastry cream, Linzertortes, Swiss apple flans, and tarte Tatin (*see Figure 16-4*).

Figure 16-4 Tartlets can be made with a variety of fillings.

Short dough is prepared quite differently than pie dough. Another name for short dough is 1-2-3 dough, which refers to its ingredients by weight: 1 part sugar to 2 parts butter to 3 parts flour (3 eggs are also added). Rather than cutting in the fat by hand, short dough is made by the creaming method.

1. Mix sugar, eggs, and butter at a low speed with a dough hook.
2. When the ingredients are just combined, add the flour, and continue to mix until the dough is smooth.
3. Let the dough rest in the refrigerator, preferably overnight, to relax the gluten and firm the butter.
4. Avoid overmixing to prevent overdevelopment of the gluten, which will result in an undesirable product.

French Short Doughs

The French have many types of short doughs that include: pâte sablée, pâte sucrée, and pâte brisée. They are all rich in butter and eggs and are mixed by hand using a method known as *fraisage* (this method can be used in the making of other doughs as well). Fraisage is done on a table to help prevent overmixing. It is prepared by sifting the flour and salt together and then rubbing in the butter using a bowl scraper. When the butter is sufficiently coated with the flour and is still in relatively large pieces, a well is made in the center of the ingredients, and the liquid (usually eggs and a little water or milk) is placed in the well. The mixture is then worked with the hands—moving from the inside of the well to the outside—until it is cohesive.

Fruit Tart with Pastry Cream

There are many variations on the classic fruit tart. Its basic components are a prebaked short-dough shell, a pastry cream filling, and fruit decoratively arranged on top and brushed with a glaze. The edge of the tart shell may be finished with finely chopped nuts that are usually toasted almonds.

Linzertorte

Actually a tart rather than a torte or cake, this pastry originated in the town of Linz, Austria. The traditional Linzertorte has a hazelnut crust and contains a filling of sweet raspberry jam. It is finished with a lattice-style top crust.

Swiss Apple Flan

This tart contains a mixture of custard and tart apples. The apples are arranged in a spiral on a semibaked crust over which the custard filling is then poured. After baking, the top is brushed with a light glaze.

Tarte Tatin

The tarte Tatin is the most classic of all tarts. This upside-down tart was created by two French sisters at their restaurant in the Loire Valley outside of Paris over 100 years ago. The tart is made with butter and sugar that caramelizes over apples and a crust. When the pastry is golden brown, the tart is turned upside down to reveal the beautifully caramelized apples.

STEP PROCESS

Making Fruit Tarts

While pies have become American in nature, tarts are particularly European. Tartlets are smaller versions.

To make a fruit tart, follow these steps:

1 Line a tart ring with short-dough pastry.

2 Use a rolling pin to trim the dough.

3 Use parchment paper and dried beans to prevent the pastry from blistering and shrinking.

4 Paper liners can be used for tartlets.

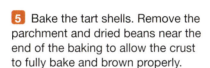

5 Bake the tart shells. Remove the parchment and dried beans near the end of the baking to allow the crust to fully bake and brown properly.

6 Allow the crust to cool and brush with melted chocolate.

7 Fill with pastry cream.

8 Arrange the fruit over the pastry cream.

9 Brush with a thin coating of warm apricot glaze.

Baking Pies and Tarts

Baking procedures and times vary with the type of pie or tart. General guidelines include:

1. When ***blind baking,*** or baking an unfilled pie crust, it is necessary to dock, or pierce, the unbaked dough to allow the steam to escape and prevent blisters from forming during baking. Placing an empty pie pan on top of the dough and then turning the pan upside down to bake, or lining the shell with parchment paper and then filling it with dried beans or pastry weights, will prevent both blistering and shrinkage.

STEP PROCESS

Making a Linzertorte

Actually, Linzertorte is a traditional Austrian tart made with a hazelnut crust. It is often assembled in a cake ring.

To make Linzertorte, follow these steps:

1 Roll out Linzer dough about 1/4 inch (6 mm) thick. Use a cake ring to cut the bottom layer to fit.

2 Roll out the crust for the top layer about 1/4 inch (6 mm) thick and cut into strips.

3 Pipe raspberry preserves on the bottom layer leaving about 1/4 to 1/2 inch (6 mm–12 mm) around the outside with no preserves.

4 Begin laying strips over the preserves to form a lattice crust.

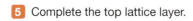

5 Complete the top lattice layer.

6 Press the cake ring over the filled dough to form the tart.

7 Roll out additional Linzer dough to make small balls.

8 Arrange the balls of Linzer dough on the tart. Brush with egg wash and sprinkle with sanding sugar.

9 Bake at 350°F (177°C) until golden brown.

STEP PROCESS

Blind Baking

Blind baking is the term used to describe the process of prebaking pie crusts before filling. Unbaked crusts can be lined with parchment paper and filled with dried beans or pie weights. For this process, two pie tins were used.

To blind bake using two pie tins, follow these steps:

1 Roll out and dock pie dough.

2 Fold pie dough into quarters and place in pie tin.

3 Unfold pie dough and press lightly into pie tin. Cover with an empty pie tin and trim excess dough.

4 Turn the pie tins upside down and bake.

5 When the pie dough has cooked and turned golden brown, flip the pie tins over and remove the empty pie tin from the top of the cooked pie crust. The cookied pie crust can be brushed with heated jam and allowed to cool, if desired, to prevent filling from softening the dough.

2. Remove the parchment paper and dried beans or weights near the end of the baking to allow the crust to fully bake and acquire proper browning.
3. Set fruit pies on a preheated sheet pan to catch drippings.
4. Bake fruit pies at high heat (400°F–425°F/204°C–218°C) to set the bottom crust so that it does not absorb moisture from the filling. For custard pies, reduce the oven temperature to 325°F–350°F (163°C–177°C) after 10 minutes to bake the pie slowly.
5. Test custard pies for doneness by shaking the pan to check for firmness or by inserting a knife near the center of the pie. The knife should come out clean when the pie is done. Fruit pies will begin to bubble toward the end of the baking process. Crusts should be golden brown. Fruit should be tender but not mushy when a knife is inserted into the pie.

Cooling and Storing Pies and Tarts

Unbaked pie crusts and unbaked fruit pies may be stored, tightly wrapped, and kept in the freezer for up to two months. Always cool a blind-baked crust before filling. Prebaked crusts will stay fresh for two to three days at room temperature. Wrapped in plastic, these crusts will keep in the freezer for up to three months. Refrigerate cream, custard, chiffon, and specialty pies, such as Boston cream, to prevent the growth of bacteria. Freezing is not appropriate for baked fruit pies, cream, custard, chiffon pies, or some specialty pies such as chocolate mousse.

CHAPTER 17

Pastry Doughs & Batters

Pastry doughs and batters are the foundation of a wide spectrum of sweet and savory classic and modern pastries. A mastery of doughs and batters provides a solid basis for the preparation of a variety of pastry items. This chapter introduces several types of pastry doughs and batters including pâte à choux, baba, savarin, hand-stretched doughs, and pourable batters.

KEY TERMS

pâte à choux
baba, or baba au rhum
savarin
phyllo dough
kataifi
strudel dough
cannoli
crêpe

Pâte à Choux

Pâte à choux (pronounced *pot-ah-shoo*), also known as éclair paste or choux paste, is a basic pastry dough and a staple in every pastry chef's repertoire. Pâte à choux is a cooked dough made of water or milk, butter, flour, and eggs. When cooked, it results in a product that is light and tender with a hollow interior and a shiny outer shell. Pâte à choux is used for many filled sweet and savory pastries including cream puffs, éclairs, gougères, and profiteroles. The name pâte à choux is French for "cabbage paste," a possible reference to the appearance of cream puffs.

Unlike most pastry doughs, pâte à choux is mixed on the stove before being baked. It has the consistency of a thick paste rather than a dough and is generally piped from a pastry bag. Many eggs are incorporated into the cooked mixture to achieve the proper consistency for piping into a variety of shapes. The dryness or wetness of the mixture when removed from the stove top determines the number of eggs required for the desired thickness. Pâte à choux is physically leavened by steam. Its structure is set by the proteins in the eggs and flour. To achieve the desired hollow interior and a firm outer shell, a strong flour (rather than the soft flour generally used to make tender pastry products) is usually used. A hollow interior is also created by baking the pâte à choux in a hot oven, so that maximum steam is generated. Pâte à choux should not be mixed ahead of time. It should be piped into the desired shapes while still warm, and immediately baked or frozen.

Pâte à Choux Products

A variety of pastry products are made with pâte à choux. These include:

Beignets

Also known as French doughnuts, beignets are deep-fried dollops of pâte à choux dusted with confectioner's sugar. Beignets are a signature breakfast item in New Orleans, where they were introduced by French colonists in the 1700s. *Beignet* is a French word that means "fritter," which implies that the dough is deep-fried. **(See Chapter 14: Quick Breads for a general discussion of fritters.)**

Churros

Similar to beignets, these Spanish and Mexican pastries are deep-fried sticks of pâte à choux rolled in cinnamon sugar.

Cream Puffs

Cream puffs are rounds of pâte à choux with a cream filling, usually pastry cream. The top is sliced off so the puff can be filled, and then the top is replaced. The top may be dusted with powdered sugar or finished with chocolate glaze.

STEP PROCESS

Making Pâte à Choux

Baked pâte à choux shells may be stored for several days at room temperature or several weeks in the freezer. After the shells are filled, the pastries should be served within several hours, before they become soggy.

To make pâte à choux, follow these steps:

1 Bring butter, salt, sugar, and water or milk to a boil. Add sifted flour all at once.

2 Stir constantly with a wooden spoon for about 5 minutes or until the mixture forms a ball that does not stick to the sides of the pot.

3 Transfer the mixture to a mixer, and blend on low speed until slightly cooled.

4 Add eggs gradually, mixing on low speed.

5 Pâte à choux mixture should be somewhat sticky and elastic.

6 Pipe into desired shapes on sheet pan lined with parchment paper.

7 Bake immediately at 425°F–475°F (218°C–246°C). After 15 minutes, reduce the oven temperature to 375°F–425°F (191°C–218°C) for the final baking time to ensure that the product becomes firm and dry but not too brown. Perfectly baked pâte à choux should have a crisp, golden-brown outer shell and a dry center when broken open.

Éclairs

Filled with pastry cream and glazed on top with chocolate, these popular pâte à choux pastries are oblong versions of cream puffs.

Gougères

Gougères are savory French cheese puffs, made by folding grated or finely diced cheese (traditionally Gruyère) into the pâte à choux immediately after the eggs are incorporated. Gougères are served warm from the oven, often as hors d'oeuvres to accompany light, acidic wine.

Paris-Brest

This pastry was developed in the late 1800s by a baker who lived along the route of a famous bicycle race that runs from Paris to Brest and back. This ring-shaped pastry designed to resemble a bicycle wheel is made by piping pâte à choux into a large ring. Sliced almonds are sprinkled on top before baking. Once baked, the ring is cut horizontally and filled with a praline-flavored cream filling.

Profiteroles

Classic profiteroles are small puffs of pâte à choux filled with ice cream and drizzled with a topping of warm chocolate sauce. Three profiteroles typically make up a serving.

Croquembouche

This tall structure makes a dramatic dessert presentation. It consists of many pâte à choux puffs filled with pastry cream, stacked into a pyramid shape and held together with caramelized sugar, then finished with spun sugar. Croquembouche is often served at Christmas celebrations, and is used in France as a traditional wedding cake.

Gâteau Saint Honoré

Named for the patron saint of pastry chefs, this classic French dessert has a circle of puff pastry as its base with a ring of pâte à choux piped on the outer edge. After the base is baked, small cream puffs are dipped in caramelized sugar and attached side-by-side on the top of the circle of the pâte à choux. This base is traditionally filled with Crème Chiboust and finished with Chantilly cream using a special St. Honoré piping tip.

Swans

These classic pâte à choux pastries are made by cutting a shell-shaped puff in half horizontally and filling one half with Chantilly cream. The remaining piece is cut in half vertically and inserted in the sides of the filling to resemble wings. An arching neck (piped and baked separately) is placed at the front of the shell.

Baba and Savarin

Baba, also known as ***baba au rhum***, is a yeasted raisin cake that has been soaked in a sweet rum syrup. The dough, which is similar to brioche, is enriched with butter and eggs, and has an open, tender crumb, perfect for sponging up the rum syrup. The classic shape of a baba is a tall cylinder, but babas can be baked in a variety of shapes, including individual cylindrical baba molds or muffin tins. As the cake rises out of the mold it develops a characteristic domed crown.

Savarin is an adaptation of the Baba that does not contain raisins and is baked in a ring mold. The savarin was developed in the mid-1800s to honor Antoine Brillat-Savarin, a celebrated French gastronome and food writer. Savarins are often soaked in a kirsch syrup, rather than rum, and can be shaped into small individual rings or large rings that are cut like a cake. The center of the ring is usually filled with fresh fruit and whipped cream.

Stretched Doughs

Three basic doughs make up the stretched dough category of pastry: phyllo, strudel, and cannoli. All three are unleavened and rolled or stretched thin to yield crisp, delicate products. Phyllo and strudel dough are extremely thin and require a careful hand-stretching method to achieve the desired thinness. Cannoli dough is rolled rather than stretched by hand, and is not as thin.

Phyllo Dough

Phyllo dough is paper-thin and bakes into a crisp, golden-brown crust of extraordinary flakiness. Phyllo ("leaf" in Greek) originated along the Eastern Mediterranean and is the basis for many popular sweet and savory Greek, Turkish, and Armenian dishes. When assembling a phyllo product, individual sheets are lightly buttered and then stacked or rolled. The butter contributes to the flakiness of the product. The steam released during baking lifts and separates the layers.

Commercially-made phyllo dough is available, so the difficult process of making phyllo dough by hand is rarely done in modern bakeshops or pastry shops. Phyllo comes frozen, in boxes of roughly 24 sheets per box. Phyllo dough should be thawed overnight in the refrigerator. Before removing it from the box, make sure that it is completely thawed. Phyllo sheets are prone to breaking if thawed too quickly or handled when frozen. Phyllo dough will dry out in less than 3 minutes if exposed to air. When working with phyllo, keep it covered with a lightly moistened cloth. Wrap unused sheets airtight and refrigerate.

Baklava

Perhaps the best-known pastry made from phyllo dough is baklava, a delicacy made of butter-coated phyllo layered with chopped walnuts, pistachios or other nuts, and sweetened with a hot rose or citrus-scented honey syrup that is poured over the baked layers. The pastry is left to rest overnight to allow the layers of phyllo and nuts to absorb the syrup.

Shredded phyllo dough, known as *kataifi*, is found in many desserts of the Eastern Mediterranean. It is commonly used as a base for various fillings and soaked in a sweet syrup, much like baklava.

Spanakopita

Spanakopita is a savory Greek spinach-and-feta pie with a crust of phyllo dough. It is at times baked in a pan and cut into serving-size pieces as is lasagna. To achieve a more elegant presentation, individual pockets can be made by folding the phyllo dough around the filling into triangular shapes. Small spanakopita triangles are popular hors d'oeuvres.

Strudel

Like phyllo, *strudel dough* is rolled out and then stretched gently by hand until it is paper-thin and transparent. Although strudel dough is traditionally not quite as thin as phyllo, the two are quite similar. Commercial phyllo dough is sometimes substituted for strudel dough.

STEP PROCESS

Working with Phyllo Pastry

Phyllo dough can be used to make an assortment of Middle Eastern desserts as well as savory dishes. Phyllo dough can also be used to make crisp, flaky pie shells, pastry cups, bases, and garnishes.

Follow these steps when working with phyllo dough:

1 Carefully spread phyllo dough out onto a work surface. Brush with butter or use pan spray if desired.

2 Sprinkle lightly with sugar. Top with another sheet of phyllo dough and repeat this process several times as desired or according to the specific formula.

3 Crisp, baked phyllo dough can be used in many applications.

Because the dough is stretched to an extreme, the dough's extensibility is paramount. The use of a strong flour, such as bread flour, the proper development of the dough, and the resting of the dough sufficiently are integral before stretching. The addition of vinegar also helps to relax the gluten, and increase the dough's extensibility.

Strudel dough is first rolled on a floured bench and is then stretched by placing the back of both hands beneath the dough and gently lifting and stretching outward from the center. The pastry chef must move around the bench, to stretch the dough evenly in all directions. If properly made and rested, the dough will stretch over an entire pastry bench. The dough should be so thin that it is almost transparent. The ability to read a newspaper through the stretched dough is the traditional test of thinness for strudel dough.

Strudel dough is lightly buttered, cake crumbs are sprinkled along one edge of the dough, the filling is placed on top of the crumbs, and then the dough is rolled tightly like a jelly roll. Common strudel fillings include apple, cheese, or nuts.

After the strudel dough has been filled and rolled, the strudel is placed on a sheet pan, brushed with melted butter and baked. A very hot oven (400°F–425°F [204°C–218°C]) is necessary to ensure a crisp, golden-brown crust and to prevent the filling from overcooking. Strudel is best served warm, sprinkled with confectioner's sugar.

Cannoli

The word *cannoli* means "pipe" in Italian, which is also a description of the shape of this pastry. A favorite dessert originating in Sicily, **cannoli** are cylinders of sweet, crispy deep-fried pastry traditionally filled with a sweetened ricotta cheese. Various pastry cream fillings can be substituted for the ricotta for a less traditional cannoli.

Cannoli dough is rolled out to a thickness of about 1/8 inch (3 mm) thick and cut into circles. The circles of dough are wrapped around metal cylinders and deep-fried until crisp and golden brown. Cannoli shells can be fried ahead of time, cooled, and stored tightly covered at room temperature for later use. The filling is piped into the shells just before service.

STEP PROCESS

Making Apple Strudel

The dough for strudel likely evolved from Byzantine and Turkish origins to form the foundation for this popular Austrian pastry made famous in the coffeehouses of Vienna.

To make apple strudel, follow these steps:

1 Cover a table with a tablecloth, dust lightly with flour and roll the prepared dough out as thin as possible.

2 Using the floured back of your hands, stretch the dough uniformly.

3 Continue to stretch the dough toward the opposite corners and the sides until the table is completely covered.

4 Trim the excess dough with a pastry wheel.

5 Brush the dough with butter.

6 Sprinkle and spread cake or bread crumbs on the dough into a long rectangle.

7 Place the fruit filling onto the crumbs.

8 Roll up, using the tablecloth to hold and guide the dough.

9 Continue using the tablecloth to form a tight roll.

10 Transfer to a sheet pan lined with parchment paper.

11 Brush with butter.

12 Bake at 400°F (204°C) for 20 minutes, or until golden brown. Slice and dust with confectionery sugar.

Pourable Batters

These simple batters yield familiar, popular products that are at their best when made to order and served hot off the griddle or immediately from the oven.

Pancakes

Pancakes are chemically leavened quick breads made with a thin, pourable batter and cooked on a hot griddle or skillet rather than baked in an oven. Because most pancakes do not hold well for extended periods of time, they are best when cooked to order and served immediately. Pancakes should be moist with a light, tender texture. To avoid tough pancakes caused by excessive gluten development, ingredients should only be mixed until incorporated.

Pancake batter can be mixed and refrigerated up to 12 hours ahead if baking powder is used as the leavener. When baking soda is used as a leavener, the batter should be mixed just shortly before cooking to ensure a satisfactory rise.

Waffles

Waffles are made with a thick but pourable batter and cooked between the two sides of a waffle iron. Crisp, slightly sweet, and a bit fancier than pancakes, waffles are standard fare at breakfast buffets and Sunday brunches. The waffle's many dimples are perfect for holding pools of melting butter, whipped cream, maple syrup, or fresh, sautéed or puréed fruits. For variety, additional ingredients such as blueberries or coarsely chopped pecans may be sprinkled on the batter after it has been ladled onto the iron, just before closing the lid. Pecan waffles are a popular variation, especially in the South.

Waffle batter is richer than pancake batter, with more fat and less liquid. For a lighter batter, separate the eggs and beat the whites until stiff, then gently fold them into the batter after the other ingredients are mixed. As with pancakes and all quick breads, avoid overmixing.

Crêpes

Crêpes are thin French-style pancakes prepared from a very thin unleavened egg-rich batter. Crêpes are usually served rolled or folded around a filling, and covered with a thin sauce. Many crêpe dishes are sweet, but savory crêpe dishes are also popular.

Prepare crêpe batter at least one hour in advance to allow the flour to absorb the liquid ingredients. Crêpes may be cooked in advance and frozen. To freeze crêpes, make sure they are tightly wrapped and that paper is used to separate each layer to prevent them from sticking together.

There are many varieties of sweet and savory crêpe dishes. The following sweet crêpes are classic desserts:

> *Crêpes Suzette* Flavored with oranges, sugar, and spices, these crêpes are flambéed tableside with orange liqueur and cognac or brandy. *See Figure 17-1.*
> *Crêpes Jacques* These crêpes are filled with sautéed bananas and seasoned with butter and spices.
> *Crêpes Empire* A dessert that features crêpes with pineapple macerated in kirsch or cherry brandy.

Figure 17-1 Classic Crêpes Suzette with a garnish of fresh raspberries.

Popovers

Popovers are small, hollow rolls made with a thin, unleavened egg batter. As their name implies, popovers are characterized by the dramatic rise they achieve when baked. Although they are made with a thin, pourable batter and often considered a type of quick bread, popovers share many characteristics with pâte à choux. Specifically, both rely sole-

STEP PROCESS

Making Crêpes

Crêpes are prepared from a thin, egg-rich batter. They can be rolled or folded with a variety of fillings both sweet and savory. Classic Crêpes Suzette are flavored with brandy and orange liqueur and often flambéed tableside.

To make crêpes, follow these steps:

1 Heat a crêpe pan over moderately high heat and add a small amount of clarified butter or oil.

2 Add a ladle of crêpe batter and tilt the pan slightly so that the batter spreads evenly on the bottom.

3 Cook until just set and the bottoms are a golden, light brown color.

5 Slide the crêpes onto a sheet pan lined with parchment paper. Continue the process until you have made as many crêpes as necessary.

4 Flip the crêpe over and cook a few seconds.

---

ly on steam for leavening and ideally achieve a hollow interior. Both also require ample protein in the flour and eggs to hold the risen structure aloft. For this reason, a strong flour and a high percentage of eggs are recommended for popover batter, and thorough mixing is required to develop the gluten. Also like pâte à choux, popovers are baked in a hot oven to generate maximum steam. Popovers can be served with sweet items such as jam or jelly or with savory items such as poultry or meat. It is important to remember that popovers are at their best when served hot from the oven.

CHAPTER 18
Custards & Cheesecakes

At first glance custards and cheesecakes may not appear to have much in common. Upon closer examination, however, many similarities between these products can be found. The products are related in terms of their ingredients and the manner in which they are baked. This chapter will examine both of these areas and focus on the ingredients in each of the products.

KEY TERMS

wet slow baking
custard
scald
water bath

crème caramel
crème brûlée
brûlée
stirred crème brûlée

flan
bread pudding
pots de crème
cheesecake

New York-style/deli-style cheesecake
French-style cheesecake
no-bake-style cheesecake

283

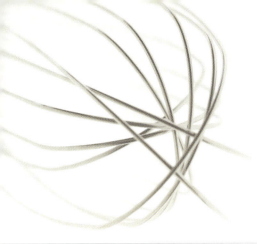

Defining the Custard and Cheesecake Family

Custards and cheesecakes are related by both their ingredients and their cooking methods. The ingredients in these products are dairy, eggs, sugar, and flavoring. A wide range of dairy products including milk, heavy cream, half-and-half, and soft, bland cheese are options. The eggs used can be whole eggs, whites, or yolks. All of the products are made with sweeteners, usually granulated sugar, and flavorings.

The dairy base for custards and cheesecakes is thickened through the coagulation of eggs, which is achieved with the application of heat. The specific form of baking used is known as *wet slow baking*. The product is placed in a water bath in the oven and baked slowly at a low oven temperature of 325°F (163°C) or below. More information on wet slow baking is found later in this chapter.

Throughout this chapter the term *custard* will be used to refer to an entire family of products that are dairy based, egg thickened, and cooked in a water bath at a low temperature. In the industry these products are referred to by their specific names (i.e., crème caramel, crème brûlée, etc.). "Custard" is used here to identify a specific category of products.

Custards

Food historians believe that cooks in ancient Rome were the first to discover the ability of eggs to bind ingredients together. Roman cooks used this knowledge to produce both sweet and savory products. Custard was used to hold together other ingredients such as vegetables, nuts, dried fruit or meat, in the product.

Custard was first eaten by itself as a dessert in the Middle Ages. It was at this time that a version of our modern-day custard was created. When Europeans immigrated to America they brought their custard formulas with them. At the end of the 1800s custard was praised for its nutritional value and was given to invalids and children to improve their health.

Basic Custard Method of Preparation

Two methods of preparation are used when making custard products: the cold and the hot method of preparation. In the cold method of preparation all of the ingredients, dairy, sugar, eggs, and flavoring, are simply whisked together and mixed just until they are completely incorporated. Whisking too vigorously is not recommended because too much foam in the base mix will result in large air pockets on the top of the finished custard. Once all of the ingredients are combined, the mixture is poured into ramekins and baked in a water bath.

There are both advantages and disadvantages in using the cold method of preparation for custard. This method of preparation is quick and easy, and convenient to use when time is a factor. Conversely, the cold method of preparation limits flavoring additives to extracts and/or liqueurs. Because there is no heat involved, fresh flavors cannot be infused into the custard base.

The second method of preparation for custards is the hot method of preparation, where dairy and some, or all, of the sugar is brought to a *scald*, or preboil (when tiny bubbles form around the side of the pan). Fresh flavorings such as mint leaves, toasted nuts, or vanilla beans can then be infused into this mixture. The infusion of flavor into the custard base results in a strong overall flavor. Fresh vanilla beans, for instance, have a much more intense flavor than vanilla extract.

A disadvantage of the hot method of preparation is the time involved in infusing flavor. Dairy must be brought to a scald and infused with the flavor before it is added to the eggs. The mixture may also have to be allowed to steep because of the potency of the item being infused. Once the flavor has completely infused, the dairy and eggs are whisked together.

Regardless of the method of preparation employed, the custard base can be made ahead of time and refrigerated for a few days before being baked. Place the custard base in a tightly wrapped container in the refrigerator and bake as needed.

The Role of Ingredients

Because of the simplicity of the ingredients in custards, it is vital to understand the contribution of each ingredient to the end product. A basic knowledge of the ingredients ensures consistent success. The type of dairy used in the custard base has a dramatic effect on the richness of the end product. A custard base utilizing 100% milk will result in a product that is noticeably less fatty than one using 100% heavy cream. Many formulas call for a combination of both.

The type of egg product used is also reflected in the structure and mouthfeel of the end product. A custard formula that uses only egg whites will result in a product with a very sturdy structure because of the protein present in the egg whites. It will not, however, have a creamy, rich mouthfeel. The custard will feel like gelatin in the mouth. A custard made with 100% egg yolks, on the other hand, will have an extremely rich mouthfeel and a very weak structure because of the large amount of fat and limited protein in egg yolks. Most custard formulas require a combination of egg whites and egg yolks.

The Importance of a Water Bath

The manner in which custards are baked is as important as the ingredients used in the base mix. Custards must be baked in a *water bath*. Ramekins are placed in a hotel pan and filled with the custard base. The pan is then placed into the oven and filled with hot water. A pan with edges at least as high as the sides of the ramekins is necessary and the hot water should reach between half and two-thirds of the way up the sides of the ramekins.

It is important to use a water bath that produces a gentle heat when making custard because as eggs cook their protein strands begin to unravel. Protein strands must unravel slowly to give custard a smooth mouthfeel. Proteins that are exposed to high heat, and heat up too quickly bond up tightly, prohibiting the hardened egg strands (proteins) from holding the surrounding liquid (the milk). Egg strands clump together and are not able to bind the liquid together, resulting in a watery milk mixture containing bits of egg. The overbaked custard looks curdled, with a flavor that becomes noticeably more eggy.

Baking in a water bath is a slow and gentle process because a water bath in a 325°F (163°C) oven will not rise above simmering, 185°F–200°F (85°C–93°C). Starting the baking process with a cold custard base also aids to slow the baking process. The gentle heat of a water bath prevents a custard from rising in the oven to produce a pale yellow product with a soft top.

When placing the ramekins in the oven, allow for ample room around each ramekin so that the hot water may freely circulate and evenly bake the custards. To test if the custard is sufficiently baked, gently shake the ramekin or mold. The center should no longer jiggle. When the entire custard appears to be firm, remove the tray from the oven. Remember that carryover cooking (the heat that is retained by a product after it is taken out of the oven) will continue to cook the custard, so it is important to remove the ramekins from the water bath as quickly as possible.

The Custard Family

Some of the more popular desserts that are included in the custard family are crème caramel, flan, crème brûlée, bread pudding, and pots de crème.

Crème Caramel

Crème caramel is a caramel infused custard. A thin layer of caramel is poured into a ramekin and then topped with a custard base. The caramel permeates the custard as it bakes. The custard is inverted out of the ramekin for service and the caramel bottom becomes the top when presented. The custard base for crème caramel must have enough structure for it to be tipped out onto a plate without falling apart. When making a crème caramel, using the correct proportion of egg yolks to egg whites is extremely important. In most formulas there is a greater ratio of egg whites to egg yolks. Crème caramel must consist of a combination of heavy cream and milk to produce an end product that has a richness that is not overpowering.

The caramel for the crème caramel should be cooked to a medium-dark stage. If the sugar is too light in color (i.e., not cooked long enough), the result is a colorless and flavorless caramel top. If the caramel is cooked too long, a strong bitter flavor is infused throughout the custard. Either the wet or the dry method can be used when cooking the sugar. *(See Chapter 28: Pastillage, Sugar Artistry, and Marzipan for more information on cooking sugar.)* Cooked sugar has a tremendous amount of carryover cooking and it is important to line up the ramekins before starting to cook the sugar. Once the sugar has reached the desired color, pour the caramel immediately into the ramekins. There is no need to spray or butter the ramekins.

After the crème caramel is baked, try to let it cool for at least 24 hours to make it easier to unmold. Do not unmold a warm custard, as it will fall apart. To keep the original round shape of the custard intact, take care when unmolding by running a paring knife or a small metal spatula around the sides of the ramekin. Pressing against the sides of the ramekin during the process will also prevent accidentally cutting into the corners of the custard. It is best to keep the crème caramel in the ramekins until ready for service. Crème caramel can be kept refrigerated for a few days.

Crème Brûlée

Crème brûlée is the custard family's richest member. Its base is made with 100% yolks and 100% heavy cream. The direct origin of crème brûlée is unknown and despite its French name, one of the earliest recipes for crème brûlée comes from England, rather than France. The recipe dates back to the 1600s, when a faculty member at Trinity College in Cambridge, England, developed this dessert for guests. Within a hundred years, the dessert appeared in America and Thomas Jefferson had a crème brûlée recipe in his collection. *See Figure 18-1.*

Crème Caramel

Using a water bath helps produce a gentle heat and gives the custard a smooth mouthfeel.

Follow these steps in making crème caramel:

1 Gather all equipment and ingredients. Preheat oven to 325°F (163°C).

2 Carefully pour hot caramel into ramekins and swirl the caramel to evenly coat the bottoms of the ramekins.

3 Put milk, cream, and half of the sugar in a saucepan over medium heat. Add vanilla or scraped vanilla seeds. Heat until scalded.

4 Whisk remaining sugar and whole eggs in a mixing bowl until well combined.

5 Pour the hot milk mixture into the egg mixture, whisking constantly.

6 Whisk slowly until smooth.

7 Strain the mixture and pour into caramel-lined ramekins.

8 Place ramekins in a hot water bath and bake approximately 45 minutes or until set. Remove the ramekins from the water bath and cool.

9 Run a paring knife or small metal spatula around the sides of the ramekin.

10 Carefully unmold the cold custard.

11 Garnish the crème caramel as desired and serve.

Figure 18-1 Crème brûlée has been a popular dessert since the 1600s.

Crème brûlée is a popular standby on many dessert menus. It is loved by customers and a favorite of many pastry chefs. Pastry chefs like crème brûlée because it is easy to prepare and therefore helps to keep down labor costs. Crème brûlée requires little finishing work, and when properly stored, will last for days in the refrigerator.

Crème brûlée is certainly the most luxurious and probably the best-known and most popular member of the custard family. James Beard, a famous American chef, said that crème brûlée was "one of the greatest desserts in the realm of cooking." A glance at its ingredients makes crème brûlée's luxurious nature clear. This custard base consists of 100% heavy cream and 100% egg yolks that result in a silky smooth and extremely rich dessert. Crème brûlée is presented in the same ramekin in which it is baked because it lacks the structure needed to maintain its shape.

The fact that crème brûlée must be served in a ramekin offers the pastry chef some interesting possibilities. A thin layer of sauce such as caramel, chocolate, ganache, or fruit coulis can be spread on the bottom of each ramekin and placed in the freezer until the layer has completely hardened. The custard base is then added to produce well-defined layers of crème brûlée and the accompanying "sauce." If the sauce or the custard base is too warm, the result will be a murky looking crème brûlée. Items such as fresh berries, diced caramelized ginger, or a fruit compote can also be added to crème brûlée with delicious results.

Crème brûlée translates from the French as burnt custard or burnt cream. The "burnt" refers to the layer of **brûlée**, or caramelized sugar on top of the custard, which is integral to the custard's success. Ideally, the sugar crust should be made directly before service, after the custard base has cooled. There are various schools of thought as to the type of sugar that produces the superior thin sugar shell. Granulated sugar works well but must be added in thin layers rather than in a single thick layer, to allow the bottom of the sugar to caramelize. When using brown sugar it can be dried by placing it on a sheet pan in an oven at a low temperature of 250°F (121°C) for approximately 15 minutes. After the sugar is dry, crumble or grind it finely before sprinkling it on top of the crème brûlée. Raw sugars also work well. Sugars such as turbinado and muscovado offer a flavorful sugar coating. The type of sugar used for the crème brûlée's crust is a matter of personal preference. Keep in mind, however, that brown and raw sugars have stronger inherent flavors than granulated sugar.

Another important component in the preparation of crème brûlée is the application of heat. Intense, direct heat produces the best caramel crust. Although a salamander or broiler can be used, pastry chefs prefer to use a propane torch. A propane torch ensures speedy results, allows the pastry chef to control the placement of the heat, and produces the crème brûlée's trademark shiny, caramelized crust.

Only a small dusting of sugar should be used on the top of the crème brûlée. Pass over the sugar with a torch to quickly melt the sugar. It is important to move the torch back and forth. The resulting sugar crust should be thin enough that it can be easily broken with a light tap of a spoon. Sugar that is too thick is difficult to cut through and unappetizing as well.

Some pastry chefs make crème brûlée entirely on the stove top. These **stirred crème brûlées** are prepared much like an anglaise that has a pourable consistency. A stirred crème brûlée is not served in a ramekin, but rather poured into a chocolate or a cookie cup and garnished with fresh berries, or used as a filling for tartlettes.

If ramekins are not available, a stirred crème brûlée offers the pastry chef an alternative. This type of brûlée can be made more quickly than a classic crème brûlée. The absence of a ramekin also offers additional plating options. A disadvantage of this product is its overall texture, which is not similar to that of the traditional baked crème brûlée. A customer desiring a classic crème brûlée may be disappointed at the thin consistency of a stirred crème brûlée.

Figure 18-2 A Swiss apple flan features a thin layer of sugared, spiced cake crumbs and ground almonds in a short-dough crust. Apple slices are arranged in a spiral pattern and custard is poured over them. After baking, the flan is glazed with apricot glaze.

Flan

Flan is the Spanish relative of crème caramel. The name "flan" is related to the French word "*flaon*," which originally comes from the Latin word meaning custard. Like crème caramel it is a custard base into which a caramel layer is infused. It is usually made with 100% whole milk and whole eggs and is typically flavored with vanilla, or variations that include fruit, almonds, and pistachios.

Flan is also seen on many culinary menus that include vegetable flans such as asparagus, sweet corn, and sweet potato. In these instances the flan is usually made in a ramekin, and inverted onto the plate as a side dish. In Mexico, Spain, and Cuba, however, flan is seen only on dessert menus. In these countries a small amount of cream cheese is sometimes added to a custard base made with a combination of evaporated, sweetened condensed, and whole milk. Traditionally flan is made in a pie-shaped tin, cut into wedges, and served with its caramel sauce on top. *See Figure 18-2.*

Bread Pudding

Bread pudding consists of two components: a custard base and a bread or bread-like product. The custard base is made with whole milk and whole eggs to guarantee an end product that is firm enough to cut out of a pan. The type of bread used also determines the flavor of the bread pudding. The custard base takes on the flavor of the bread used in the formula. Enriched breads work well in bread puddings. Breads such as challah and brioche, and Danish and croissants add a richness to bread pudding that is not found in its custard base ingredients. Cake scraps, chocolate, gingerbread, and dried fruits are just a few ingredients that can contribute to creating exciting bread puddings. Génoise sponges and other light cakes will fall apart if added to the custard base.

Once the custard base has been made, the dried bread or cake scraps are added. Allow the bread to completely absorb the custard before baking. Dried scraps work best because they absorb more liquid than fresh scraps that contain more moisture.

Pots de crème

When translated from the French, *pots de crème* means pot of cream. The name refers to both the custard base as well as the utensil in which it is made. The ramekins classically used for pots de crème are taller and a bit smaller in diameter than the traditional ramekin. Sometimes the small "pots" have two small handles and may have a lid with a small hole. The ingredients in the pots de crème formula require that the custard be served in the container in which it is baked.

In terms of overall richness, pots de crème lies somewhere between crème caramel and crème brûlée. The dairy product used for pots de crème is half-and-half. The end result is richer than crème caramel and a bit lighter than crème brûlée. Pots de crème uses fewer egg whites than crème caramel. The final product has less richness than a crème brûlée yet with slightly more structure. The relatively loose structure of the pots de crème does not allow it to be removed cleanly from the ramekin in which it is baked. Chocolate pots de crème is the traditional flavor.

Cheesecake

Cheesecake is a member of the custard family. It consists of a dairy base or soft, bland cheese, bound by eggs, and baked in a water bath. Cheesecake is not a modern invention. Its history dates as far back as Ancient Greece where cheesecakes were given as prizes to athletes competing in some of the first Olympic games. A recipe from as far back as A.D. 200 calls for a "well crushed cheese" to which an egg and some wheat flour are added and then placed in a loaf pan and allowed to slowly cook in a hot fire.

Today, cheesecake is mostly associated with soft, bland cheeses. There are numerous types of cheesecakes made throughout the world that use a variety of cheeses. Cheesecakes in America are made with cream cheese, in Italy with ricotta, and in Germany and Austria with quark.

The Creaming Method

Regardless of the type of cheesecake, the method of preparation consists of the same basic steps. The handling of the cheeses is perhaps the most important step. The cheese must be lump free. This can be easily accomplished by using a room temperature cheese. If cold cheese is used extra care must be taken when scraping the bowl, as the cold cheese will stick to the bottom and sides.

The mixing method for all cheesecakes is the creaming method. Cream the cheese with a paddle on a low speed. Continue to cream until the cheese is lump free as this is vital to the success of the end product. Do not mix the cheese on a high speed because it will incorporate too much air into the batter and cause the cheesecake to rise and then fall during baking. The cheesecake will also crack and give an undesirable appearance. It is important to scrape the bowl well and often throughout the creaming process. Do not add the eggs or any additional dairy until the cheese is completely smooth and has become slightly lighter in texture.

Once the cream cheese is lump free, other ingredients can be added. Continue to scrape the bowl carefully and thoroughly using a plastic bowl scraper to ensure that all of the cheese is scraped off the sides and the bottom of the bowl.

Types of Cheeses

There are many types of cheese that can be used when making a cheesecake. They are, for the most part, relatively bland cheeses that are pale in color.

Cream Cheese

Originally made in Philadelphia and still referred to as Philadelphia cream cheese, this cheese was America's try at imitating the French Neufchâtel cheese. It is rich and creamy with a slightly acidic tang.

Neufchâtel Cheese

Originally made in France, this is a soft, white cheese that is often used as a lowfat alternative to cream cheese due to its lower content of butterfat.

Baker's Cheese

This is a relatively bland, white cheese that contains about 75% water. When added to cream cheese the result is a cheesecake with a lower percentage of overall fat. Baker's cheese lacks the slightly acidic tang of cream cheese, and is rarely used by itself. An extremely bland tasting cheese, it is commonly used in cheese fillings for items such as Danish and coffee cakes.

Ricotta Cheese

This cheese is usually used in both Italian baking and cooking. Its slightly grainy texture differs from the smooth richness of cream cheese. Ricotta based cheesecakes associated with the Easter holiday are a Southern Italian tradition. A classic Italian cheesecake, Cassata alla Siciliana, is made with a base of ricotta cheese. This "cheesecake" is really more of a layer cake. The cheese filling is mixed with candied fruit and nuts (either pistachios or pignolia) and flavored with cinnamon and Strega (an herb- and flower-based Italian liqueur). The cheese filling is placed between two layers of sponge cake and the entire cake is then covered with a chocolate frosting.

Mascarpone Cheese

Another Italian favorite, mascarpone is cream cheese's rich cousin. Unlike cream cheese, mascarpone starts with a base of heavy cream that results in a smooth, luxurious cheese with a slightly tan color. Mascarpone is most often used in tandem with cheeses that are lower in fat content such as cream cheese or baker's cheese.

Chèvre or Goat Cheese

Chèvre and goat cheese are made with a base of goat's milk rather than cow's milk. The result is a slightly dry, white cheese with an almost unidentifiable pungent flavor. The very distinct flavor of chèvre is not to everyone's liking and can easily overpower other items on a plated dessert. For these reasons, it is best that goat cheese be used with a delicate hand and a light touch. Chèvre seems to work best when paired with other more bland cheeses.

Quark

This is a cheese not frequently seen in America. It is the European equivalent to America's cream cheese, and is most often found in Germany, Austria, and Switzerland. Quark is pale white in color with a bit less fat than Philadelphia cream cheese. It is a soft, unripened cheese.

Types of Cheesecake

Cheesecakes are a universal dessert. Most countries throughout the world have their own version. In the United States, the primary types of cheesecake are New York-style/deli-style, French-style, and the no-bake-style.

New York-Style/Deli-Style

New York-style/deli-style cheesecakes are rich in flavor and fairly dense in texture. Usually baked on top of a graham cracker or cookie crust, they are occasionally served with a fruit topping. This cheesecake consists of cream cheese, eggs, and sugar and is cheesecake in its purest form. Some formulas call for a bit of additional dairy to help loosen the cream cheese. This addition gives the cheesecake its trademark creamy texture. The addition of a few supplementary egg yolks also increases the cake's overall richness.

Like custard, New York-style/deli-style cheesecake is baked in a water bath to give the end product a creamy texture, and to prevent it from rising and/or cracking. The ideal cheesecake is pale tan in color with a smooth unbroken top.

To achieve a lump-free batter, start with room temperature cream cheese. Once the cheese has softened and is lump-free, add sugar. Take care to scrape the bowl regularly. Flavoring and eggs are then added. Incorporate the eggs slowly for best results and continue to cream the mixture. Once all of the ingredients are fully incorporated, the mixture is poured into a prebaked cookie crust. The cheesecake is placed in a hotel pan and in the oven. Once in the oven, fill the pan with hot water. The water should rise about halfway up the sides of the cheesecake pan. Bake at 325°F (163°C). To test for doneness, lightly shake the cheesecake pan. The center should be firm and not wiggle. Touch the center of the cheesecake. Although it will not have the same resilience as a leavened cake, such as a génoise sponge cake, it should be firm to the touch. Remove the cheesecake from the water bath and cool in the cake pan. When the cake is completely cool to the touch, wrap tightly and place in the refrigerator. Do not try to remove a warm cheesecake from the pan as it will fall apart. For best results, the cake must be chilled completely before being cut.

French-Style

Like the New York-style/deli-style cheesecake, the **French-style cheesecake** base consists of cream cheese thickened with eggs, and baked in a water bath in a low oven. As with the New York-style/deli-style, the cream cheese must be lump-free before additional ingredients are added. French-style cheesecake is quite a bit lighter in texture than the New York-style/deli-style cheesecake. When making a French-style cheesecake the eggs must be separated. The yolks are added to the cream cheese mixture and the whites are whipped with a portion of the sugar and folded into the cream cheese mixture just before it is baked. The addition of a meringue gives this cheesecake a light, airy texture. Instead of a cookie crust, the French-style cheesecake utilizes a layer of génoise sponge cake.

No-Bake-Style

No-bake-style cheesecake is an unbaked cheesecake, set with gelatin and lightened with the addition of whipped cream. No-bake-style cheesecakes vary from New York-style/deli-style cheesecakes in both texture and in their method of preparation. The cheesecake is much lighter than both the New York-style/deli-style and French-style cheesecake. There are no eggs in the no-bake-style cheesecake and no application of heat. Rather than thickening the cream cheese through the coagulation of eggs, gelatin is used for firmness. The no-bake-style cheesecake adds whipped cream that results in an overall light texture and mouthfeel.

As with all other cheesecakes, it is vital that the cream cheese for the no-bake-style cheesecake be creamed until it is completely lump-free. While the cream cheese is being creamed, whip the heavy cream to a medium peak and place in the refrigerator. Bloom and dissolve the gelatin. Once the cream cheese is lump-free, temper the gelatin into the cream cheese mixture and quickly fold in the whipped cream. Immediately place the mixture into a piping bag to fill the desired molds as quickly as possible. Place the molds in the refrigerator or freezer until the cheesecake is firm enough to unmold. The cheesecake can be placed on a crust made of graham cracker crumbs, sponge cake, or even a thin sugar cookie.

The no-bake-style cheesecake offers a lighter alternative in terms of mouthfeel and texture for those customers who do not want a heavy dessert. The cake, however, is not lighter in terms of calories and or fat content. Flexipans work extremely well for making individual no-bake-style cheesecakes.

Flavoring Options

There are many ways in which to flavor a cheesecake batter. One option is to flavor the entire batter; pumpkin, lemon, and chocolate cheesecakes are some examples. Another option is to flavor a portion of the cheesecake batter and to swirl it into the remaining

STEP PROCESS

Making a New York-Style/Deli-Style Cheesecake

This cheesecake is often called a deli-style cheesecake. In this example, the use of chocolate is demonstrated to create a marbling effect, and as a way of decorating a plain cheesecake with chocolate curls.

To make a New York-style/deli style cheesecake, follow these steps:

1 Gather equipment and ingredients. Preheat oven to 375°F (191°C).

2 Prepare graham cracker crust and press it into the cake pans.

3 Mix cream cheese and lemon rind until smooth.

4 Add sugar and continue to mix, scraping the sides and bottom of the bowl and paddle.

5 Add heavy cream slowly while mixing; scrape often.

6 Slowly add eggs, incorporating well after each addition. Scrape bowl and paddle well.

7 Add sour cream, half-and-half, and vanilla extract; scrape well and mix until all ingredients are blended. Pour the mixture into prepared cake pans.

8 To make a marble cheesecake, pour chocolate ganache into a small amount of batter and blend well. Add the chocolate batter in circles.

9 Swirl the chocolate batter to create a marbling effect.

10 Place cake pans on a sheet pan and add water to make a water bath. Bake approximately 1 1/2 hours, or until the center is firm. Remove cheesecakes from water bath when done and cool completely.

11 Cheesecakes are often served with fruit toppings or decorated with chocolate curls.

batter. To do so, prepare the crust and cover it with approximately 3/4 of the cheesecake batter. Flavor the remaining 1/4 of the batter with either extracts, liqueurs, jams, or chocolate. The flavored batter is swirled into the remaining batter using the point of a paring knife or a skewer. The flavor can also be piped or poured onto the remaining batter forming swirls, concentric rings, or polka dots. Bear in mind that this is added flavor to the overall cheesecake, not just the addition of color. Be generous and make sure that each slice has a liberal amount of the flavor swirl.

Many items can be used to flavor cheesecakes such as: fruit curds, fruit compotes, fresh berries, fruit purées, chocolate chunks or chips, cookie pieces, candy pieces, pieces of candied fruits, citrus zest, and fruit preserves. Cream cheese has a relatively neutral taste that lends itself well to the addition of almost any other flavor combination. Remember that items such as fruit preserves and fruit compotes often contain a lot of sugar and some alterations to the original cheesecake formula may be necessary when adding these types of flavorings.

Crust Options

There are many options for cheesecake crusts. The type of crust chosen depends upon the type of cheesecake. Always consider the flavor of the cheesecake when deciding upon the type of crust.

Crumb or Cookie Crusts

These crusts consist primarily of cookie crumbs. Cookies containing little fat, such as graham crackers, biscotti, and gingersnaps, are often used for these crusts. In most instances the richness of the cheesecake does not require a crust that is equally rich. Cookies must be well pulverized before assembling the crust. Using a food processor ensures good results. After the crumbs are processed, they are mixed with melted butter and some sugar, if desired. Just enough butter is then added to hold the crumbs together. If too much butter is added the crust will be too fatty, with an undesirable mouthfeel. Because cookies have varying amounts of fat and moisture, the amount of butter added will vary. Once the ingredients for the crust have been combined, prepare the cake pan. Spray the pan with pan spray or brush lightly with butter. Place a parchment circle on the bottom of the pan to make the cheesecake easier to remove once it has been cooled.

Best results are achieved when the crumb crust is tapped into the pan. Use your hands or some kind of weight to compress the crumbs. Place the cake pan on a sheet pan and blind bake until the crust is dry to the touch. Oven time will vary according to the type of cookies used.

Cookie Crust Variations

Some variation on crumb crust can include nuts and other flavored cookies or spices in graham crackers for an additional flavor component. Consider a pumpkin cheesecake on top of a gingersnap crust or a peanut butter cheesecake on top of an Oreo cookie crust.

Short dough or 1-2-3 dough can also be used as a cheesecake crust. Simply roll out the dough to about 1/4 of an inch thick and place in a papered and sprayed pan. When baking a short-dough crust use blind baking. Parchment paper or aluminum foil weighted with dried beans can be used to line the pan when blind baking the crust. This step is not required for a crumb crust.

Individual cheesecakes can also be presented on crusts that have been baked separately. In this case the short dough is rolled out and cut into the appropriate size and the crusts are then baked on a sheet pan. Just before service, the cheesecake is inverted on the cookie. This technique works particularly well with no-bake-style cheesecake. The cookie should not be more than 1/8 of an inch larger in diameter than the base of the cheesecake and should be thin enough to be easily broken with a fork.

Sponge Cake as a Crust

There is no limit to the types of sponge cakes that can be used when making a French-style cheesecake. This variety can include such diverse flavors as vanilla, chocolate, nut, or lemon. Remember that the sponge cake layer is acting as a crust for the cheesecake and should not be too thick. As with cookie crusts, consider the flavor of the sponge cake as an integral part of the overall flavor of the cheesecake.

Finishing Cheesecake

Cheesecake can be topped with a variety of items such as fruit toppings or a rich chocolate glaze. The cheesecake should be chilled before finishing begins.

Traditional Fruit Toppings

American New York-style/deli-style cheesecake is traditionally served with a fruit topping. The topping usually consists of soft fruits such as strawberries or cherries. Regardless of the fruit used, it is important that it be flavorful, ripe, and if possible, in season. If fresh fruit is unavailable, frozen or canned fruit can be substituted. Check the flavor of the fruit for its sweetness level as additional sugar may or may not be required.

The fruit topping for cheesecake is prepared in the same manner as for a cooked-juice or cooked-fruit-and-juice pie filling. *(See Chapter 16: Pies and Fruit Tarts for more information on this procedure.)* Once the topping has been made it is placed on top of the cheesecake. The cheesecake is then refrigerated until the fruit is completely chilled and firm enough to cut.

Chocolate Glaze

Perhaps the most popular chocolate glaze is ganache. Its luster can certainly add an elegant touch to a cheesecake. Best results are achieved when the cheesecake is completely chilled. A chilled cheesecake holds its shape when covered with a warm glaze. A warm cheesecake, on the other hand, will begin to melt as it is being covered with a glaze.

Fruit or Mirror Glaze

A mirror or a fruit glaze works well with the tangy, light nature of a French-style or a no-bake-style cheesecake. *(See Chapter 25: Buttercreams, Icings, and Glazes for more information on glazing cheesecakes.)* The bright color of a fruit glaze complements a cheesecake's otherwise relatively bland exterior.

Unmolding Cheesecake

Cool the cheesecake completely while it is still in the pan. Then tightly wrap it in plastic wrap and place it in the refrigerator until it is firm. Ideally, the cheesecake should stay in the refrigerator overnight. The next day, unwrap the cheesecake and run a small paring knife around the rim, pressing the knife against the side of the cake pan to help maintain the cake's round edges. To release the cheesecake, gently warm the bottom of the pan over a burner. This should take no more than 20–30 seconds. Cover a cardboard cake round with plastic wrap to prevent the cake from sticking to the cardboard. Place the cardboard round on top of the cake, invert, and tap on the bench to release the cake. Quickly place another cake board on the bottom of the cake and invert again. The cake is now ready to be sliced and served. Use a hot, wet knife to cut cheesecake, and wipe the blade between each cut.

CHAPTER 19

Creams & Mousses

Cream forms the base of many bakeshop and pastry shop products. This chapter discusses the composition of cream, the various types of creams used in the bakeshop and pastry shop, and the many variations of cream-based desserts.

KEY TERMS

whipped cream
Chantilly cream
crème parisienne
protease enzymes

Bavarois/Bavarian cream
charlotte royale
charlotte russe
mousse

panna cotta
pâte à bombe
ribbon
sabayon

pastry cream
diplomat cream
crème Chiboust
Gâteau St. Honoré

Basic Creams

The most basic of all creams is heavy cream. In the bakeshop or pastry shop, there are a number of ways in which heavy cream is used.

Whipped Cream

Whipped cream is simply heavy cream that has been whipped to a thickened foam. No additional ingredients are added to the cream as it is whipped. Whipped cream is used as an ingredient in formulas to lighten items such as mousses, Bavarian creams, and frozen soufflés. Like meringue, whipped cream is a type of foam. **(See Chapter 20: Meringues and Soufflés for more information on meringues.)**

Heavy cream consists of small butterfat globules that are suspended in milk. It is the butterfat that stabilizes the cream as it is whipped. The butterfat coats each air bubble as it is formed in the whipped cream, to give the resulting foam its structure. Heavy cream that contains more than 30% butterfat produces the best foam.

Hints and Tips for Whipping Cream

The structure and stability of whipped cream depends upon the temperature of its butterfat. Heavy cream should be kept as cold as possible because of the butterfat's low melting point. The colder the butterfat in heavy cream, the more stable the whipped cream. When working in a hot environment, try to keep the utensils and tools used for whipping cream, as well as the cream itself, as cool as possible. Whip the cream quickly. The longer the cream is whipped, the warmer it becomes. Cream can be very easily overwhipped. When this happens, the cream begins to look grainy. Soon after, the cream separates, with the butterfat parting from the surrounding liquid. Once this happens, the process cannot be reversed.

Like meringue, cream can be whipped to either a soft, medium, or a stiff peak depending upon its intended use. When cream is folded into other ingredients, such as a mousse, it is generally whipped to a soft peak. The very action of folding the cream into a base will continue to whip the cream. If a stiffly whipped cream is used, there is a greater chance that it will become overwhipped during the folding process.

Chantilly Cream

Chantilly cream is whipped cream with the addition of sugar and vanilla. Chantilly cream is used primarily to finish work. It is used to cover cakes, garnish plated desserts, and to top cream pies.

Chantilly cream stands alone; it is not combined with other ingredients. It is sweetened and flavored to bring out its unique rich flavor. A number of sweeteners such as powdered sugar, granulated sugar, or simple syrup are used for Chantilly cream. Vanilla is the traditional flavoring for Chantilly cream although various liqueurs, fruit purées, spices, and/or zest can also be used.

Crème Parisienne

Crème parisienne is essentially a chocolate-flavored Chantilly cream. Add sugar to heavy cream and bring to a boil. Pour the hot mixture over chopped dark chocolate and stir gently until the chocolate is completely melted. Place the chocolate cream in the refrigerator and chill for at least 12 hours. After that time, whip as you would Chantilly cream. Crème parisienne is also used for finishing work.

Cream-Based Desserts

Many desserts fall under the category of cream-based desserts. They are related by their light consistency and smooth textures. Some of these cream desserts form one component of a dessert, such as the Bavarian cream filling in a charlotte royale, while others serve as main items, as in a chocolate mousse. Cream-based desserts can be further divided into those that are stabilized by gelatin and those that are stabilized by eggs.

Creams Stabilized with Gelatin

Gelatin is an important component in a number of desserts. The nature of gelatin must be understood to have consistently successful products. Gelatin is an animal protein derived from bones, cartilage, tendons, and other tissues of beef and veal. Much of the commercial gelatin used in the bakeshop or pastry shop is a by-product of pigskin. Gelatin comes in two forms: powdered or sheets. Both forms have the ability to turn a liquid into a solid, as long as the ratio of gelatin to liquid and the temperature of the liquid and the gelatin are appropriate. If too little gelatin is added, the liquid will not set, and if too much gelatin is added the product will have a rubbery texture.

Gelatin must be bloomed to work. Blooming, or rehydrating, the gelatin is the first step in any formula in which gelatin is used. Bloomed or rehydrated gelatin can absorb up to five times its weight in water. Always bloom the gelatin in a cold liquid. When using powdered gelatin, it should be carefully stirred into a cold liquid to moisten all of the granules. Gelatin is fully bloomed when it is no longer dry to the touch. After gelatin is bloomed, it must be dissolved or melted. Gelatin can be melted in a number of ways: in a microwave, over a double boiler, or in a hot component of the formula.

When using sheet gelatin the sheets must be covered with cold water until they soften. Squeeze out the excess water and stir the softened sheets into the hot liquid until completely dissolved. A liquid with gelatin begins to set at 68°F (20°C) and melts at 86°F (30°C). Timing is important when working with gelatin, and the necessary molds or forms should be ready before beginning a formula that uses gelatin. Do not allow a gelatin-based mixture to gel in the mixing bowl as the result will be a lumpy end product.

Caution is necessary when adding gelatin to a mixture containing raw exotic fruits. Many fruits, such as papayas, pineapples, guavas, kiwis, mangoes, passion fruit, and figs contain *protease enzymes* that break down gelatin. Thus, when gelatin is added to a raw mango mousse, for instance, it will not set. Always cook exotic fruit before using it in a gelatin-based dessert. Cooking the fruit to 165°F (74°C) will kill the protease enzymes.

Bavarian Cream/Bavarois

Bavarois is the French name for Bavarian cream. *Bavarian cream* consists of a mixture of crème anglaise, gelatin, and whipped cream. Bavarian cream is used as a filling in the classic charlotte royale and charlotte russe, and as an entrée item on plated desserts.

The method of preparation for Bavarian cream begins with crème anglaise. Once the crème anglaise has reached "nappé," the bloomed gelatin is added. Stir to make sure that

STEP PROCESS

Making Bavarian Cream

Bavarian creams are basically custard-like creams, with the addition of gelatin for setting and whipped cream for lightness.

To make Bavarian cream, follow these steps:

1 Bring milk and sugar to a boil. Slowly add the hot milk to lightly beaten egg yolks stirring constantly.

2 Continue to temper the egg yolks by adding the hot milk until the egg yolk mixture is warm.

3 Add the tempered egg yolk mixture into the hot milk.

4 Stir and cook until the mixture reaches the nappé stage. Do not allow the mixture to boil. The mixture should coat the back of a spoon or spatula.

5 Strain the mixture into a container set in an ice bath to stop the cooking and cool the mixture slightly.

6 Softened gelatin sheets are added to the warm (120°F/49°C) mixture and stirred until completely dissolved.

7 Add flavorings such as chocolate and blend to combine.

8 Fold whipped cream into the chocolate mixture. Pipe into molds immediately.

STEP PROCESS

Making Charlotte Royale and Charlotte Russe

Charlotte royale is a Bavarian cream filling lined with a jelly roll sponge cake filled with jam or preserves and rolled. Charlotte russe is a Bavarian cream filling surrounded by ladyfingers.

To make charlotte royale and charlotte russe, follow these steps:

1 Slice jelly roll into thin slices.

2 Line the mold with the jelly roll slices.

3 Fill the lined molds with Bavarian cream.

4 Top with a slice of jelly roll and chill.

5 Once unmolded, apricot jam can be used to glaze the outside of the charlotte royale.

6 To make charlotte russe, pipe Bavarian cream into a mold lined with ladyfingers.

the gelatin is completely dissolved. If using sheets, it is necessary to squeeze out any excess water from the sheets before adding them to the crème anglaise. **(See Chapter 23: Sauces and Syrups for more information on crème anglaise.)** Once the mixture starts to thicken (it should be no warmer than 80°F [27°C]), gently fold in the whipped cream and pipe into molds immediately. An ice bath can be used to control the temperature during production.

Charlotte Royale

Charlotte royale utilizes a Bavarian cream filling. A traditional dome mold is lined with plastic wrap followed by thin slices of jelly roll. The Bavarian cream is then poured in and covered with a thin layer of plain génoise sponge cake. After the Bavarian cream has set, the dome is inverted and the mold and plastic wrap removed. The outside of the charlotte royale is brushed with a thin layer of apricot jam.

Charlotte Russe

Bavarian cream is the filling used in the classic dessert ***charlotte russe***. A mold is lined with ladyfingers and then filled with the Bavarian cream, and pieces of fruit may be added to the Bavarian cream filling. The Bavarian cream may also be piped into flexi molds or individual ring molds as an entrée on plated desserts.

Chocolate Mousse

Considered by many customers to be the height of luxury, chocolate mousse can range from the light and delicate to the sinfully rich. Regardless of the end result, all chocolate mousses begin with a base of chocolate. When dark chocolate mousse is made, the cocoa butter and cocoa powder in the couveture works to stabilize the mousse and gelatin is not needed. Melted butter is also sometimes added to chocolate mousse to increase its stability and richness. Whipped cream lightens the mousse while the butterfat in the heavy cream helps to stabilize it. The addition of a French/common meringue also contributes to an overall light mouthfeel.

When making a chocolate mousse, the timing and temperature of the ingredients is critical to success. If the chocolate is too warm it may melt the whipped cream, and result in a thin mousse. If the chocolate is too cold, the mousse will have chocolate chunks in it. To prevent either of these scenarios, it is important to prepare each component before the mousse is assembled. The cream should be whipped and the meringue made, and the chocolate and butter (if used) melted before any of the components are put together. Correctly folding the cream or meringue into the chocolate base is essential in making a mousse. Use a rubber spatula or a bowl scraper to fold as the flat edges of these utensils are more effective at folding together ingredients than a whisk. Start by folding a small amount of cream into the base chocolate mix to lighten the mix. This will also make it easier to fold in the remaining portion. Folding the whipped cream or meringue in thirds allows for the least amount of volume loss.

White Chocolate Mousse

The formulas used for white and dark chocolate mousse are not the same due to the differences in the manufacturing of each type of chocolate. Unlike dark chocolate, white chocolate does not contain cocoa solids. The lack of cocoa solids requires the addition of a stabilizer to white chocolate mousse formulas. White chocolate mousses call for the addition of gelatin. Once it has been bloomed and dissolved, the gelatin is added to the melted white chocolate. Then proceed using the same method of preparation for a dark chocolate mousse.

Fruit Mousse

The components of a fruit mousse can be as simple as fruit purée, gelatin, and whipped cream. Both the gelatin and the butterfat in the whipped cream help to stabilize the mousse. Tropical fruits should be heated to 165°F (74°C) to kill the protease enzymes. Failure to do so will result in a watery mousse.

Mousse

Mousse is another category of cream that is stabilized with gelatin (with the exception of chocolate mousse). Mousse translates in English to "foam." Mousses are mixtures that are lightened with whipped cream or a meringue. Unlike many of the other items in this chapter, mousses are largely uncooked and should be made with pasteurized yolks and whites. A large variety of bases are used to make mousses. They range from sabayon, fruit purée, curd, crème anglaise, pâte à bombe, and chocolate.

STEP PROCESS

Making Chocolate Mousse

Mousses are very popular and versatile; they can be served as an individual dessert or as a filling for tortes and pastries.

To make chocolate mousse, follow these steps:

1 Melt chocolate and butter over a double boiler.

2 Blend pasteurized egg yolks into the melted chocolate mixture.

3 Add liqueur or flavorings.

4 Fold in whipped pasteurized egg whites.

5 Fold whipped cream into mixture.

6 Do not overfold but make sure the components are fully combined.

7 Portion mousse for service.

Panna cotta consists of a sweetened and flavored dairy base set by the addition of gelatin. Panna cotta is made in individual forms and then unmolded. The ideal panna cotta contains just enough gelatin to hold it together. The end product should melt instantly on the tongue with no rubbery mouthfeel.

Various dairy products can be used to make a panna cotta; milk and cream are only two possibilities. Using buttermilk or yogurt can add a wonderful tangy element to panna cotta. Dairy products may be heated and infused with flavorings such as citrus zest, vanilla bean, or fresh mint. The bloomed gelatin is then melted in the warm dairy mixture. The mixture is chilled over an ice bath to allow the mixture to just slightly thicken before it is poured into molds and refrigerated.

Creams That Are Egg Thickened

Not all creams are stabilized by gelatin. Some creams are thickened by eggs and are only stabilized with the application of heat. Many of these creams have a rich mouthfeel due to a high yolk content.

Pâte à Bombe

Pâte à bombe derives its name from the bombe mold in which it is traditionally made. The bombe mold is a demi-sphere or dome mold. The ingredients in a pâte à bombe are egg yolks and sugar. The yolks are heated so that pasteurized yolks are not necessary. The yolks and sugar are placed over a double boiler and whisked continually. The yolks *ribbon*, thicken, and turn a light lemon color.

Pâte à bombe is used as the base for mousses. The richness of the pâte à bombe adds an elegant mouthfeel to the product in which it is incorporated. Chocolate should only be added when the pâte à bombe has reached its full volume. Pâte à bombe is also used in frozen parfaits and buttercreams.

Sabayon

Sabayon is a variation of pâte à bombe. It is simply pâte à bombe with the addition of wine or liqueurs. The French sabayon uses white wine, the Italian zabaglione uses Marsala wine, and sherry is used in Spain to make sabañon. A sabayon may also include liqueurs or fruit purées for equally flavorful results. Sabayon made with Amaretto, Frangelico, Grand Marnier, and Kahlua are just some of the possibilities.

Sabayon is classically poured over fresh berries in a gratin dish and lightly bruléed directly before service. Sabayon can also be a flavorful dessert in its own right. Fold whipped cream into a cooled sabayon, or pipe the mixture into a glass or cookie cup and serve with fresh fruit. Sabayon may be molded in a flexipan or used as a cake filling. In both of these instances gelatin, as well as whipped cream, is added to the sabayon mixture. Sabayon is also sometimes used as a sauce. See Figure 19-1.

Figure 19-1 Sabayon can also be poured over fresh berries and served in an attractive glass.

Pastry Cream

Pastry cream is crème anglaise that includes whole eggs as well as egg yolks and has been thickened with a starch. The type of starch used in pastry cream can vary from formula to formula. Cornstarch is sometimes used for the luster and shine it contributes, while flour is used for strength. Some formulas call for a combination of both cornstarch and flour. The addition of either starch prevents eggs from curdling, and allows the mixture to be brought to a boil. When flour is used to thicken the product it should be brought to a boil for at least 5 to 6 minutes. Boiling the product activates the starch, eliminates the starchy flavor, and brings a shiny luster to the product.

Many pastry cream formulas use a combination of egg yolks and whole eggs. The addition of albumin in pastry cream gives the cream its structure so that it can be easily piped into doughnuts, pâte à choux, and tartlet shells.

Pastry cream begins in much the same way as does crème anglaise. Like crème anglaise, pastry cream can be flavored with extracts, liqueurs, or infusions. Bring 2/3 of the dairy product, sugar, and infusion to a scald. Whisk the remaining dairy with the eggs and starches until lump-free. Temper the scalded dairy mixture in the egg and starch mixture. Whisk until well incorporated and return to the heat. Continually whisk the pastry cream while it is on the stove to prevent lumps from forming. The presence of starch in the pastry cream allows it to be boiled for 3 minutes.

Once the pastry cream has boiled, remove it from the heat and add butter. If an extract or liqueur is used to flavor the pastry cream, it should be added at this time. Pour the pastry cream into a hotel pan or on a sheet pan that is lined with plastic wrap or parchment paper. Cover the pastry cream completely with plastic wrap or parchment paper to make sure that no air bubbles form between the top of the pastry cream and the plastic wrap or parchment paper. The plastic wrap or parchment paper will prevent the formation of a skin on top of the pastry cream as it cools. Place immediately in the refrigerator.

Traditionally, pastry cream is used as a filling for fruit tarts and tartlettes. It can, however, also be used to fill Boston cream pie, doughnuts, and pâte à choux. It is enjoyed as a topping on Danish and used as a base for soufflés.

Pastry Cream Variations

Pastry cream serves as a base for the following creams. These variations of pastry cream are lightened through the addition of whipped cream or meringue.

DIPLOMAT CREAM *Diplomat cream* consists of pastry cream and whipped cream. The addition of whipped cream results in a light product. Diplomat cream is classically used as a filling for cream horns and Napoleons. Gelatin should be added to the diplomat cream when it is used as a filling in products that are sliced for service such as tortes or Napoleons.

CRÈME CHIBOUST *Crème Chiboust* is pastry cream that includes Italian meringue. As in diplomat cream, gelatin is added to crème Chiboust when it is used in a product that is cut or sliced. The bloomed gelatin can be dissolved simply by whisking it into the hot pastry cream. After the gelatin is completely dissolved, 1/3 of the Italian meringue is whisked into the mixture to lighten it. The remaining meringue is then gently folded in.

Ganache

One of the most versatile creams in the bakeshop or pastry shop is chocolate ganache. Ganache consists of boiled heavy cream poured over chocolate, and then stirred until a shiny chocolate cream results. There are many variations of ganache ranging from hard to soft. Each type has its specific uses in the bakeshop and pastry shop. The ratio of cream to chocolate determines the consistency of the ganache. When using dark chocolate, the ratio of chocolate to cream for a hard ganache is 2 to 1. The ratio changes to 2 1/2 to 1 when using a milk or white chocolate. More chocolate is used in a milk or white chocolate ganache because each of these chocolates contain fewer cocoa solids and cocoa butter than dark chocolate. (**See Chapter 27: Chocolate for more information on chocolate-and ganache-based confections.**) Soft ganaches usually have a greater percentage of cream to chocolate.

To make ganache, boil the cream and pour over chopped chocolate. Stir until the two ingredients are completely incorporated. Flavors can then be added into ganache. Add an extract, liqueur, or fruit purée to the finished ganache or infuse the cream while it comes to a boil with fresh flavorings such as mint leaves or citrus zest.

Ganache is extremely versatile. A hard ganache can be used for truffles and/or confections, and softer ganaches may be used as glazes, icings, or fillings. See Figure 19-2.

Figure 19-2

To make chocolate ganache, pour boiled cream over chocolate.

An immersion blender can be used to combine the chocolate and the cream to make a smooth ganache.

The classic use for crème Chiboust is as a filling for **Gâteau St. Honoré**. A praline (roasted hazelnut) flavored Chiboust is used for this famous dessert. A praline paste is added to the warm pastry cream and the process used for a traditional crème Chiboust is followed.

St. Honoré is the patron saint of pastry chefs and this famous dessert was designed in his honor. A Gâteau St. Honoré starts with a blind-baked puff pastry ring. Small pâte à choux puffs are filled with pastry cream, dipped in caramelized sugar, and attached to the outer ring of the puff pastry disc. Crème Chiboust is piped in the center of the puff pastry ring using a special decorating tip that gives the dessert its distinctive finish.

STEP PROCESS

Making Pastry Cream

The procedure for making pastry cream begins in much the same way as crème anglaise. Starch and butter are used to make a thicker product that can be used as a filling or a topping.

Follow these steps to make pastry cream:

1 Gather ingredients for pastry cream.

2 Combine the starch, half the sugar, and salt, if using, in a bowl. Add a small portion of milk, and blend.

3 Add eggs to milk mixture.

STEP PROCESS

4. Beat eggs and milk until combined.

5. Put remaining milk into a saucepan with the remaining sugar, and scald over moderate heat.

6. Temper hot milk into egg mixture.

7. Combine the tempered egg mixture and hot milk in the saucepan.

8. Heat until mixture thickens and there is no recognizable starch flavor.

9. Add vanilla.

10. Add butter, and blend until combined.

11. Pour into a stainless steel hotel pan or sheet pan lined with parchment paper or plastic wrap.

12. Cover with parchment paper or plastic wrap, label, date, and refrigerate.

CHAPTER 20

Meringues & Soufflés

Like so many other food items, the exact origin of meringues is unclear. However, there is an account of a Swiss pastry cook who prepared a meringue for the wedding feast of a king as early as the 1700s. The name *meringue* is thought to refer to the Swiss town of Meringen. There are three main categories of meringues: French/Common, Swiss, and Italian.

All three meringues consist of sugar and beaten egg whites, and vary only in their methods of preparation and their varying degrees of stability. Stable meringues do not easily deflate or fall apart and are also the heaviest meringues. Light meringues are considerably more fragile. This chapter explores the world of meringues. Categories of meringues will be discussed as will their methods of preparation. A discussion of soufflés is also included.

KEY TERMS

meringue
French/common meringue

Swiss meringue
Italian meringue

Japonaise
soufflé

frozen soufflé/soufflé glacé

Meringues

Meringues consist of only two ingredients, egg whites and sugar. Considering the simplicity of these ingredients, meringues are extremely versatile and are responsible for the success of a large number of products. Such products include macaroons, buttercreams, mousses, soufflés, and pies that are topped with meringue such as lemon meringue pie.

The Science of Meringues

Raw egg whites and sugar are the only ingredients needed to make a meringue. When egg whites are beaten, they can increase by as much as seven times in volume because of their protein. Egg whites consist of protein and water. When egg whites are whisked their proteins unravel. As whisking continues, the proteins rise to the surface of the bubbles that are being created and latch on to the other proteins that have also risen to the surface. A flexible film is formed. This film is the meringue.

Tips for Success

A successful meringue relies upon an understanding of the components found in egg whites.

Fat and Meringues

When egg whites are whisked, proteins rise to the surface and begin to coagulate and form a thin film. The introduction of any fat interferes with the protein's ability to form this film and prohibits the egg whites from reaching a peak or forming a secure structure. Meringues must be made with equipment that is free of any type of residual fat. Because egg yolks contain a great deal of fat, it is important that the egg whites used in making a meringue are completely yolk-free. Frozen egg whites are a clean and convenient product that is ready for whipping.

Overwhipping Egg Whites

It is possible to overwhip egg whites. When this happens the whites continue to enlarge until they eventually collapse. As they collapse the water held by the coagulated proteins escapes and produces hard clumps of meringue floating in water. The addition of sugar can help to prevent this from happening.

When sugar is dissolved, a thick syrup that is heavier than the original water contained in the whites is produced. This syrup gradually drains from the cells as they are forming. Sugar also helps to prevent overwhipping by interfering with the proteins as they try to coagulate.

Coagulation of these proteins resembles a chain reaction. Once sugar is introduced to the whites it interferes with the speed at which the reaction occurs. The addition of sugar extends the time needed to establish the meringue's flexible film. This slowing down of the coagulation process reduces the possibility of overwhipping the meringue. Adding a pinch of salt or cream of tartar can also produce the same results.

Additional Tips for Success in Making Meringues

Use a thin, flexible whisk to whip the egg whites. A thin whip allows for the incorporation of more air into the meringue because the meringue itself is relatively fragile. Room temperature egg whites should also be used because they foam better than cold egg whites. To get the chill out of refrigerated whites, gently heat them over a pot of simmering water at 185°F–200°F (85°C–93°C). Continuously stir the whites so that they do not coagulate.

The sugar used to make a meringue must be clean and free from any impurities such as bits of flour or other matter as these impurities will prevent the meringue from reaching a medium or stiff peak. Also use a large bowl because egg whites will increase by as much as seven times in volume when they are whipped.

Stages of Meringues

As egg whites are whipped they go through three stages:

Soft, Medium, and Stiff Peaks

In all three stages the meringue should be glossy. A dry looking meringue has been overwhipped. When a meringue is at a soft peak, it stays on the whisk when lifted from the bowl. A medium peak meringue forms soft peaks when the whisk is raised. When a whisk is removed from a stiff peak meringue, the meringue holds the peak with a stiff, pointed top. Different formulas call for meringues beaten to various peaks. A soft or medium peak meringue is usually used for soufflés and cake batters, whereas buttercreams require stiff peak meringues. Because a stiff peak meringue holds its shape so well when piped, it is used for meringue cookies as well as for garnish work. See Figure 20-1. **(See Chapter 24: Cookies and Petits Fours for more information.)**

Figure 20-1   Two different effects: Meringue browned in an oven (background) will be lighter than meringue browned using a blowtorch (foreground).

Categories of Meringues

There are three major categories of meringues: French/common, Swiss, and Italian. Each of these meringues has specific applications in the bakeshop or pastry shop.

French/Common

French/common meringue is sometimes referred to as cold or common meringue. This meringue is made without any application of heat and requires the use of pasteurized whites if it is to be used as a component of a product that is not cooked or baked, such as mousse. It is the simplest meringue to make and the lightest and

STEP PROCESS

Making French/Common Meringue

Because this method of preparing meringue requires no heat, pasteurized egg whites must be used if the meringue is not a component of a baked or cooked product.

To make French/common meringue, follow these steps:

1 Whip egg whites until frothy.

2 Continue whipping the egg whites and gradually add sugar.

3 When the egg whites reach the soft peak stage, the meringue will stay on the whip when lifted from the bowl.

4 At the medium peak stage, the meringue will stay on the whip when lifted from the bowl and form a soft, flexible peak.

5 Meringue reaches the stiff peak stage when the whip is lifted and the peaks point upward or just bend slightly.

most delicate of the three types of meringues. It is the least stable, and is used to lighten mousses. It can be piped and slow baked for decorations, or edible containers, such as swans. It is also used for base layers in tortes such as a Japonaise. *See Figure 20-2.*

The method of preparation for a French/common meringue is fairly straightforward. Before beginning, make sure that your bowl is large enough to handle the increase in volume of the whites. Use a thin, flexible whisk and make sure that the whisk, bowl, and whites are completely clean and free from any residual fat or grease. Using room-temperature whites, beat at high speed until the whites have tripled in volume. Reduce the speed to medium and begin to add the granulated sugar. The granulated sugar should be incorporated in a slow, steady stream. Adding the sugar too quickly, or all at once, will destroy the thin film you are trying to create. Continue to beat the whites until the meringue is at a stiff peak. A meringue is at a stiff peak if it holds its shape when piped and does not fall out of a bowl when tipped. The meringue should be shiny and glossy looking, and extremely light and fluffy. A French/common meringue must be used immediately or it will begin to deflate and separate if allowed to sit.

STEP PROCESS

Making Swiss Meringue

Because Swiss meringue is very stable, it forms the base for Swiss buttercream and is also used to make macaroon cookies.

To make Swiss meringue, follow these steps:

1 Heat egg whites and sugar in a double boiler and stir the mixture until it reaches 110°F (43°C).

2 Transfer mixture to a mixer and whip.

3 Whip until cool and egg whites have achieved full volume.

Figure 20-2

Use a plain tube to make necks and heads for meringue swans.

Pipe the wings for the swans using a star tip.

A plated meringue swan filled with raspberry sorbet and finished with both crème anglaise and chocolate sauce.

Swiss

A ***Swiss meringue*** is more stable than a French/common meringue and lighter than an Italian meringue. Its stability allows it to accept the addition of butter without collapsing, as in a Swiss buttercream. Swiss meringue is also used to make macaroon cookies. The method of preparation for a Swiss meringue is a bit more time consuming than that used in preparing a French/common meringue.

Place the egg whites and the granulated sugar in a bowl and gently heat over a double boiler. Stir the mixture so that the egg whites on the sides of the bowl (which are the hottest) do not coagulate. When the temperature of the egg white and sugar mixture has

STEP PROCESS

Making Italian Meringue

Italian meringue is the strongest type of meringue. It has many uses including providing a base for Italian buttercream.

To make Italian meringue, follow these steps:

1 Prepare a simple syrup using ⅔ the sugar in the formula and all the water. Cook until the syrup reaches 238°F–242°F (114°C–117°C).

2 When the syrup reaches that temperature, begin whipping egg whites and the remaining ⅓ sugar until frothy.

3 When the sugar syrup reaches 250°F (121°C) begin adding it to the whites.

4 Continue whipping the whites to full volume.

reached 110°F (43°C) and the granulated sugar is completely dissolved, remove the mixture from the heat. Pour the egg white/sugar mixture into a mixing bowl and whip at medium speed until it has cooled.

Italian

The strongest type of meringue is the **Italian meringue**. It is made by adding a hot simple syrup to a French/common meringue. It is the most time consuming of the three types of meringues and requires a bit more skill. An Italian meringue has many uses, such as lightening a parfait mixture or making an Italian buttercream. Italian meringue is not baked and its stability results in an extremely hard end product.

The method of preparation for an Italian meringue also begins with room-temperature egg whites, and clean, grease-free equipment. The two components of an Italian meringue are the hot simple syrup and a French/common meringue. Prepare

Japonaise

The classic *Japonaise* is a meringue variation consisting of a meringue base into which ground nuts are folded. Japonaise is used for cake layers in classic French tortes. This meringue is baked following a circular pattern. Trace the outline of a cake pan on a piece of parchment paper for a pattern. Place the pattern under a clean piece of parchment. Using a large, plain tip, pipe the Japonaise mixture in a spiral pattern starting from the center of the circle pattern and moving outwards. Continue to pipe until the circle pattern is filled. Place the parchment paper on a sheet pan and bake. *See Figure 20-3.*

Figure 20-3

Ingredients	French/Common	Swiss	Italian	Japonaise	Basic Meringue Formulas
Egg whites	1 pint	1 pint	1 pint	1 pint	
Granulated sugar	2 pounds	2 pounds	2 pounds	1 pound, 8 ounces	
Cornstarch	—	—	—	2 ounces	
Water	—	—	4 ounces	—	
Nuts	—	—	—	1 pound	
Uses	Cake layers Soufflé glacé Cookies	Fillings Buttercreams	Fillings Buttercreams	Cookies Torte layers	

the simple syrup by cooking at high heat ⅔ of the sugar in the formula, and all of the water. When the sugar syrup reaches 238°F–242°F(114°C–117°C) begin to make a French/common meringue with the egg whites and the remaining ⅓ of the sugar. Continue to cook the syrup as the eggs are whipping. By the time the sugar reaches 250°F (121°C), the whites will be at full volume. Remove the sugar mixture from the stove.

Turn down the mixer to medium speed and carefully stream the cooked sugar syrup into the French/common meringue. When adding the sugar to the whites be careful not to splash the hot syrup. Aim the syrup so that it hits the whites between the whisk and the side of the bowl. Continue to whip the whites on medium speed until the meringue is cool.

Baking and Storing Meringues

Baked meringues are used for cookies, torte layers, or as edible containers. Meringues should be baked at low temperatures, anywhere from 210°F–220°F (99°C–104°C) to ensure that they will dry out without caramelizing or browning. Sometimes, meringues are placed in the oven at the end of the day's production, although the ovens are turned off so that the remaining heat continues to dry out the meringue. In the days of wood burning ovens, the meringues were also placed in the oven at the end of the day so that the residual heat of the embers could dry them out. Store the baked meringues in an airtight container to prevent the meringue from absorbing moisture and becoming soggy. *See Figure 20-4.*

STEP PROCESS

Piping Meringue Nests

Meringue nests make attractive containers for ice creams, mousses, and fruits. They are baked at very low temperatures or placed in hot ovens that are turned off at the end of the day and allowed to dry out overnight.

To pipe meringue nests, follow these steps:

1 Using a template, draw a circle on parchment paper and turn the paper over. Use a pastry bag to pipe a base.

2 Continue in a circular motion to pipe the sides of the nest.

3 A star tube can be used for a more decorative nest.

4 The meringue nests are baked at a very low temperature and allowed to dry out without caramelizing or browning. The nests can be filled with a variety of fillings.

Figure 20-4 Meringue mushrooms can be made by piping round caps and pointy stems and then baking the meringues at low temperature until they dry out. They can then be assembled for garnishes.

Soufflés

A *soufflé* is a light dessert leavened with meringue. The term *soufflé* translated from the French literally means to puff or blow up. Seeing a soufflé it is easy to see just how appropriate this name is. Baked soufflés are made with a variety of bases, that all share one trait. They all owe their height to the leavening power of meringue.

The Rise and Fall of Soufflés

Egg whites contain about 90% water. When water is heated, it turns to steam, which expands and rises. When soufflés are placed in the oven the water in the meringue turns to steam and rises, pushing the surrounding structure and forcing it to rise. The structure of soufflés comes from the protein in the egg whites. After the air in the cells has expanded, the protein that surrounds each cell begins to coagulate. Once the protein has coagulated, the soufflé is considered to be "set." This delicate structure and the cells that make it up are extremely fragile.

Just as air is responsible for the soufflé's rise, it is also responsible for its collapse. Air constricts as it cools. The air inside of a soufflé starts to cool as soon as it is removed from the oven. As the cool air contracts, the surrounding cell walls collapse.

Soufflé Bases

There are four main types of soufflé bases consisting of either a fruit reduction, pastry cream, roux, or egg yolks. The base used determines the soufflé's ultimate stability, richness, and overall texture.

Fruit Reduction

A soufflé made with a fruit reduction base is extremely flavorful and produces a fairly stable end product. The base is made from a reduction of fruit pulp. The type of fruit used dictates the length of time needed to reduce the pulp. Some water must be removed from the fruit so that the resulting soufflé batter is not too thin. Care must be taken that not too much liquid is removed or the pulp will be too thick and will collapse the whites as they are folded together.

An Italian meringue is used when making fruit reduction based soufflés. The stability of this type of meringue holds up well to the density of the fruit reduction. Once the fruit base has been made and cooled, simply fold it into the meringue.

Pastry Cream

Pastry cream is somewhat similar to an elegant pudding. It consists of a dairy base that is thickened with eggs and starch. When used as a soufflé base, pastry cream produces a fairly stable soufflé, due in part, to the presence of pastry flour and cornstarch in the pastry cream base. This type of soufflé is made by simply folding the meringue into a slightly warm pastry cream. The pastry cream's richness gives the soufflé an overall rich mouthfeel and texture. Traditionally, soufflés made with a pastry cream base are flavored with liqueurs such as Grand Marnier and Frangelico.

Roux

A roux is traditionally used in the culinary world to thicken sauces. It consists of flour and butter that are cooked on the stove top. The butter is melted and the flour is slowly added and whisked in. Continue to whisk the mixture for 3–6 minutes to prevent lumping and burning. This procedure produces a blond roux with a golden-brown color. Hot milk is then slowly drizzled in the mixture and cooked for 5–7 minutes more until the mixture thickens.

STEP PROCESS

Making Soufflés Using a Roux Base

Soufflés are made up of two basic parts, the base and the egg whites. The base contains the elements that give the soufflé its flavor. The following hazelnut soufflés are made using a butter and flour base known as roux.

To make soufflés using a roux base, follow these steps:

1 Place butter in a saucepan and melt. Add flour stirring constantly and cook 3–6 minutes on low heat until the mixture begins to smell like toasted nuts. The cooked roux should be a light tan color, not dark brown.

2 Add boiling milk to the mixture and cook 5–7 minutes. The mixture will be quite thick.

3 The roux and milk mixture will pull away from the sides of the pan as it cooks.

4 Remove from the heat, cool slightly, and fold in nuts and egg yolks.

5 Fold a small amount of meringue into the base.

6 When combined, fold in the remaining meringue.

7 Fill soufflé cups that have been buttered and sprinkled with granulated sugar almost to the top with the mixture.

8 Bake at 375°F–400°F (191°C–204°C) for 15 minutes without opening the oven door. Soufflés are ready to serve when they bake out of the cups and are slightly firm to the touch.

STEP PROCESS

Making Soufflés Using an Egg Yolk Base

Egg yolks form the base for this chocolate soufflé. After the soufflé was removed from the oven, the top was cracked open and apricot sauce was poured in.

To make soufflés using an egg yolk base, follow these steps:

1 Whip egg yolks and sugar until they become light in color and form ribbons.

2 Melt chocolate and stir the chocolate into the egg yolk mixture.

3 Blend in cream and liquor such as rum or brandy. Instant espresso can also be used.

4 Carefully fold meringue into the chocolate mixture.

5 Pour the mixture into soufflé cups that have been buttered and sprinkled with sugar.

6 Bake at 375°F–400°F (191°C–204°C) until done. Serve immediately.

Temper the egg yolks if desired with a bit of the hot roux and return the mixture to the saucepan. Continue to whisk until well incorporated. Remove the mixture from the heat and allow the base to cool to room temperature before folding in the meringue. This is an extremely strong soufflé, and heavy ingredients such as ground nuts can be folded into the mixture without causing its collapse. Savory soufflés are usually roux based. Some common savory soufflé variations add grated cheese, shredded seafood, or cooked spinach.

Egg Yolk

An egg yolk based soufflé consists of a foundation of ribboned egg yolks. Chocolate soufflés are most commonly made with an egg yolk base. Warm, melted chocolate is folded into the ribboned yolks and sometimes a bit of espresso or brandy is added as well. Once the egg yolk mixture is at room temperature, the meringue is folded in. The resulting soufflé is rich and luxurious and a perfect environment for chocolate.

Tips and Hints for Making Successful Soufflés

Regardless of the type of soufflé the following hints will help to guarantee a successful product. Take care when folding the meringue into the flavoring base. Although a certain amount of effort is required to incorporate a light meringue into a thick soufflé base, too much folding will cause the meringue to lose a great deal of volume. If the meringue sacrifices too much volume the final soufflé will not rise. To prevent this from happening, take a small amount of the meringue and carefully mix it into the soufflé base. The base will, consequently, become a bit lighter and looser in texture. The remaining meringue can then be folded in without losing too much volume.

After the soufflé has been assembled, bake it at once in a hot oven at 375°F–400°F (191°C–204°C). Avoid opening the oven door throughout the baking process. Bake the soufflé for an additional 2–3 minutes, even after it has reached its maximum volume to allow the proteins time to completely set. If the proteins are not entirely set, the soufflé will collapse the moment it leaves the oven.

Prepping Ramekins Prior to Baking Soufflés

The correct preparation of the ramekin is vital to the soufflé's success. A straight-sided ramekin is used to lend support to the soufflé as it begins to rise. The striations on the sides of the ramekin are not just decorative; they work to equally disperse the heat throughout the ceramic dish. Butter the ramekin evenly, using either softened or melted butter for the best results. Dust the buttered ramekin with sugar to support the soufflé as it starts to rise. Make sure, however, that the outer rim of the ramekin is clean, with no overfill of butter and/or sugar. A clean outer rim gives the soufflé its beautiful top crust, or cap.

Serving Soufflés

Few products in the bakeshop or pastry shop are as dependent upon proper timing as are soufflés. Communication with the waitstaff, from the moment the soufflé is ordered until the moment it is served, is vital. Soufflés are always served immediately, or à la minute. Sometimes the waitstaff serves the soufflé tableside, breaking open the soufflé's cap and pouring in a chocolate sauce, fruit coulis, or sauce anglaise.

Frozen Soufflé/Soufflé Glacé

Frozen soufflé/soufflé glacé is made to visually resemble a baked soufflé. It is, however, not baked. An assortment of bases are used to make these soufflés, ranging from a fruit or chocolate mousse to a Bavarian cream or pâte à bombe. ***(See Chapter 19: Creams and Mousses for more information on all of these products.)*** When making a frozen soufflé, the base mix is lightened with additional whipped cream or a meringue to allow the frozen soufflé to acquire an airy nature. The frozen soufflé glacé is often sprinkled with a dusting of cocoa powder or powdered sugar directly before service. Because the frozen soufflé will not deflate at room temperature, timing is not as crucial to the service of this dessert as it is for the baked soufflé.

Figure 20-5 Soufflé glacé.

Regardless of the base used, the frozen soufflé is assembled to look like a baked soufflé. The mixture is frozen in a ramekin to give the appearance that the frozen soufflé has "risen" and a paper collar that extends approximately 2 inches over the top is then fitted around the outside of the ramekin. With the collar in place, the ramekin is filled above its rim. Just before the frozen soufflé glacé is served, the paper band is removed and the resulting soufflé has the appearance of a traditional baked soufflé. *See Figure 20-5*.

CHAPTER 21

Frozen Desserts

There are many popular and classic frozen desserts that are appreciated throughout the world. In modern foodservice, however, ice cream, sorbet, and sherbet are perhaps the most frequently used by pastry chefs as components of other plated desserts. These frozen desserts provide contrasts in temperature, flavor, texture, and color to the dessert item with which they are served.

Frozen desserts are served year-round, and are an important part of every pastry chef's repertoire. House-made ice creams in flavors that reflect the changing seasons are versatile dessert components that can be easily dressed up or down, while refreshing sorbets are ever-popular intermezzos to cleanse the palate and to transition between courses.

KEY TERMS

ice cream
French-style ice cream
gelato
American-style ice cream (a.k.a. Philadelphia-style)
frozen yogurt
serum solids
churn-frozen dessert
overrun
hardening
coupe
sundae
ice cream bombe
still-frozen dessert
parfait glacé
soufflé glacé
semifreddo
ices
Baumé scale
sorbet
spoom
sherbet
granita

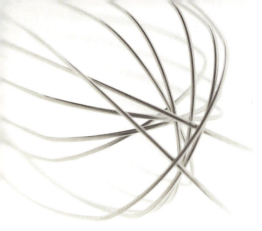

Ice Cream

Ice cream is a smooth frozen dessert made by simultaneously churning and freezing a base mixture of milk, cream, sugar, egg yolks (optional), and flavorings. Ice cream has a universal appeal and is found in numerous cuisines. So many cultures have a long history of making ice cream that its origin is unclear. A record that dates as far back as four thousand years ago does include a description of a dish of frozen milk and seasoned rice that was prepared in China.

By the 1700s, ice cream was common in England and the colonies, but its popularity exploded in 1846 when a dairymaid in Philadelphia named Nancy Johnson invented a portable hand-cranked ice cream freezer. This invention allowed for the incorporation of air through the constant churning during the freezing process, and made the texture of ice cream much smoother. This improvement also made ice cream one of America's favorite desserts.

Types of Ice Cream

There are two main types of ice cream: rich custard-based ice cream that contains egg yolks, and a lighter style that does not use eggs. Both are flavored with a wide variety of flavoring components, but have differences in body, richness, and texture. The two types also have different methods of preparation.

French-style ice cream is a custard-based ice cream made with milk, cream, sugar, egg yolks, and flavoring. The ingredients and preparation of the base mixture are the same as those used for crème anglaise. Because of its egg content, French-style ice cream must be cooked prior to freezing.

Gelato is an Italian custard-based ice cream, similar to the French-style except that gelato contains a higher percentage of egg yolks. Gelato is also made exclusively from milk rather than milk in combination with cream.

American-style ice cream, also known as *Philadelphia-style ice cream*, does not contain eggs, only milk, cream, sugar, and flavorings. American-style ice cream need not be cooked prior to freezing.

Frozen Yogurt

Frozen yogurt was developed as a lowfat alternative to ice cream. The ingredients used are similar to those in ice cream except that lowfat yogurt is substituted for the milk, cream, and eggs, and stabilizers such as gelatin may be added to improve smoothness and body. Frozen yogurt is made by the same method as that used for American-style ice cream.

STEP PROCESS

Making French-Style Ice Cream

The richness of ice cream depends on the ingredients used and the manner in which it is made. French-style ice cream starts with a base of crème anglaise.

To make French-style ice cream, follow these steps:

1 Prepare the crème anglaise and chill it in an ice bath.

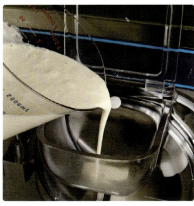

2 Pour chilled base into the ice cream maker.

3 When the ice cream freezes, it is ready to be extracted from the ice cream maker.

4 Fold in fruits if using.

5 Cover, label, date ice cream and place in freezer to harden.

Because frozen yogurt was developed as an ice cream substitute, the flavors and varieties of frozen yogurt generally mimic those of ice cream. The popularity of frozen yogurt peaked in the early 1990s, but began to wane in the new millennium as the lowfat diet craze lost ground to a "foodie" culture that fosters an appreciation of high-quality cream and butterfat.

Ingredients in Ice Cream

Ice cream today is made from a variety of ingredients including milk, cream, eggs, sweeteners, and flavorings such as fruit purée, infusions, chocolate, spirits, and liqueurs.

Milk and Cream

The butterfat found in milk and cream is the chief contributor to the texture, body/firmness, and richness of ice cream. Butterfat percentages usually range between 14% and 16%. The total butterfat should not exceed 22%–25% or the excess butterfat will freeze separately, resulting in undesirable grains of butterfat throughout the ice cream. Excess butterfat also inhibits the mixture's ability to whip, or incorporate and trap air.

Solids found in the milk and cream other than the fat are referred to as *serum solids* and should total at least 20%. Milk solids contribute to whipping capabilities, but too much can cause the lactose in the milk to crystallize, resulting in a sandy texture.

Eggs

Eggs contribute to the smoothness, body/firmness, richness, flavor, color, and fat content of ice cream. Egg yolks provide smoothness because they contain lecithin, an emulsifier that controls the separation of butterfat during the freezing process. When eggs are used in ice cream, they should be prepasteurized or the custard base should be heated to at least 145°F (63° C) to prevent salmonellosis.

Sweeteners

In addition to enhancing flavor, sweeteners contribute to the smoothness of ice cream. Although sugar is a freezing inhibitor, the total sugar content should be between 14% and 16%, and not exceed 18% or it will interfere with the body/firmness of the ice cream. A percentage of granulated sugar can be replaced with invert sugars such as corn syrup or glucose. Invert sugars inhibit the growth of ice crystals to further enhance smoothness.

Flavorings

Ingredients in the base mixture as well as added ingredients such as fruit purées, nuts and nut pastes, flavor infusions, chocolate, and liqueurs or other spirits contribute to the flavor of ice cream. Natural flavor extracts and compounds are also options. When using these types of flavorings in a custard-based ice cream the mix must be heated and the flavorings are added after the mix is removed from the heat. When flavoring ice cream, always adjust the flavor intensity to complement the flavors of the other components in a dessert. Ice creams with novel or multiple flavors are best enjoyed by themselves.

Ice cream requires more flavoring than most nonfrozen products because its cold temperature decreases the sensitivity of taste buds, reducing their ability to perceive flavors. Some of the flavor is also captured in the butterfat content.

Fruit Purées

Thick fruit purées promote smoothness and firmness, and add fresh fruit flavor and color to ice cream. The sugar in fruit purées acts as a freezing inhibitor, and must be considered in the overall formula to avoid an overly soft product. A high water content in a fruit purée will also promote an undesirable iciness.

Infusions

Flavor infusions can be made of ingredients such as ground coffee beans, tea leaves, or spices. The flavor is infused in the milk and sugar mixture, then strained before proceeding.

Chocolate

Cocoa butter promotes smoothness and increases the body/firmness of ice cream. The sugar content of some types of chocolate acts as a freezing inhibitor, and must be evaluated in the overall formula.

Spirits/Liqueurs

Alcohol is another freezing inhibitor. When using liqueurs to flavor ice cream, the freezing ability is inhibited by both the alcohol and the sugar content. So while Kirschwasser,

Kahlua, and other liqueurs offer many tempting flavors, ice creams flavored with liqueurs and other alcoholic beverages are usually softer and may not freeze satisfactorily.

Additives in Ice Cream

Additives are used in some commercial ice cream products to create body and a smooth texture, and to cut the cost of expensive ingredients. Lowfat ice cream varieties that omit or reduce ingredients such as cream and eggs rely on additives to enhance body and smoothness. Gums, soy lecithin, gelatins, and chemical additives are also used to stabilize ice cream. Pastry chefs making ice cream for in-house service rarely include additives and usually rely on quality ingredients to create smoothness and body.

Qualities of Ice Cream

Smoothness of texture, body/firmness, richness, and flavor are the main evaluative criteria for grading ice cream. The quality and quantity of the butterfat in the milk products used are the main factors that affect these qualities.

Smoothness of Texture

Smoothness of texture is primarily determined by the butterfat content. A higher butterfat content within the recommended range produces a smoother ice cream, while a lower butterfat content yields an icier texture. Egg yolks also promote smoothness, as do thick fruit purées, sweeteners, constant churning, and rapid freezing.

Inadequate churning promotes the growth of ice crystals, as does the water content of ingredients such as fruit purées. The growth of ice crystals on products has a detrimental effect on their smoothness.

Body/Firmness

Egg yolks, the butterfat content of the milk and cream, and the fat content of flavoring ingredients such as chocolate all increase the body/firmness of ice cream. Sugar and alcohol, on the other hand, are freezing inhibitors that decrease firmness to yield a softer product.

Body or firmness is also affected by the amount of air that is incorporated during churning and freezing. Incorporation of air increases the total volume to create a lighter and smoother product. The percent by which the volume increases during churning is referred to as overrun. As overrun increases, the body/firmness decreases. Overrun is desirable to a point, but too much incorporated air creates an overly soft, foamy product. An overrun of 20% produces an ice cream with good, firm body.

Richness

Richness is determined by butterfat and egg yolk content, and by certain rich flavor ingredients such as chocolate.

Flavor

The quality of each ingredient, including the specific "flavoring" ingredients, contributes to the overall flavor of ice cream. Care should be taken in selecting fresh, high-quality eggs, milk, cream, and flavoring ingredients. Freshly ground, aromatic spices or coffee beans, vibrant, and flavorful fruit purées, the highest quality cocoa or chocolate, and natural flavor extracts or compounds yield the best results.

The Production of Ice Cream

The production of ice cream consists in preparing the base, aging, freezing and overrun, and hardening.

Preparing the Base

There are two methods used to prepare an ice cream base. The first is used for custard-based ice creams (French-style or gelato) that include eggs, while the second is for American-style ice creams that are produced by merely mixing together the base ingredients.

Aging

The ice cream base should be chilled for 4 to 12 hours at approximately 40°F (4°C) before freezing. Chilling the ice cream for this period of time at a food safe temperature allows the proteins to fully absorb the water, preventing the formation of ice crystals, and producing a smoother finished product.

Freezing and Overrun

Ice cream is known as a ***churn-frozen dessert*** because it is churned continuously during the freezing process. Two important things happen during this process: (1) the mix is frozen, and (2) air is incorporated to make the mix lighter (less dense) and to increase its volume. The freezing process occurs at 0°F (−18°C).

The percent by which an ice cream base increases in volume as air is incorporated into it is referred to as the ***overrun***. An insufficient amount of air causes ice cream to be heavy and feel very cold in the mouth. In contrast, the inclusion of too much air produces an unpleasant foamy texture. As the percentage of overrun increases, the quality of the ice cream decreases. The industry standard for the highest quality ice cream is 20% overrun. Some supermarket brands incorporate a 70%–100% overrun.

Factors that affect overrun include:

- the freezing equipment used.
- how full the machine is at the start (For maximum overrun, the freezer should be filled only halfway).
- the length of time the base is frozen and churned (While a rapid freezing process is beneficial in inhibiting the growth of ice crystals, the overrun is compromised if the process is too rapid. Air can only be incorporated while a portion of the ingredients remain in a liquid state).
- the total amount of solids in the mix (Total solids including fat, sugar, nonfat milk solids and egg solids should be approximately 40%).

Ice cream comes out of the freezer at a temperature between 22°F (−6°C) and 26°F (−3°C) and has the consistency and texture of soft serve ice cream. The product is most flavorful at this stage and can be served immediately, or stored for up to one week in a subzero hardening cabinet. It must be packed airtight to prevent ice crystals from forming on the surface and to prevent the absorption of odors from the freezer. Fluctuations in temperature during storage can cause coarseness of texture and shorten the shelf life of the product.

Hardening

Hardening refers to the time during which packed ice cream remains in the subzero hardening cabinet. The hardening cabinet has a constant temperature of 6°F (−14°C) to 10°F (−12°C). Ice cream should be tempered several hours before service by placing

> **Sanitation Standards for Ice Cream Production**
>
> An ice cream base provides an excellent environment for the growth of bacteria. Strict sanitation guidelines must be followed when making and handling ice cream. Do not touch any ice cream mixtures with bare hands. Properly clean and sanitize machines and equipment after each use and keep storage units sanitary. Many machines have automatic sanitizing systems.

it in a serving cabinet with a temperature between 8°F (−13°C) and 15°F (−9°C) to soften it appropriately.

Plating Frozen Desserts

Frozen desserts should be served in chilled serviceware. When plating a frozen product, some form of anchor should be used to prevent the product from rolling or sliding on the plate as it is taken from the kitchen to the guest. Examples of anchors include a cup of tuile, almond lace, fruit or tempered chocolate. Toasted ground nuts or cake crumbs sprinkled on the plate also allow for ice cream to be placed securely on top. *See Figure 21-1.*

Ice Cream Products and Classic Desserts

The variety of ice cream products is extensive. Descriptions of the most popular products, and their most common specific presentations follow.

Coupes and Sundaes

Coupe is the French name for the American dessert called a *sundae*. A sundae is a scoop (or scoops of one or more flavors) of ice cream served in a glass or dish, with sauce, and topped with fruit or nuts spooned on top or layered with the ice cream. An endless variety of ice creams, sauces, and topping combinations exist for coupes and sundaes, although some combinations have become classics. Caramel, chocolate, hot fudge, and fruit are popular sauces used for coupes and sundaes, while fresh or poached fruit, nuts, candies, whipped cream, and maraschino cherries are frequently used toppings.

Peach Melba

Peach Melba is a coupe made of poached or blanched peaches served with vanilla ice cream and Melba sauce. Auguste Escoffier created this classic dessert in London in 1892 for a celebration honoring the great Australian opera diva Nellie Melba. The original "Peach Melba" was served in an elaborate carved ice swan and topped with spun sugar. A few years later Escoffier added a sauce made from red currant preserves and raspberry coulis—Melba Sauce—and the dessert has been a classic ever since. Today in the United States, Melba sauce is commonly made from raspberry preserves thinned with simple syrup. The

Figure 21-1 Raspberry sorbet.

Figure 21-2 Peach melba.

modern presentation of Peach Melba usually includes a slice of sponge cake with peaches, ice cream, and Melba sauce. *See Figure 21-2.*

Poire Belle Hélène

Poire Belle Hélène is a classic French coupe. A firm pear such as Bosc is poached in vanilla syrup then served upright on a bed of vanilla ice cream and topped with chocolate sauce. The dish was created in 1864 and named for a popular operetta at the time, Offenbach's La Belle Hélène. The operetta tells the story of the abduction of Helen, the queen of Sparta, by the Trojan prince, Paris, which prompted the Trojan War.

Ice Cream Bombes

An *ice cream bombe* is a molded frozen dessert that classically consists of a parfait filling encased in concentric layers of ice cream in a dome-shaped mold, with an optional sponge cake base. The outer layer is enrobed with a variety of finishes such as chocolate glaze, chopped nuts, or meringue. Today, any dome-shaped frozen dessert is referred to as a bombe. Bombes are also served in wedges that reveal the contrasting layers within.

To assemble a bombe, chill the mold in the freezer until ready to use. Ice cream or sorbet taken directly from an ice cream machine has the perfect soft, spreadable consistency for a bombe. Hardened ice cream must be used softened at room temperature to the correct consistency for spreading. Fruits or other garnishes may be folded into the softened product at this time. Pack the mold with successive layers until it is filled, placing the mold back in the freezer to harden each layer before adding the next. A sponge cake base may be added if desired. Cover with plastic and freeze until hard. To serve, dip the mold in warm water briefly to facilitate unmolding, then turn the bombe out onto a chilled serving platter and decorate as desired.

Contemporary bombes are made with any variety of outer layers and fillings, but some combinations, such as those used in the Baked Alaska, are considered classic.

Baked Alaska

This extravagant ice cream bombe features the contrast of a baked exterior encasing a frozen center. The technique of baking ice cream inside an insulating layer of pastry dough may have been developed in China. The French chef and author Honoré de Balzac is said to have learned this process when visiting Chinese chefs in the mid-1800s. Thomas Jefferson also served ice cream encased in hot pastry at his home, Monticello, after spending time as the United States ambassador to France.

The famous New York restaurant Delmonico's popularized Baked Alaska using meringue and named it "Alaska/Florida" to highlight the temperature contrasts. Meringue has become a popular alternative to pastry as the outer layer for baked ice cream dishes, because the many trapped air cells in meringue make it an excellent insulator. Fannie Farmer is credited with naming the dessert "Baked Alaska" in 1896 to celebrate America's purchase of Alaska. Other names for Baked Alaska include "Norwegian Omelet" and "Omelet Surprise."

The modern version of Baked Alaska is often prepared in individual servings and consists of vanilla, chocolate, and strawberry ice cream layered in a bombe with a base of sponge cake and an outer layer of meringue. Baked Alaska is often lightly doused with a liqueur and flambéed.

STEP PROCESS

Making Baked Alaska

Baked Alaska is an ice cream bombe encased in sponge cake and decorated with meringue. Originally, the finished dessert was browned in an oven to caramelize the exterior. It is more common now to use a blowtorch or a salamander. Baked Alaska is often served flambéed.

To make Baked Alaska, follow these steps:

1 Line a bowl with sponge cake and add soft ice cream. Insert a bowl into the center to form a well. Remove the bowl once the ice cream has set.

2 Fill the well with ice cream or sorbet.

3 Top with a layer of sponge cake which will form a base when unmolded.

4 Invert the bombe onto a platter and pipe with meringue.

5 Brown the meringue with a blowtorch and flambé. A small cup or eggshell is often inserted into the meringue to hold the flambé liquid.

6 Slice Baked Alaska for service.

Still-Frozen Desserts

Frozen products that are not churned during the freezing process are called *still-frozen desserts*. Air is incorporated into still-frozen products by whipping the base or folding in whipped ingredients such as egg whites or whipped cream. Parfait glacés, soufflé glacés, and semifreddos are examples of still-frozen products.

Parfait Glacé and Soufflé Glacé

Parfaits are made from a pâte à bombe or a crème anglaise base that has a lightly whipped cream folded into it. Sauces and fruits are often layered with the cream. *Parfait glacés* are frozen in glasses or molded in terrines. *Soufflé glacé* also uses a pâte à bombe or

STEP PROCESS

Making Soufflé Glacé

Soufflé glacé is a still-frozen dessert made to resemble a traditional baked soufflé.

To make soufflé glacé, follow these steps:

1 Place egg yolks, whole eggs, and granulated sugar in a bowl and heat over a double boiler; heat to 145°F (63°C) whipping constantly.

2 Remove from heat and whip to a medium peak; cool. Flavor as desired.

3 Whip heavy cream to a soft peak stage and fold one-third of the cream into the egg mixture. Fold in the remaining cream in two stages.

4 Fill a pastry bag with the mixture.

5 Pipe into ramekins fitted with a paper collar. Freeze.

6 Dust tops with cocoa if desired and remove the collar.

7 Soufflé glacé plated for service.

crème anglaise base folded with a lightly whipped cream, but it is served in ramekins. The ramekin is fitted with a tall collar before being filled with the glacé. When frozen, the collar is removed to reveal a frozen parfait that rises high above the rim of the ramekin, to resemble a lofty baked soufflé.

Semifreddo

Semifreddo means "half cold" in Italian, and refers to a dessert that consists of partially to completely frozen ice cream, mousse, or custard, piped or packed in a mold, and sometimes layered with cake. It can be frozen in rectangular molds similar to loaf pans, then turned out and served in slices. Semifreddos are often garnished with a warm sauce. *See Figure 21-3.*

Figure 21-3 Chocolate semifreddo prepared in a pyramid mold.

Ices

Ices are frozen desserts made from sugar-syrup bases with the addition of fruit purée, fruit juice, and sometimes an alcoholic beverage. Pasteurized egg whites, gelatin, or other stabilizers may also be added.

Ices differ from ice cream in the following ways:

- Ices, with the exception of sherbet, do not contain milk or cream. This lack of butterfat and other serum solids results in a coarser texture.
- The sugar content of ices is much higher than that of ice cream, between 25% and 35% (the sugar content of ice cream is generally no more than 18%).
- Ices usually have a tart flavor due to a high fruit-acid content.
- Ices have a greater cooling characteristic when consumed due to their coarser texture and lower melting point.

Special Problems with Ices

Three common problems sometimes encountered in the production of churn-frozen ices are: the separation of the sugar syrup, the crystallization of sugar and ice (a grainy texture), and a coarse or crumbly texture. Tactics for avoiding or correcting these problems follow.

Separation of the Sugar Syrup

The concentrated sugar syrup may separate if the sugar content is too high, or if there is excess overrun and insufficient total solids to bind the mixture. Reduce the amount of sugar in the formula, and add stabilizers in the form of gelatin, agar-agar, pectin, or egg whites to the base mixture. The usual amount of gelatin necessary is 2 teaspoons per quart of the base mixture. Fruits high in pectin, such as apples, citrus fruits, blueberries, and cranberries, should not require the use of a stabilizer.

Crystallization of Sugar and Ice (Grainy Texture)

The type of sugar used to make the syrup affects the texture of the finished product. A grainy texture can be reduced by substituting part of the granulated sugar with an invert sugar such as honey or glucose.

Exposure to air in the freezer and fluctuation of storage temperatures promote ice crystallization, so storage in airtight containers and a stable temperature in the hardening cabinet are recommended.

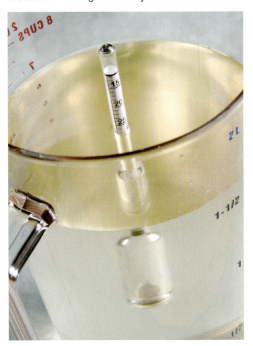

Figure 21-4 A Baumé scale is used to measure the sugar density of a solution.

Coarse or Crumbly Texture

Low sugar content in the base mixture, inefficient freezing methods, and too much overrun can result in an undesirable coarse or crumbly texture in the final product. If necessary, adjust the sugar content, use more efficient freezing techniques, and/or reduce overrun.

Measuring Sugar Density

A *Baumé scale*, or hydrometer, is a device used to measure sugar density (the concentration of sugar) in a sugar syrup or other solution. *See Figure 21-4*. This measurement is expressed on a Baumé scale in units of degree (°Bé). The denser the syrup, the higher the reading. The syrup must be at room temperature for the scale to read accurately.

If a Baumé scale is not available, a rough measurement of sugar density can be taken by floating a sanitized whole, unbroken egg in the tepid syrup. For sorbet, the area of the egg that rises above the surface of the syrup should be between the size of a dime and quarter.

Types of Ices

Sorbet, spoom, and sherbet are churn-frozen ices made using the American-style (uncooked) method of ice cream preparation. Granita is a still-frozen ice with an intentionally coarse, grainy texture.

Sorbet

Sorbet is an ice made with a sugar syrup and a base infused with fruit, wine, or fresh herbs. As America has become increasingly interested in quality fat-free dessert options, sorbet has gained in popularity. The Density Reading of sorbet should be between 14°Bé and 18°Bé. *See Figure 21-5*.

Figure 21-5 Measuring sugar density for sorbet.

Spoom

A *spoom* is a light, frothy variation of sorbet, made by folding an uncooked (pasteurized) Italian meringue into the sorbet base during the final stages of freezing. Spooms are typically made with champagne or white wine. The Baumé reading for spoom should be between 14°Bé and 18°Bé.

Sherbet

Sherbet is a fruit-based ice made from sorbet syrup and a base containing milk or cream and/or egg whites. Butterfat and other solids from the milk or cream make sherbet smoother and richer than sorbet and other ices. Sherbet does not contain as much milk or cream as ice cream, so sherbet is sometimes considered a lowfat substitute for ice cream.

The Baumé reading of sherbet should be between 14°Bé and 18°Bé.

Granita

Granita (French *granité*) is a coarsely textured Italian ice made from a sweetened, flavored water base and is still-frozen rather than churned. Its name is derived from the Italian word, *grana*, which means "grainy." Coffee, tea, wine, or fruit juice can be used as a base for granita.

Figure 21-6

When making granita, the flavored water base is first stirred before it is completely set.

Scrape the frozen ice with a fork to create the "grainy" texture.

The sugar content of granita is lower than that of other ices, and the Baumé reading is 10°Bé to 12°Bé. The low sugar content and the freezing method used for granita contribute to its coarse texture. Granita is still-frozen in a hotel pan. During the freezing process the granita is stirred occasionally to break up and distribute the ice crystals, creating light flakes that are spooned, without packing, into serving glasses or dishes. See Figure 21-6.

CHAPTER 22

Fruit Desserts

Fresh, seasonal fruit that is perfectly ripe, aromatic, and attractively presented is a light but elegant dessert, and a satisfying way to end a rich dinner. Unadorned seasonal fresh fruit was served as a sweet finish to Persian and other Near Eastern cuisine and is a popular choice today, throughout the world. In the West, consumers increasingly seek dishes made from locally grown produce in season, including fruit-based desserts. Fresh fruit offers the pastry chef an almost limitless range of options for meeting this demand. In winter when fresh fruit is not in season, warm fruit desserts such as compotes, cobblers, and crisps are welcome choices.

Many pastry chefs alter their menu offerings nightly based on the availability of products found that day at the farmers' market. Menu items are designed to include the freshest fruit seasonally available in a geographic region. Developing a relationship with local farmers and vendors is advantageous, and pastry chefs today often choose to include the name of the local orchard or farm in the name of the dessert listed on the menu to let consumers know that the fruit is local. This acknowledgement also creates good will by promoting local food producers.

KEY TERMS

compote
fruit salad
fruit salsa

fruit chutney
cobblers
grunts

crumbles
crisps
Betty

pandowdy
flambéed desserts

Fruit Desserts

Most fruit-based desserts are fairly simple creations, designed to showcase the natural flavor and color of the fruit itself. While fruit desserts can be easy to prepare, they need not be considered plain. With careful ingredient choices and straightforward preparation fruit desserts can be appropriate for even the most regal occasion.

Poached Fruit

Poaching infuses fruit with the flavors of the poaching liquid, gently softening the fruit while allowing it to retain its shape. The poaching liquid also stains the fruit to add color and appeal. Served warm or chilled, whole poached fruit can make a striking plated dessert, or may be sliced for use in pastries or to accompany other dishes.

Firm fruits such as pears, apples, and peaches may be poached because they are able to withstand the necessary heat and cooking time while remaining intact. Fruit is generally poached whole, so the fruit's good shape and appearance are important to consider. Wash, peel, and core the fruit as needed, but consider leaving the stem on fruits such as pears and apples for visual appeal.

Figure 22-1 Poaching pears in a flavor-infused liquid.

The poaching liquid can be any syrup made from wine, water, liquor, or a liqueur—usually four parts liquid to one part sugar—and flavored with herbs, spices, and other flavoring ingredients such as citrus peel or vanilla bean. The liquid is kept at a low temperature (160°F–185°F/71°C–85°C) so that the fruit poaches gently. This slow, gradual cooking helps to retain the shape of the fruit.

To poach fruit, combine the ingredients for the poaching liquid in a deep saucepan that is large enough to hold the fruit in a single layer. Bring to a boil, reduce the temperature, and place the fruit so that the liquid covers it. To ensure even cooking and color, weigh the fruit down with cheesecloth or a plate to keep it submerged. Poach the fruit until a small knife can easily pierce the fruit.

To optimize flavor and color, allow the fruit to steep in the liquid while cooling (preferably overnight). The poaching liquid may be reused for poaching, or it can be reduced to make a sauce or glaze to accompany the fruit. It may also be used as a base for other fruit desserts or sorbets. *See Figure 22-1*.

Roasted and Grilled Fruit

The dry heat of an oven or grill concentrates flavor and caramelizes the sugar in fruit to enhance its flavor and to add an interesting flame-broiled appearance. Fruit is usually roasted or grilled quickly, just to the point where the sugar caramelizes and before the fruit's structure begins to break down. Pineapples, apples, peaches, and grapefruit are excellent choices for these methods of preparation. Roasted or grilled fruit can be presented alone with an accompanying sauce, or as a topping for another dessert.

Moisture content, sugar content, and the texture of the fruit determine the type of preparation and the outcome of the grilling. Firmer fruit can be cut into smaller pieces (as for skewering) without losing structure. High moisture content prevents fruit from drying out and allows for a greater concentration of flavor as the moisture evaporates. For caramelization, fruit with a high sugar content is desirable.

Select ripe, firm-textured fruit for grilling and roasting. Peel, core, or pit the fruit as necessary, and cut or slice it to the desired form. Rub the fruit with butter or oil and a sweetener, or brush it with simple syrup, a glaze made of honey, a liqueur, or lemon juice. Roast in a baking pan at 325°F–425°F (163°C–218°C), or grill on an oiled grill until it is tender and caramelized. *See Figure 22-2.*

Figure 22-2 Roasting concentrates the flavors of fruits.

Compote

Compote is a stewed fruit dessert prepared by simmering fruit in syrup until it begins to lose its shape and fall apart. Compote is served warm or cool, alone or accompanying another dessert.

Compote is often made with an assortment of dried fruits, and a simmering liquid made up of any flavorful combination of water, liquor, wine, liqueur, or fruit juice. Sugar, honey, and spices are added to taste. If fresh fruit is used, cut or dice the fruit before cooking. In addition to dried and fresh fruit, canned and frozen fruit may also be used for compote. Stir the simmering compote periodically as it cooks, and continue cooking until the fruit is completely soft.

Compotes can be left overnight to absorb flavor and color, or they can be used immediately. Compote is often chosen as a light alternative to richer, sweeter desserts, and is traditionally considered an aid to digestion after a heavy meal.

Fruit Salads and Salsas

Fruit salads and salsas use fresh fruits at the peak of their ripeness. *Fruit salads* are a simple combination of complementary fruits, decoratively cut, and often lightly sugared to encourage maceration. Fruit salads may also be tossed with a liqueur, a floral water such as rose or orange blossom water, or a light dressing. Always dress fruit salad lightly so that the dominant flavors and aromas come from the fruit. Fruit salads can be served alone or used to garnish another dessert.

When choosing ingredients for a fruit salad, consider texture, color, flavor, and aroma. Achieve contrast and balance using a variety of fruits from different families. Fruit salads are often garnished with fresh mint leaves.

Fruit salsas have become increasingly popular and are used to accompany desserts as well as entrées such as fish and poultry. The fruit is finely diced or julienned and combined with herbs, spices, and other flavorings that add zest and balance to the flavor and aroma of the fruit.

Although fresh fruit is usually used to make fruit salads and salsas, dried, frozen, or canned fruits may also be employed as long as the appearance and texture of the fruit is acceptable. Poached fruit may be sliced or diced, and served chilled.

Figure 22-3 Clockwise from the left: Fruit chutney, salsa, and salad.

Fruit Chutney

The *fruit chutney* that is familiar to us is a sweet and spicy cooked fruit-based condiment with roots in the exotic cuisine of India. The word *chutney* is derived from the Hindi word *chatni* used to describe flavorful fruit relishes. British colonists living in India developed a taste for chutney, and soon adopted and incorporated it into Western cuisine. Depending on the spices and ingredients used, the flavor of chutney can range from sweet to tart, spicy or hot. Chutney is traditionally made with bruised or excess local fruit that is in season. Although any fruit can serve as a base for chutney, mangoes, apples, pears, and peaches are most commonly used.

To make chutney, the fruit is diced, combined in a saucepan with raisins, chopped onions, vinegar, brown sugar and spices, and simmered until the desired consistency and thickness is achieved. Because of a high sugar and vinegar content, chutney can keep almost indefinitely. Served hot or cold, fruit chutney adds an interesting flavor and color to a dessert. *See Figure 22-3.*

Fruit Gratins

Fruit gratins are simple but elegant desserts that can be made from almost any fruit at hand. Figs and berries are excellent choices for fresh fruit gratins because they are tender at the outset and require no extra cooking time to tenderize the fruit.

Like savory gratins, fruit gratins are often baked in individual ramekins. The fruit is placed inside, sometimes on a thin layer of genoise or other sponge cake, and covered with a custard-like sauce. The gratins are baked in a hot oven or beneath the broiler just until the custard is set. Fruit gratins are often garnished with a dusting of powdered sugar, sometimes broiled again briefly to caramelize the sugar, and then garnished with a complementary fruit sauce and served hot or at room temperature.

Traditional Fruit Desserts

Rustic or fancy, fruit pies, tarts, and flans are classic creations that feature seasonal fruit. (**See Chapter 16: Pies and Fruit Tarts for more information.**) Other traditional baked fruit desserts in the classic American repertoire include cobblers, grunts, crisps, crumbles, Betties, and pandowdies.

Although historically rustic and informal, some of these traditional fruit desserts, in the hands of creative pastry chefs, become elegant plated desserts. Others, such as grunts, Betties, and pandowdies with their peculiar, old-fashioned names still evoke a more homespun feel.

Cobblers

Cobblers are deeper versions of fruit pies and are traditionally baked in square pans. Lattice pastry crust or sweet biscuit dough is baked on top of the fruit fillings and a bottom crust is optional. When biscuit dough is used it can either be spread as a single solid layer across the top or dropped in individual "cobbles." Cake batter is sometimes substituted for the biscuit dough.

Fruit extracts were popular in the 19th and early 20th centuries.

STEP PROCESS

Making Individual Peach Cobblers

Fruit cobblers are usually made in deep, square pans. Here is a rustic version prepared in small cast iron skillets presented in a more upscale style.

To make individual peach cobblers, follow these steps:

1 Sprinkle fruit with sugar and toss to coat evenly.

2 Fill individual baking dishes with fruit.

3 Top with biscuit dough.

4 Brush with egg wash and sprinkle with sanding sugar.

5 Sliced peaches and vanilla bean ice cream are perfect accompaniments.

Grunts

Grunts are similar to cobblers in that their biscuit dough is dropped on top of the fruit. Grunts are steamed on the stove top rather than baked to make the dropped biscuit dough resemble dumplings. The name "grunt" is thought to be a reference to the sound that the fruit makes as it stews and bubbles up through the dough.

Crumbles and Crisps

With sugary streusel-like toppings and warm fruit fillings, crumbles and crisps are comforting, familiar desserts. The distinction between the two is generally made based on

Blueberry flan

Figure 22-4 Apple crisp.

whether the topping contains brown sugar and oatmeal (a ***crumble***), or white sugar and no oatmeal (a ***crisp***). Both toppings contain flour, butter, and spices, and are granular like streusel. Apples and peaches are popular fillings for crumbles and crisps. If firm fruit is used, the filling is precooked to tenderize the fruit. *See Figure 22-4.*

Crumbles and crisps are usually baked in individual ramekins. The precooked filling is divided among the ramekins, the topping is sprinkled liberally on top, and the dessert is baked until heated through. Crumbles and crisps are commonly served warm, with vanilla ice cream, or with a small pitcher of heavy cream. Crumbles have origins in Britain, while crisps are considered American. The two are very similar and the names are often interchanged.

Betty

Similar to crumbles and crisps, a ***Betty*** consists of spiced fruit that is baked with a granular topping made of buttered bread crumbs or stale cake crumbs, combined with brown sugar. The topping of a Betty is layered with the fruit in a shallow baking dish before baking. "Apple Brown Betty" is the most well known Betty.

Pandowdy

Pandowdy is essentially a deep-dish double crust fruit pie that has had its crust broken and mixed with the filling for the final minutes of baking. Pandowdy is baked in a casserole rather than a pie pan. Just before it has finished baking, its crust is broken into large pieces and gently folded into the fruit filling using a fork and spoon to lift the bottom pieces and to push the top pieces under. Pandowdy is served warm in bowls to accommodate the juicy filling.

Dried fruits make attractive, flavorful, and crisp garnishes.

Flambéed Fruits/Tableside Desserts

Flambéed desserts are ignited with rum, cognac, or a fruit liqueur immediately before serving. Their preparation is usually done tableside, in view of the guests. Beyond the drama of leaping flames, however, the burning off of the alcohol serves to change and intensify the flavor of the sauce. The dramatic presentation of flambéed desserts makes them appear elaborate, but most are fairly simple in both ingredients and method of preparation. Flambéed desserts consist of fruit that is flambéed in a sweet sauce and served over vanilla ice cream. There are three classic flambéed desserts.

Crêpes Suzette

This elegant dessert consists of crêpes immersed in a hot sauce of orange, sugar, and butter. It is then flambéed with an orange liqueur such as Grand Marnier or Cointreau and folded into quarters. The origin of this dish is credited to an error made by Henri

A classic presentation of Crêpes Suzette garnished with raspberries.

Charpentier, a young waiter working at the Café de Paris in Monte Carlo in the 1890s. As Henri prepared a dessert for the Prince of Wales and his companion Suzette, flames from the chafing dish ignited cordials at the table. Unsure of what to do, the waiter served the dish anyway. The prince enjoyed the flavor immensely, and requested the dessert be named for his companion, Suzette.

Cherries Jubilee

This dessert was created in 1887 by the French chef Auguste Escoffier to celebrate the Golden Jubilee of Queen Victoria's reign. Bing cherries simmered in a syrup of their own juice are flambéed with Kirsch and served over vanilla ice cream.

Bananas Foster

Bananas Foster is a classic New Orleans dessert of banana sections flambéed in a sweet sauce with banana liqueur and rum, and served over vanilla ice cream. Bananas Foster was created in 1951 by a chef at Brennan's in New Orleans, and named after a frequent patron at the establishment whose name was Foster. At the time, bananas were imported to this country mostly through the port of New Orleans, and this recipe was intended to showcase the popular imported fruit.

STEP PROCESS

Making Banana Flambé

Banana Flambé is often prepared tableside. This version is a variation of the famous dessert Bananas Foster that originated at Brennan's in New Orleans.

To make Banana Flambé, follow these steps:

1 Melt butter in skillet and add granulated sugar.

2 Cook, stirring frequently until lightly caramelized. Add bananas.

3 Cook bananas briefly and coat with caramel sauce.

4 Remove the pan from the flame and add brandy. Return to the flame and ignite.

5 Shut off the heat and add banana liqueur.

6 Serve bananas and sauce over vanilla ice cream.

CHAPTER 23

Sauces & Syrups

Sauces are important components of plated desserts because they add flavor and color, and enhance the overall visual appearance of the plate. Dessert sauces include fruit coulis and various reductions, as well as classics such as crème anglaise, caramel and chocolate sauce, and sabayon. Flavored and unflavored simple syrups, on the other hand, are basic ingredients that have many applications in the bakeshop and pastry shop.

KEY TERMS

coulis
fruit sauce
preserve-based sauce

reduction
crème anglaise
caramel sauce

chocolate sauce
sabayon

simple syrup
dessert syrups

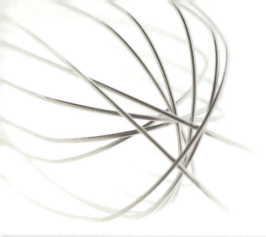

Sauces

The dessert sauce has a tremendous impact on the overall success of a plated dessert. While the flavor of a sauce is paramount, the sweetness, color, texture, temperature, and consistency are all important factors that must be considered as well. A sauce should complement and contrast with the main dessert item in all these areas to create a balanced presentation. The sweetness of the sauce should complement the sweetness of the main dessert, and its intended application or decorative design should determine its consistency. More than one sauce may be used on a plate, but sauces should never be added merely for color. In general, 1 to 2 ounces of sauce is appropriate, although the moistness of the main item, the richness of the sauce, and the size of the plate should determine the amount of sauce needed. *See Figure 23-1.*

Figure 23-1 When multiple sauces are married together and feathered into a design, they should both be of the same consistency.

Types of Dessert Sauces

Coulis, fruit sauce, preserve-based sauce, reductions, crème anglaise, caramel sauce, chocolate sauce, and sabayon are all basic sauces commonly used for plated desserts.

Coulis

Fresh, canned, or frozen fruit that is puréed may be used to make a *coulis*. The procedure is simple: purée fruit to the desired consistency, adjust the sweetness and flavor, and strain through a fine-mesh sieve to produce a smooth sauce. Granulated sugar is often used to sweeten coulis, but simple syrup or other sweeteners also work well. Citric acid such as fresh lemon or lime juice is used to balance the flavor and sweetness of the coulis and to accent the fruit's natural flavor. Liqueurs or other spirits are optional flavoring ingredients that may be added to complement the flavor of the fruit. A dessert coulis is uncooked and contains no added thickeners.

Fresh, seasonal fruit is the best choice for flavor, texture, sweetness, and price, although IQF (individually quick frozen) fruits and frozen purées also produce good results. A smooth purée of cooked fruit is sometimes used as the base for coulis. Fruits vary in flavor, sweetness, texture, and water content, so adjustments must be made to accommodate these variables when preparing coulis. Sample each fruit in its raw state to evaluate the balance of sweetness needed to complement the main item.

Fruit Sauce

A *fruit sauce* is a base of puréed fresh, frozen, or canned fruit, thickened with a thickening agent such as cornstarch, arrowroot, or instant starch, and briefly cooked. The brief cooking activates the thickener, decreases the starchy flavor, and regains the original vibrant color that is temporarily clouded by the thickener. (This method is the same as the "cooked-juice method" used for making a fruit pie.) Fruit juice adds flavor and is a cost-effective way to stretch a fruit sauce. Some fruits must be sweetened after they are heated to retain the true flavor of the fruit.

Preserve-Based Sauce

A *preserve-based sauce* is a fruit sauce made by thinning fruit preserves with simple syrup. Melba sauce is a well-known raspberry preserve-based sauce. Its name is derived from the dessert Peach Melba, of which this sauce is a component. Peach Melba consists of vanilla ice cream served with poached peaches and Melba sauce. The classic version of Melba sauce is made with raspberry coulis and red currant preserves. The modern version, often used in the United States, consists of raspberry preserves thinned with simple syrup.

Reductions

A *reduction* is a sauce made from a sweet liquid that is thickened by cooking it uncovered, over low heat. The sauce thickens as water is lost through evaporation. The reduction process intensifies the flavor and sweetness, and alters the color of the sauce.

Any liquid that contains sugar can be reduced, but reductions used as dessert sauces are generally made from fruit purée or coulis, fruit juice, poaching liquid, wine, or fortified wine. Additional sweeteners (usually granulated sugar or simple syrup), flavorings, spices, and aromatics can also be added to modify the flavor. To reduce any sauce, cook uncovered in a wide, shallow pan over low heat.

Crème Anglaise

Crème anglaise, also called "vanilla sauce," is a classic vanilla custard sauce made from egg yolks, milk or cream, and sugar. This versatile sauce is a staple in pastry shops. It has many applications as an accompaniment to pastries, cakes, dessert soufflés, and fruit dishes, and is used to decorate plated desserts. If cream is used rather than milk, crème anglaise can be churned in an ice cream machine and served as French vanilla ice cream.

Crème anglaise is made by incorporating scalded milk or cream into whipped egg yolks and sugar, and gently cooking it in a bain-marie, stirring constantly, until the mixture is thick enough to coat the back of a wooden spoon. Vanilla flavoring is added by infusing the milk or cream with a fresh vanilla bean or by scraping the seeds into the milk or cream. If vanilla beans are not available or if additional vanilla flavor is desired, vanilla extract can be added after the mixture is removed from the heat.

In addition to vanilla, crème anglaise can be flavored with liqueurs, other flavor extracts, or infusions such as ginger, orange, or coffee. Chocolate crème anglaise is a common variation, made by stirring melted chocolate into the sauce while it is still warm.

Caramel Sauce

Caramel sauce is made from melted, caramelized sugar that is thinned to the proper consistency with water or other liquids and then made richer with the addition of butter and cream. Citrus juice, liqueur, or fruit juices may be used to enhance flavor. There are three main types of caramel sauce.

TRADITIONAL CARAMEL SAUCE This rich sauce contains butter and heavy cream. This is the caramel sauce that tops many familiar desserts including traditional American ice cream sundaes.

STEP PROCESS

Making Crème Anglaise

Take care not to boil the final mixture, or it will curdle. The final product should be smooth and thick.

To make crème anglaise, follow these steps:

1 Put milk, heavy cream, half the granulated sugar, vanilla beans, and salt in a pot, and bring to a boil.

2 Whip egg yolks slightly with the remaining granulated sugar.

3 Temper the egg yolks with boiling milk.

4 Add the tempered mixture to the hot milk.

5 Cook the mixture to 185°F (85°C), stirring constantly. Never let the temperature go above 190°F (88°C). The mixture should coat the back of a spoon.

6 Strain the mixture into a container set in an ice bath. Cool, stirring constantly.

CLEAR CARAMEL This is the most basic caramel sauce, made by thinning caramelized sugar with water. For flavor variations, other liquids such as coffee can be used in place of the water.

FRUIT CARAMEL Fruit caramel is caramelized sugar combined with fruit purée, fruit juice, or a fruit reduction.

Chocolate Sauce

Chocolate sauce can be made in a variety of ways, but the basic method is to simply incorporate melted chocolate and/or cocoa powder in sugar syrup. The addition of butter makes a richer and more luxurious sauce. As with all products made from chocolate, high-quality chocolate is vital for creating the finest chocolate sauce.

The consistency of the chocolate sauce should be adjusted according to its intended use. To thicken, add more melted chocolate, and to thin, add water. Remember to consider the temperature at which the chocolate sauce will be served, because the viscosity changes dramatically when a chocolate sauce is heated or chilled.

STEP PROCESS

Making Traditional Caramel Sauce

The process for making a traditional caramel sauce starts by caramelizing sugar and then enriching it with the addition of butter and cream.

To make caramel sauce, follow these steps:

1 Combine sugar and water in a saucepan. Heat over medium to high heat and bring to a boil.

2 As the sugar cooks, crystals may form on the sides of the pan. Dissolve the crystals with a brush dipped in clean water.

3 Continue cooking the mixture, wiping down the sides of the pan as needed, until the desired color is achieved. Do not stir.

4 Take the saucepan off the heat and add butter.

5 Blend in the butter.

6 Add the cream and stir to make a creamy, rich sauce.

Making Caramel

To make caramel, place sugar in a saucepan and add just enough water to make the mixture appear gritty. Wipe down the sides of the saucepan with a brush. Place on high heat until the water evaporates and the sugar begins to caramelize. Keep the sides of the saucepan clean by periodically wiping them down with a brush to prevent crystallization. Do not stir.

The color of caramel darkens and the flavor increases and becomes more bitter the longer caramel is cooked. Color change indicates a change in flavor as well. When the desired color/flavor stage is reached, the saucepan is removed from the heat and the butter and heavy cream are stirred in. Avoid the dangerous splashing of hot sugar by warming the liquids before adding them to the caramel sauce.

See Figure 23-2.

Cocoa powder is sometimes added to or used in place of chocolate to intensify the flavor. When using cocoa powder, first make a paste by combining the cocoa with a small amount of the hot sugar syrup. Blend until smooth, and then add the remaining syrup.

Figure 23-2 Three stages of sugar caramelization.

Sabayon

Sabayon is a classic dessert sauce of egg yolks, sugar, and white wine. Champagne or Marsala is often substituted for the white wine. When Marsala is used, sabayon is referred to by its Italian name, *zabaglione*.

Classically, sabayon is spooned over a flat dish of fresh fruit and placed under a salamander or broiler until lightly caramelized. Although considered a sauce, sabayon is sometimes served by itself as a dessert.

To prepare sabayon, vigorously whisk the ingredients over a water bath until the mixture turns a pale lemon color and thickens to a light peak. The sauce should be cooked to at least 140°F (60°C) for 3 1/2 minutes and the consistency should be thick and fluffy.

Syrups

Simple syrup is a concentrated solution of sugar and water, usually in a one-to-one ratio (i.e., equal parts sugar and water). Simple syrup is used as a base for a variety of dessert syrups, poaching liquids and sorbets, and is a fundamental ingredient in every bakeshop and pastry shop. Simple syrup is used in sugarwork and candy making, and to modify the consistency of fondant and chocolate syrup.

To make simple syrup, combine sugar and water in a saucepan and stir to make sure that all the sugar is moistened. Add several lemon slices to help prevent recrystallization of the sugar. Bring to a boil and cook, stirring occasionally, until the sugar is dissolved. Remove from heat and let cool. Store covered in the refrigerator until needed. The lemon slices may be left in the syrup during storage unless a lemon flavor is not appropriate for the syrup's intended use.

The ratio of sugar to water in simple syrup may be adjusted to accommodate the intended use and the sweetness of the other components with which the syrup is to be combined. Keep in mind that adjusting the ratio of sugar to water changes the concentration (the sugar density) of the syrup. The length of time the solution boils on the stove top also affects the concentration of simple syrup; sugar density increases as water evaporates. These two factors should be closely monitored when the sugar density is important to the success of the final product. Use a Baumé scale, or hydrometer, when a precise sugar density is required. **(See Chapter 21: Frozen Desserts, for more information on using a Baumé scale to measure sugar density.)**

Invert sugars such as corn syrup and glucose are sometimes used as a portion of the sugar in simple syrup to help prevent recrystallization of the sugar, and to improve the smoothness of the finished product. When the syrup is used in sugarwork or candy making, or as a base for sorbet or other frozen desserts, this combination is employed.

Dessert syrups are flavored simple syrups that are used to add moisture, sweetness, and flavor to a variety of desserts such as baklava, Baba, Savarin, and many sponge cakes. Dessert syrups are also used to decorate or garnish plated desserts. Common flavorings include rum, brandy, Kirschwasser, Amaretto, orange, coconut, and vanilla. In Middle Eastern and North African cuisines, dessert syrups are frequently flavored with floral waters such as rosewater and orange blossom water.

Liqueurs, liquors, floral waters, and flavor extracts should only be added after the syrup has cooled to prevent the evaporation of the volatile flavors. Flavorings such as citrus wedges and whole or ground spices may be added to the sugar and water solution while it is still hot. Strain syrup as necessary to remove the flavoring agents.

STEP PROCESS

Making Zabaglione

Zabaglione can be both a sauce and a dessert. It can be poured over ladyfingers, cubed sponge cake, or fruit. Both Venice and Tuscany claim to have originated this dessert, referred to as sabayon in France.

To make zabaglione, follow these steps:

1 Place the egg yolks and sugar in a bowl; place over a double boiler.

2 Whip constantly, allowing the mixture to thicken.

3 Whip until the mixture reaches 180°F (82°C). It will become very thick, pale yellow, and fluffy.

4 Blend in Marsala, white wine, or Champagne. In some preparations, the wine can be added in the beginning before the mixture is cooked.

5 Continue whipping until well blended.

6 Warm zabaglione can be poured over fresh fruit.

7 Zabaglione can also be served in a flat dish with fresh fruit. The dessert is often glazed with a blowtorch or placed under a broiler.

CHAPTER 24
Cookies & Petits Fours

While they may at first appear to be unrelated, cookies and petits fours are quite similar. Both products come in a number of varieties and are handheld foods that are consumed without utensils. The word *cookie* comes from the Dutch word *koeke*, which means "little cake," while bite-sized petits fours are elaborately decorated iced cakes. The French also use the term *petit four* to describe small, fancy cookies. This chapter examines the many varieties of cookies and petits fours.

KEY TERMS

- crisp-textured cookies
- chewy-textured cookies
- soft/cake-textured cookies
- bagged/pressed cookies
- bar cookies
- dropped/deposit cookies
- icebox/refrigerator cookies
- molded cookies
- rolled cookies
- sheet cookies
- wafer cookies
- petits fours
- petits fours secs
- petits fours demi-secs
- petits fours glacés
- petits fours variés

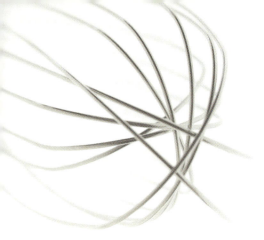

Cookies

Every country has its own specialty cookie: Germany has a nut-filled "kekse," Scandinavia, the spice-filled "lebkuchen," and America, the Toll House chocolate chip. Cookies are often associated with specific events such as the everyday after-school snack, afternoon tea, or as a special treat linked to a holiday. The primary identifying characteristic of any type of cookie is its texture. The texture of a cookie is the result of both the method of preparation and the ingredients used in its production.

Types of Cookies

There are three categories of cookie textures that range from crisp to soft.

Crisp-Textured Cookies

Crisp-textured cookies are cookies with a crunchy texture. They have a relatively low moisture content and are usually high in fat and sugar. Crisp-textured cookies often contain more flour than other cookies. Because of their large ratio of sugar, crisp cookies tend to spread a lot during baking. As they spread, they become thin and crisp.

Chewy-Textured Cookies

Chewy-textured cookies are relatively soft in texture and are high in moisture. They have a high ratio of eggs, or other liquids, to dry ingredients and usually contain less fat than crisp cookies. The development of gluten during the mixing stage contributes to their chewy texture. Gluten's elasticity provides the cookie with its chewy characteristic. Although pastry flour is appropriate for making most cookies, a combination of cake flour and bread flour creates a cookie with a chewy texture.

Soft/Cake-Textured Cookies

Soft/cake-textured cookies have a texture that resembles that of a dense cake. Like chewy-textured cookies, these cookies are relatively low in fat but high in moisture. Batter for soft/cake-textured cookies contains a high ratio of liquid to fat and sugar. Formulas for soft/cake-textured cookies often call for corn syrup, molasses, or honey in addition to granulated sugar. These liquid sweeteners help the cookies to retain moisture after baking, thus contributing to their cake-like texture.

Ingredients

Most cookies contain the same list of basic ingredients: sugar, fat, flour, liquid, and occasionally a leavening agent. The combination and ratios of these ingredients affects how much a specific cookie will spread in the baking process. The spread of a cookie, in turn, determines its final texture.

Sugar

The finer the grain of sugar used, the less the cookie spreads. Finer grains absorb more moisture from the batter than coarse grains resulting in less spread. A coarser grain of sugar is used when greater spread is desired.

Fat

Cookies contain a great deal of fat in comparison to other baked goods. Butter, with its low melting point, creates greater spread in a cookie. Shortenings with a higher melting point cause cookies to spread less. Many formulas call for a combination of both butter and shortening.

Flour

The structure of a cookie is created by the flour used in its production. Flour that has a high gluten content helps to maintain the cookie's structure and offset its high fat content. Used by themselves, strong flours develop more gluten than is needed for cookies. Many formulas therefore call for blends of hard and soft flours. It is a good idea to experiment with such blends until the desired spread is achieved.

Liquid

Eggs are the chief source of moisture in cookies. Cookie batters made with a large amount of liquid spread more than batters made with less liquid. Eggs also help to give cookies some structure.

Leavening Agents

Many cookie formulas call for the inclusion of chemical leavening agents such as baking soda or baking powder. Baking soda helps to relax gluten and contributes to the overall spread of a cookie, while baking powder is most often used in soft/cake-like cookie batters to both create a rise and to help the cookie spread. The cookie products in which leavening agents are used must contain strong air cells. *(See Chapter 26: Cakes, for more information on leavening and air cells.)* The high amount of fat and sugar in cookies does not allow for the development of these types of air cells.

Methods of Preparation for Cookies

Three methods of preparation are used when making cookies: creaming, blending, and whipping.

Creaming Method

Creaming is the method of preparation most often used for cookies. Formulas utilizing the creaming method contain a solid fat that should be used at room temperature. The fat and sugar are creamed together on medium speed with a paddle attachment until the mixture lightens in color and texture. Mixing at higher speeds creates friction and reduces or destroys the batter's air cells. During the creaming method the batter increases in volume as small air cells are formed and incorporated into the mixture. Over- or undercreaming the fat will dramatically affect the finished product. Overcreaming causes cookies to spread too much, while undercreaming prevents them from spreading enough.

    A variety of factors can affect creaming. These include the temperature of the fat, the type of eggs used, the climate in the bakeshop or pastry shop, and the speed at which the

STEP PROCESS

The Creaming Method

The creaming method is a procedure that is common for producing both cookies and cakes. The following procedure demonstrates the use of the creaming method in the production of chocolate chip cookies.

To apply the creaming method, follow these steps:

1 Combine butter and/or shortening and sugar in a mixing bowl. Spices may be added at this time if desired.

2 Mix at moderate speed using a paddle until the mixture is smooth. For less dense cookies and cakes, mix longer until creamy and light.

3 Add eggs gradually, mixing on low speed until combined.

4 Scrape down the bowl and paddle attachment.

5 Add the sifted dry ingredients, and mix only until ingredients are just combined. If more liquids are needed, alternate the dry and the liquid ingredients.

6 Add additional ingredients such as nuts and chocolate chips to finish.

mixer is set. It is important not to use too high a mixer speed to cream. The fat should also be at 70°F (21°C) to achieve the best results. Cold fats require too long a mixing time for maximum creaming, and warm fats (above 75°F/24°C) are not able to hold the proper amount of air cells.

Once the mixture of fat and sugar is soft and lump free, the eggs are slowly added. The gradual addition of eggs allows them to emulsify properly into the fat and sugar mixture. Egg yolks contain the natural emulsifier lecithin, which coats the surface of the air cells formed during creaming and holds the liquid without curdling. Curdling occurs when there is more liquid than the fat-coated cells have the capacity to retain. It is vital to scrape the bowl frequently throughout the creaming process to ensure proper incorporation of all ingredients.

In the last stage of the creaming process, all of the dry ingredients are added and mixed on low speed. It is important not to overmix the dough at this point to prevent too much gluten development. Some formulas also call for the incorporation of additional liquids to the mixture at this time. If more liquids are used in the mixture, dry ingredients and liquid ingredients are added alternately, beginning and ending with the dry. Flour and other dry ingredients absorb excess liquid to help prevent the mixture from curdling.

Blending Method

The blending method of mixing is used when the formula calls for a liquid fat. Tuile or tulip cookies use this method of preparation. In the blending method, the liquid fat and the sugar are whisked or mixed together. The eggs are added and then the dry ingredients. As in the creaming method, it is important to not overmix the batter once the flour is added. *(See Chapter 14: Quick Breads, for more information on the blending method of mixing.)*

Whipping Method

The whipping method of mixing involves whipping whole eggs, egg yolks, or egg whites with sugar into a thick foam, and gently folding sifted flour into the mixture by hand. Cookies made with the whipping method of preparation include ladyfingers and meringues. *(See Chapter 26: Cakes, for more information on the whipping method of preparation.)*

Varieties

Cookies are defined by the method of preparation of their doughs, as well as how their dough is formed before being baked. There are several popular cookie varieties including bagged/pressed cookies, bar cookies, dropped/deposit cookies, icebox/refrigerator cookies, molded cookies, rolled cookies, sheet cookies, and wafer cookies.

Bagged/Pressed Cookies

Bagged/pressed cookies are shaped soon after the batter is made. Macaroon and spritz cookies are two well-known examples of this type of cookie. The batter is pressed out through a pastry bag and deposited onto paper-lined sheet pans. When making bagged/pressed cookies it is important that the batter is firm enough to hold its shape, yet soft enough to be piped. See Figure 24-1.

Figure 24-1 Bagged/pressed cookies.

Bar Cookies

Bar cookies are baked in a bar shape. Some variations of bar cookies are baked, sliced, and then baked again. Italian "biscotti," French "biscotte," German "zwieback," and Jewish "mandelbrot" are all examples of this type of twice-baked bar cookie. When making twice-baked bar cookies it is important to slice the bar while it is still slightly warm. Do not bake the cookie too long upon its return to the oven, or the resulting product will be too hard to be enjoyably eaten. See Figure 24-2.

Dropped /Deposit Cookies

Dropped/deposit cookies are made from a relatively soft dough that is simply dropped or deposited onto the sheet pan. Classic American cookies such as

Figure 24-2

Biscotti are examples of bar cookies. They are baked into a loaf and sliced while still warm.

Biscotti are baked a second time. This formula calls for a light toasting and they are turned over for consistent color.

chocolate chip, peanut butter, and oatmeal raisin are all examples of dropped/deposit cookies. Using an ice cream scoop to deposit the dough will yield cookies that are uniform in size. *See Figure 24-3*.

Icebox/Refrigerator Cookies

When a dough is made in advance and then chilled it is often referred to as an *icebox/refrigerator cookie.* In high-production bakeshops or pastry shops large batches of dough are made in advance and then baked as needed. The dough is scaled into uniform-sized pieces and then formed into cylinders. The cylinders are wrapped and then either refrigerated or frozen. When cookies are needed the dough is unwrapped, sliced into the desired thickness, and baked. This method of storing and baking cookies saves the baker/pastry chef time and ensures fresh cookies. *See Figure 24-4*.

Molded Cookies

Molded cookies are molded into a desired shape before baking. They can be either molded by hand or by a cast or a stamp. Crescent cookies are molded by hand, while Springerles are molded using a cast or a stamp.

Figure 24-3 Dropped/deposited cookies.

Figure 24-4 Slicing ice box/refrigerator cookies.

STEP PROCESS

Making Checkerboard Cookies

Checkerboard cookies are an example of an icebox/refrigerator cookie. Two doughs are rolled out using a template to maintain a perfect rectangular shape and are then stacked, sliced, and restacked to make the checkerboard pattern.

To make checkerboard cookies, follow these steps:

1 Prepare both doughs and use a straightedge and template to make even rectangles of dough. Begin stacking layers. Brush each layer with egg wash before adding a new layer.

2 It helps to use a spatula when stacking layers.

3 Slice stacked cookie strips lengthwise.

4 Restack strips, brushing with egg wash as you stack.

5 Finish stacking the strips of cookie dough and brush with egg wash.

6 Roll out another sheet of cookie dough and wrap the stacked layers.

7 Slice the cookie dough and bake on a parchment-lined sheet pan.

Figure 24-5

Rolling out rolled cookies.

Rolled cookie dough is then cut into individual shapes.

Rolled Cookies

Rolled cookies are rolled with a pin or a sheeter. They are made with a stiff, chilled dough that can withstand the physical pressure of being rolled out. In high-production bakeshops or pastry shops these cookies are rolled out using a sheeter. The dough is then cut into individual shapes. Sugar cookies, shortbread, and gingerbread cookies are some examples of rolled cookies. Rerolling scraps of the dough will further develop the gluten and result in tough cookies. It is therefore important to cut the cookies as close together as possible to eliminate waste. *See Figure 24-5.*

Sheet Cookies

Sheet cookies are made in a sheet pan and then cut into individual pieces for serving. There are a wide variety of sheet cookies including brownies, pecan diamonds, and Swiss "leckerli." This type of cookie is often used in busy bakeshops or pastry shops because it is less labor intensive than other varieties such as rolled cookies.

Wafer Cookies

Wafer cookies are light, crisp cookies that have a characteristic waffle pattern. A thin batter is poured between hot plates that are then pressed together. If desired, the resulting cookie can be molded. Wafer cookies can be served plain, filled with praline cream or jam, rolled up like a cigarette, or shaped like a fan.

Baking

Regardless of the type of cookie, the best results are achieved when each product is equal in size and shape. The cookies should be placed in uniform rows on papered sheet pans to allow for even baking throughout the product. The high quantity of sugar content causes cookies to brown quickly, especially on the bottom, so double-panning the sheet pans can help to prevent the bottom of the cookies from burning. The use of two pans also helps to distribute the heat evenly throughout.

Most cookies are baked between 350°F–375°F (177°C–191°C). A temperature that is too low causes cookies to spread excessively and to become hard and dry. If the temperature is too high, the cookies will burn and fail to spread adequately. Cookies should be removed from the oven when their edges and bottoms are a light golden brown. There is a certain amount of carryover baking that occurs with cookies, so it is important to remove the cookies from the oven when they are slightly underbaked.

Cooling and Storing

Cookies should be cooled completely before wrapping. Failure to do so will cause crispy cookies to become soggy. Once the cookies are cool they can be tightly wrapped in plastic or placed in airtight containers. Cookies should not be refrigerated, although they can be frozen.

Petits Fours

Petits fours are small, bite-sized pastries. Petit four, in French, literally means "little oven," thus all items within this category are baked. Petits fours are extremely versatile and constitute a major part of any pastry chef's repertoire. They are served at high tea, formal coffee service, buffets, presented with the check after a meal, or used as part of a hotel's turndown or amenity service. The petit four category offers the pastry chef a chance to showcase his or her skills.

Producing Petits Fours Assortments

When planning a petit four assortment for a buffet or tea service the following points should be considered:

Taste and Flavor

If possible, do not repeat the same flavor in more than one petit four. Instead choose a variety of ingredients including chocolate, nut, and various fruit flavors.

Texture

Make sure the products contain a variety of textures, soft, moist, crunchy, and creamy.

Shape

Each petit four should be of a distinct shape and size. For instance, they should not all be square or round. If it is not possible to vary shapes, focus on creating items that have contrasting colors. Remembering these guidelines can help in preparing a petit four display that has a great visual impact.

Figure 24-6 Petits fours secs and demi-secs.

Types of Petits Fours

There are four primary categories of petits fours: petit four sec, petit four demi-sec, petit four glacé, and petit four varié.

Petits Fours Secs

Sec is the French word for "dry." Petits fours in this category are not glazed or frosted. *Petits fours secs* are always small, delicate, bite-sized cookies. Cookies such as tuile, lace, meringue, shortbread, biscotti, macaroons, madeleines, or petit palmiers are all included in this group.

Petits Fours Demi-Secs

Petits fours demi-secs are petits fours that are only slightly dry. A shortbread cookie becomes a petit four demi-sec if it is dipped in chocolate. Petits fours demi-secs often consist of two dry cookies sandwiched together with a filling. Some possible fillings include marmalades, buttercreams, praline pastes, and ganache. Like the petit four sec, these cookies should be dainty and elegant in appearance. *See Figure 24-6.*

Figure 24-7 Petits fours glacés.

Petits Fours Glacés

Petits fours glacés are small, glazed cakes. Classically this type of petit four starts with a cake base. Franzipan, also called frangipane, is the cake most often used because it is a dense cake that does not crumble when cut or glazed. Pound cake can also work well. Thin layers of the cake are sandwiched together with jam or marmalade and a thin layer of marzipan is placed on the top layer. The marzipan ensures that the top of each petit four will be perfectly smooth. The cake is then cut into 1-inch squares, rectangles, or triangles and glazed. The classic glaze for the petit four glacé is fondant, although ganache, chocolate, or praline pastes are also sometimes used. The perfect fondant glazed petit four is totally enrobed in a thin layer of a glaze that remains shiny even when set. ***(See Chapter 25: Buttercreams, Icings, and Glazes, for more information on glazing procedure.)*** Once the petits fours have been glazed they are then garnished, often with a delicate chocolate filigree design, small piped rosebuds, or candied violets. *See Figure 24-7.*

Petits Fours Variés

In French *variés* means "assorted." **Petits fours variés** are a miscellaneous assortment of bite-sized baked goods that do not fit into any of the three previous categories. Bite-sized pastries such as mini éclairs, cream puffs, or tartlettes are examples of petits fours variés. Almost any pastry when made in a miniature form can be considered a petit four varié. With some imagination and organization a pastry chef can assemble petits fours variés from items already on hand in the bakeshop or pastry shop. Lemon curd that is left over from large tarts can be piped into small phyllo cups and topped with a fresh raspberry. Extra chocolate mousse can be frozen in a small dome mold, covered with ganache, and placed on top of a small short-dough cookie base for a perfect petit four. Cakes made in sheets can be assembled, filled, garnished, and then cut into small, bite-sized portions. Essentially any regularly sized dessert can be made into a miniature petit four varié with some forethought.

STEP PROCESS

Making Petits Fours Glacés

Petits fours glacés are small, glazed cakes. Franzipan, often called frangipane, was the cake used in the following process.

To make petits fours glacés, follow these steps:

1 Bake franzipan in a half sheet pan. Let cool, then trim the franzipan.

2 Cut the franzipan in half.

3 Cut the franzipan into quarters.

4 Spread 1 ounce of raspberry purée on one quarter layer and top with another quarter layer. Spread another ounce of raspberry purée on the top layer.

5 Roll out marzipan using metal yardsticks as a guide for thickness.

6 Drape the rolled marzipan over the rolling pin and cover the franzipan layers.

7 Trim excess marzipan.

8 Cut the petit four cake into 1-inch strips and then cut into 1-inch squares. Other shapes such as rectangular diamonds may be cut as well.

9 Heat fondant to 100°F (38°C). Adjust consistency with simple syrup.

10 Coat each petit four with fondant using a fondant funnel or pastry bag. Alternately, the fondant can be poured directly from a bowl.

11 Remove the petits fours from screen and decorate.

CHAPTER 25

Buttercreams, Icings & Glazes

Buttercreams, icings, and glazes are all included in the frosting category. Frostings, regardless of the type, all serve the same function in the baking and pastry industry. They moisten and protect the cakes they cover, add flavor to the end product, and can greatly increase the visual appeal of the product to which they are applied. It is difficult to imagine a birthday cake without frosting! Although they share the same functions, buttercreams, icings, and glazes are made with different ingredients and are produced using various methods of preparation. This chapter further explores these similarities and differences.

KEY TERMS

- buttercream
- American buttercream
- German buttercream
- Swiss buttercream
- Italian buttercream
- French buttercream
- icings
- fudge icing
- cream cheese icing
- flat icing
- royal icing
- décor
- fondant
- glazes
- enrobing
- mirror glaze

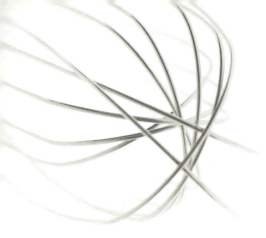

Buttercreams

Buttercreams are frostings that have a butter base that allows them to become aerated as they are creamed. The resulting frosting is lighter and fluffier in both texture and mouthfeel than icings. Buttercreams are applied to cakes with a spatula or a pastry bag. Buttercream should be applied in thin layers to allow the flavor of the cake to stand out. When using cold or refrigerated buttercream first allow it to come to room temperature. To expedite the warming process, place the buttercream in a bowl with a paddle and beat on low speed. Cream the buttercream until it is smooth, lump-free, and of a spreadable consistency.

Buttercream Functions

The primary function of buttercream is to extend the shelf life of cakes. It traps the moisture in the cake to seal it from any outside air and improves the overall mouthfeel and texture of the cake. Buttercreams also add additional sweetness and flavor to cakes, and improve their visual presentation. Artfully applied buttercream can greatly increase the appearance of any cake, whether it is used to add birthday greetings and roses, or to provide a rich chocolate layer to a devil's food cake.

Types of Buttercream

There are five types of buttercreams that are utilized in the bakeshop or pastry shop: American, German, Swiss, Italian, and French. Each are distinguished by their unique flavor, taste, and texture.

American Buttercream

American buttercream is made with a combination of vegetable shortening, powdered sugar, and flavoring. A combination of vegetable shortening and butter or margarine may also be used. The use of vegetable shortening, with its high melting point, prevents the frosting from melting or breaking in warm temperatures. When 100% vegetable shortening is used, the buttercream has a long shelf life. This inexpensive type of buttercream is commonly used in high-production bakeshops such as those found in grocery stores. *See Figure 25-1.*

German Buttercream

German buttercream is made with a combination of pastry cream, butter, and flavorings, and is considerably richer than any meringue-based buttercreams. The addition of pastry cream reduces the shelf life of German buttercream beyond that of both Swiss or Italian but-

Figure 25-1 American buttercream (foreground) typically has little, if any, butter and is often used in high-production bakeshops.

tercream. Because the egg yolks in the pastry cream impart a pale yellow color to the buttercream, German buttercream is not a good choice when a white buttercream is desired.

Swiss Buttercream

Swiss buttercream consists of a Swiss meringue with the addition of butter. *(See Chapter 20: Meringues and Soufflés, for more information on making meringues.)* Sugar and egg whites are heated over a double boiler and are constantly whisked to prevent the whites from coagulating. The mixture is heated until the sugar is completely dissolved and the whites reach 120°F (49°C). At this point the sugar and egg whites are removed from the heat, placed in a mixing bowl, and whipped on high speed. As the whites cool, a glossy meringue results. Softened or creamed butter is incorporated into the meringue a little at a time and flavorings such as vanilla or almond extract are also included. Once all of the butter has been added, the buttercream appears to be a bit curdled. Additional whipping will produce a luscious and smooth buttercream.

Italian Buttercream

Italian buttercream is made of an Italian meringue with butter. The sugar and water are first cooked to 250°F (121°C), and whites are then whipped to a medium to stiff meringue stage. The hot sugar syrup is added in a slow and steady stream. When adding the hot syrup to the meringue, carefully pour it between the whisk and the sides of the bowl to ensure that all of the syrup is incorporated into the whites without sticking to the sides of the mixing bowl. This method also helps to prevent the possibility of getting burned from splattered syrup. After all of the sugar syrup is added, continue to whip the meringue. Once the meringue has completely cooled, the butter is added. As with Swiss meringue, Italian buttercream turns out best when the butter is creamed separately and then added to the meringue. The mixture will initially appear to be broken or curdled, but continued whipping will result in a beautiful, smooth buttercream.

STEP PROCESS

Making Swiss Buttercream

Swiss buttercream is made with Swiss meringue and butter.
To make Swiss buttercream, follow these steps:

1 Place sugar and egg whites in a bowl of a double boiler using a whip to combine.

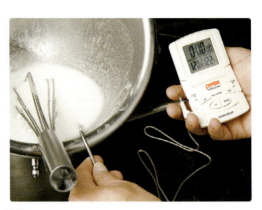

2 Continue whipping the mixture and cook to 120°F (49°C).

3 Transfer the mixture to a mixer and whip egg whites to full volume. When egg whites are completely cooled, blend in butter a little at a time.

4 Continue whipping until the mixture is light and fluffy.

French Buttercream

French buttercream consists of a pâte à bombe and butter. As with German buttercream, the addition of eggs makes this buttercream pale yellow in color and extremely rich. French buttercream contains butter rather than a meringue. Whole eggs and egg yolks are whipped until thick and lemon in color. A sugar syrup mixture is heated to 250°F (121°C) and then slowly added to the eggs. As the eggs are whipping, the butter should be creamed in a separate bowl. Once the egg and sugar mixture is cooled, the butter is added and then any desired flavorings such as vanilla, raspberry, or chocolate. Because

STEP PROCESS

Making Italian Buttercream

Italian buttercream is one of the more common buttercreams used in commercial bakeshops and pastry shops.

To make Italian buttercream, follow these steps:

1 Boil water and 2/3 of the sugar in a pot to 250°F (121°C).

2 Whip egg whites until stiff, adding remaining sugar.

3 Pour hot syrup into egg whites.

4 Add soft butter to the egg whites and whip until fluffy.

of its pâte à bombe base, French buttercream is extremely rich. Its rich nature makes it ideal as a filling between cake layers but overpowering when used to ice the outside of the cake. *See Figure 25-2.*

Buttercream Hints

- Butter should be at room temperature. Best results are achieved if the butter is creamed before it is added to the meringue or pâte à bombe. It is important to scrape the bowl often when making buttercream.
- The flavors and colors of the entire cake should be considered when flavoring and/or coloring buttercream.
- Small portions of a large batch of plain buttercream should be flavored and colored as needed.
- When frosting cakes, best results are achieved using room-temperature buttercream.

Figure 25-2

To make pâte à bombe, heat sugar, corn syrup, and water in a pot to 250°F (121°C).

Whip whole eggs and egg yolks until thick and lemon in color. Slowly blend in sugar syrup while continuing to whip. To make French Buttercream, creamed butter is added to this mixture.

Flavoring Buttercreams

There are many flavoring possibilities for buttercreams. Let your palate and the flavors of the cake be your guide.

Extracts

The essence of an extract can sometimes be lost when added to a rich buttercream. Extracts contain alcohol and adding too much can break the buttercream.

Compounds

Compounds are concentrated extracts. Their flavors are suspended in a thick paste rather than in alcohol. The advantage of using a compound is that less is needed, and there is no alcohol content to possibly break the buttercream.

Preserves

When using preserves such as raspberry or apricot, caution should be taken to assure that they are free of seeds, pits, or rinds.

Cocoa Paste

This ready-made product with concentrated chocolate flavor has a consistency similar to that of a loose peanut butter or praline paste. It does not need to be melted or tempered into the buttercream.

Chocolate

Couverture chocolate should always be used to flavor buttercream. The couverture should be melted and allowed to cool slightly. The buttercream will melt if the chocolate is too hot. The chocolate is first added into one-third of the buttercream, and once completely incorporated, the remaining buttercream is added.

Liqueur

Flavoring buttercream with liqueur can be tricky for a number of reasons. The flavors of liqueurs are light, so a large quantity must be used to achieve flavorful results. Using a lot of liqueur in turn, means adding a lot of alcohol to the buttercream, which can cause it to break. Liqueur-based compound pastes are preferable to straight liqueurs because these concentrated liqueur flavors contain no alcohol. These compounds are available in flavors such as Amaretto, Kahlua, and Grand Marnier.

Icings

Most *icings* do not contain fat and are therefore not aerated during production. Icings are the simplest members of the frosting family. They are fairly dense and are usually used on products such as brownies, cupcakes, or Danish. They are applied by dipping, pouring, or brushing.

Types of Icing

Bakeshops or pastry shops typically use five varieties of icing: fudge, cream cheese, flat, royal, and fondant.

Fudge Icing

Fudge icing is made by cooking sugar syrup and adding it to dry ingredients, usually consisting of a mixture of sugar and cocoa powder. Perhaps the most well known type of icing, fudge icing is relatively dense, stiff, and somewhat grainy in texture. Simple baked goods such as brownies, cupcakes, and cookies are often covered with fudge icing.

Cream Cheese Icing

Cream cheese icing consists of cream cheese, butter, and powdered sugar. It is commonly used on carrot cakes and is the only icing that is aerated. Best results are achieved when the cream cheese is at room temperature. The cream cheese is paddled until it is lump-free. Once it is smooth and light in texture, butter and sifted powdered sugar are added. The mixture is beaten until it is light and of good spreadable consistency.

Flat Icing

Flat icing is a combination of powdered sugar and water. It is sometimes referred to as a water or simple icing. A small amount of corn syrup may be added to flat icing to prevent it from crystallizing. Flat icing is used to finish Danish, cinnamon rolls, and other breakfast pastries. It is often brushed on baked goods immediately after they come out of the oven to provide a protective barrier that also imparts shine to the product. See Figure 25-3.

Figure 25-3 Flat icing is often used to finish Danish and breakfast pastries.

Royal Icing

Royal icing is made of powdered sugar, egg whites, and cream of tartar. Unlike fudge icing or flat icing, royal icing is used solely for decoration. Although royal icing contains egg whites, its method of preparation differs greatly from that used for meringue-based icings. Because this icing is not cooked, it is important to use pasteurized egg whites. When making royal icing, the egg whites are incorporated into the powdered sugar with a paddle. The icing is then mixed until a smooth paste is formed. Royal icing dries very quickly when exposed to air so it should always be covered with a damp towel. The damp towel prevents a dry skin from forming on top of the icing. Any dried icing will result in lumps, making it difficult to pipe. Royal icing should be stored in the refrigerator.

Because royal icing becomes hard and brittle when dry, it is the perfect icing to use for decorating. While it should not be used to cover entire cakes, it is perfect for piping intricate, delicate decorations. ***Décor,*** small piped decorations and filigrees, made with royal icing, can be produced ahead of time. The decorations are piped on parchment paper or a non-stick baking sheet and allowed to dry at room temperature. Once dried, décor should be stored in airtight containers. Royal icing should be colored immediately after it is mixed or by airbrushing the dried décor pieces.

Fondant

Fondant is essentially cooled sugar syrup. To make fondant, sugar syrup is cooked to 240°F (116°C) and then quickly cooled by pouring it onto a marble slab and moving it constantly with a bench scraper. Alternatively, the syrup can be poured in a mixing bowl and paddled until it cools. It is important that fondant be kept in constant motion to produce its characteristic white shine.

Many bakeshops and pastry shops buy fondant premade to ensure consistent results. Ready-made products are also convenient and can reduce labor costs. Before it is used, fondant must be warmed over a double boiler. Perhaps the single most important consideration when using fondant, whether handmade or ready-made, is its temperature. It must be warm enough to flow easily. If it is not heated enough, it will result in a thick layer.

Fondant should also never be heated above 105°F (41°C). If the fondant is heated too much it will result in a translucent glaze rather than one that is opaque, and will not fully cover the product. The glaze will be too thin and will have a dull, crystallized appearance when it dries. Fondant that is still relatively thick may be thinned with corn syrup, liqueur, warm water, simple syrup, or pasteurized egg whites. *See Figure 25-4.*

Figure 25-4 Fondant is commonly used to glaze petits fours.

There are two types of fondant: poured and rolled. Poured fondant is perhaps the best-known of all of the icings. It is used for most petits fours glacés. Rolled fondant is used to cover both dummy and edible cakes; and as a modeling paste for small showpieces. For more information on rolled fondant, see Chapter 28: Pastillage, Sugar Artistry, and Marzipan.

Fondant must always be warm before adding flavor or coloring to allow the flavor or color to fully incorporate. Food color pastes work well with fondant, as do liquid colorings. Liquid coloring, however, may affect the viscosity of the fondant. When coloring icings it is important to avoid using extremely vibrant colors as well as black and deep blue, because these colors make products look unnatural and unappetizing. Light pastel shades are traditionally used to color both fondant and royal icing. The following flavorings or pastel colorings may be added to fondant.

Extracts   There are two advantages to using extracts to flavor fondant. First, the essence of the extracts' flavor is clearly evident when added to the neutral flavor of the fondant, and secondly since extracts are colorless, they do not affect the stark whiteness of the fondant.

Preserves   When using preserves, such as raspberry, apricot, or orange, they should be free of seeds, pits, or rinds. It is also important to remember that the addition of preserves colors the fondant.

Cocoa Paste   This concentrated chocolate product is purchased ready-made. Due to its relatively loose consistency, it need not be melted before it is added to the fondant.

Chocolate   Always use couverture chocolate when flavoring fondant. Melt the couverture and stir it into warmed fondant.

Liqueur   The subtle nature of liqueurs works well with the neutral flavor of fondant. The colors of some liqueurs such as Chambord or Kahlua also serve to color the fondant.

Glazes

Glazes are sweetened and flavored liquids that are used to *enrobe,* or cover, baked goods. They serve to add shine and to enhance the color of a product. They also contribute sweetness and help to extend the shelf life of products such as Danish and fruit tarts by sealing the product from outside air. Glazes can be applied with a pastry brush, a glaze sprayer, or a paint gun. Application with a pastry brush is the most traditional method used, although it is more time consuming. Using a glaze sprayer or paint gun allows for a more even flow of glaze that results in a more uniform product. Glaze sprayers also reduce waste and labor costs, and are commonly used in high-volume bakeshops and pastry shops.

When enrobing a product, the layer of glaze should form a thin covering or shell around the item. To assure a thin covering, the temperature of the glaze is important. If the glaze is too cold, the covering will be thick and lumpy.

Types of Glazes

There are three primary glazes used in the bakeshop or pastry shop: chocolate, mirror, and commercial.

Chocolate Glazes

There are a couple of types of chocolate glazes. The traditional chocolate glaze is chocolate ganache, which is a combination of chocolate and cream. The cream is boiled and then poured over chopped chocolate. After a few minutes the mixture is slowly mixed until all of the chocolate is melted. When ganache is used as a glaze there is a higher ratio of cream to chocolate than when it is used in candy or truffle-making. **(See Chapter 27: Chocolate, for more information on ganache.)** Ganache can be used to enrobe both cakes and individual pastries.

A thinner chocolate glaze can also be made from a mixture of cocoa powder, couverture, water, and gelatin. There are many variations but the result is the same, a shiny, thin chocolate glaze that stays soft even when refrigerated. These glazes are most commonly used to enrobe cakes.

Mirror Glazes

Most *mirror glazes* consist of fruit purée and gelatin. Mirror glazes are not typically used to enrobe cakes. They are usually applied to a molded cake or to a dessert as a top layer. Mirror glazes are primarily used with cream-based cakes, such as charlottes, mousses, Bavarian creams, and cheesecakes.

A mirror glaze is applied to the top of the cake while it is still in its mold. The cake is then chilled to set the glaze before the form or mold can be removed. *See Figure 25-5.* The fruit purée and gelatin are heated and then cooled before being applied to the cake. Corn syrup is sometimes added to the glaze for additional sheen. When using fruit purées remember that the protease enzyme found in many exotic fruits, such as kiwi, pineapple, and papaya will interfere with the gelatin's ability to set. When using these fruits, the purée must be cooked thoroughly before the gelatin is added.

Commercial Glazes

Although commercially bought glazes may cost more than those made in-house, they make up for their price in convenience and consistency. Some commercially available glazes contain gelatin to assure that the product sets properly. Brushing a fruit tart with a gelatin-based glaze helps to hold the fruit in place and make the tart easier to cut. Glazes also seal in the fruit's moisture and prevent the fruit from oxidizing to extend its shelf life. Items such as bananas need a glaze to maintain their color and taste. Commercial glazes must be heated before they are used. They can then be brushed or sprayed on the product. Glazes are available in a wide variety of flavors, although raspberry and apricot are the two most popular. Flavorless glazes made with a combination of sugar and pectins are also available. Commercial glazes can be reused and reheated as needed.

Figure 25-5 Tilting a torte to evenly distribute the mirror glaze.

Tips for Glazing Cakes

Before glazing cakes, a thin layer of buttercream or jam should be applied to ensure that loose crumbs do not get into the glaze. The application of a thin layer of marzipan also guarantees that the cake will be a crumb-free, flat surface onto which the glaze can be poured. To glaze a cake, set a cooling rack on top of a parchment paper-lined sheet pan, and then place the cake or product to be glazed on the cooking rack. If the cake is covered with a crumb coat of buttercream, make sure that it is completely chilled before covering it with a warm glaze. The glaze must also be at the correct temperature. A glaze that is too cool will be lumpy and thick, while one that is too warm will be translucent and allow the cake layers to show through.

An adequate amount of glaze should be poured over the cake so that it falls easily down the sides. Using a large offset or straight spatula, the glaze is spread evenly over the top of the cake with as few strokes as possible. It is important to work quickly with the glaze because a thin layer placed on a chilled cake begins to harden very quickly. If any bubbles appear on the surface of the cake, they may be removed by quickly passing the flame of a butane torch over them.

CHAPTER 26
Cakes

Cakes are the building blocks upon which bakers and pastry chefs build much of their products. Wedding cakes, petits fours, and tortes all depend upon a cake that is properly mixed and baked. Knowledge of their ingredients and mixing methods is vital to a successful product. This chapter examines various types of cakes and discusses their ingredients and specific methods of preparation.

KEY TERMS

layer cakes
sponge or foam cakes
pound cakes
high-ratio cakes

emulsified shortening
blending/mixing method
two-stage mixing method

whipping method
genoise sponge cake
warm foaming method
cold foaming method

chiffon cake
two-step foaming method of mixing
angel food cake

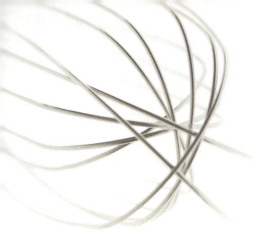

Cakes

There are two general categories of cakes: layer cakes, and sponge or foam cakes. *Layer cakes* are those that are primarily chemically leavened. Pound cakes and high-ratio cakes are examples of layer cakes. *Sponge or foam cakes* depend on the air whipped into the eggs or egg whites for leavening. Genoise, chiffon, and angel food cakes are examples of sponge or foam cakes.

All cakes rely on the development of air cells that expand during the baking process to trap the gases produced by the chemical leavener. Sponge or foam cakes incorporate a large amount of air in their cells that force the cake to rise.

There are many factors that contribute to the success of a cake. If any one of these factors is not considered, the product will be deficient. Proper scaling of ingredients is paramount to a cake's success: too much or too little of any one ingredient will adversely affect the end result. The oven must also be properly preheated. An oven that is too hot or too cold will produce an inferior cake. For each type of cake, the proper method of preparation must be applied. Improper mixing techniques may result in a cake that rises and then falls or, conversely, does not rise at all.

There are three main objectives in proper cake mixing. The first objective is to uniformly and completely mix all of the ingredients in the formula. The second is to form and incorporate air cells. The number of air cells produced in a cake depends upon the method of mixing employed. The last objective of proper cake mixing is to develop a desirable grain and texture in the cake.

Layer Cakes

Layer cakes are generally moist cakes made by creaming or blending. They contain chemical leaveners (either baking soda or baking powder). Both pound cakes and high-ratio cakes are examples of layer cakes.

Pound Cakes

Pound cakes get their name from the pounds of ingredients traditionally used in these cakes. Pound cakes were made of one pound of butter, one of sugar, one of flour, and one of eggs. Assorted fruits, nuts, and spices were also added for flavoring. The creaming method is used to combine these ingredients. Pound cakes are leavened by both the air cells that are incorporated through the creaming method as well as by the addition of baking powder or baking soda. Baking powder or baking soda should be sifted with the dry ingredients for the best results. The leavener must also be evenly distributed throughout the mix to prevent the formation of holes in the cake's texture. Overmixing the batter once the dry ingredients have been added can produce too much gluten and cause the top of the cake to rise into a dome and crack. A perfectly mixed pound cake will be dense and moist with a fine, rather than coarse crumb.

STEP PROCESS

The Creaming Method

In the creaming method, the sugar and fat are blended together first and then "creamed" by mixing. During this stage small air cells are formed and incorporated into the mix. The mix takes on added volume and becomes softer.

To make pound cake using the creaming method follow these steps:

1 Add butter and sugar to the mixing bowl.

2 Cream on medium speed until light and fluffy, approximately 10 minutes.

3 Add eggs and liquid such as vanilla and lemon extract in stages. Scrape the bowl between additions.

4 Remove the butter mixture from the mixer and fold in the sifted, dry ingredients; incorporate well.

5 When ingredients are completely incorporated, scale and bake.

High-Ratio Cakes

High-ratio cakes have a high ratio of liquid and sugar to flour to produce a cake that is more moist and tender than a pound cake. Formulas for high-ratio cakes call for emulsified shortening. An *emulsified shortening* is a shortening that has been specially formulated to absorb large quantities of liquids and sugars. When a formula calls for emulsified shortening, the mixing method is used. In the *blending/mixing method* most of the ingredients, including at least half of the liquid, are placed in the mixing bowl at the same time.

Proper steps for the blending method include:

1. Placing the sifted dry ingredients in a mixing bowl.
2. Adding the emulsified shortening and half of the liquid ingredients.
3. Mixing on low speed until blended and then mixing on medium speed until the mixture is lightened (usually about 4–5 minutes).
4. Gradually adding the remaining liquid ingredients and blending until smooth.

STEP PROCESS

Assembling a Layer Cake

Careful assemblage and meticulous attention to detail allow a baker to create a beautiful finished product.

When assembling a layer cake:

1 Use a serrated knife to level and trim the top of the cake.

2 Slice the cake into three layers, keeping as level as possible.

3 Transfer each layer to a cardboard cake circle.

4 Brush the bottom layer with simple syrup. Spread filling over the bottom layer, and top with the middle layer.

5 Brush the middle layer with simple syrup, and then spread the filling on the top of the second layer. Top with the third layer. Refrigerate several minutes to set the icing.

6 Sometimes a pre-coat or skim coat is applied first to prevent the appearance of crumbs in the icing. If using a skim coat, refrigerate the cake before applying the final coat of icing.

7 Ice the top of the cake, and then the sides.

8 Smooth icing on top of cake by working from the sides inward, holding the spatula at a 45-degree angle. Finish icing and smoothing the entire surface.

9 Pipe rosettes spacing evenly to mark individual portions.

10 Use an appropriate garnish for each cake. Candy mocha beans are being used for this cake.

STEP PROCESS

The Blending Method

The blending method of mixing is primarily used for high-ratio cakes. Air cells formed during this process are not as large as the ones formed when creaming, so the final grain of the cake layer will be tight and firm.

To use the blending method, follow these steps:

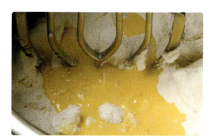

1 Place the sifted, dry ingredients in a bowl. Add the shortening and half the liquid ingredients.

2 Mix on low speed until blended and then mix on medium speed for 4–5 minutes. Add remaining liquid ingredients.

3 Blend until smooth. Scale and bake.

The blending method is sometimes referred to as the *two-stage mixing method* because the liquid ingredients are added in two stages. Although air cells are incorporated during the blending, they are smaller and fewer in number than those in cakes made with the creaming method. The resulting cake has a tight grain, which is particularly desirable to support cakes such as wedding cakes that have multiple layers and tiers.

Sponge or Foam Cakes

All sponge or foam cakes are leavened by whipping, which is the incorporation of air. In the *whipping method*, eggs are whipped and air cells are formed and incorporated into the mix. Cakes that fall under the sponge or foam cake category include classic Genoise, chiffon, and angel food.

Keys to Success for Sponge or Foam Cakes

When making sponge or foam cakes it is important to remember that once the full volume has been achieved, any further whipping will reduce the volume of the cake batter. Dry ingredients should be sifted prior to folding, and folding should be done gently and by hand.

Genoise Sponge Cake

Genoise sponge cake is leavened solely by the air whipped into eggs. Genoise is traditionally mixed using the warm foaming method. In the *warm foaming method* eggs and sugar are heated over a double boiler to 110°F/ 43°C, and the mixture is constantly stirred to prevent the eggs from coagulating. Once the mixture has reached the correct temperature it is placed in a bowl and whipped until cool. After the mixture has reached its maximum volume the sifted flour is folded in. Folding should be done gently and in stages. Lastly, melted butter is folded into the batter. To maintain as much air as possible in the batter, the butter must be incorporated quickly. This can be accomplished by tempering the butter into the batter. To temper the butter, whisk it into a small amount of batter and carefully fold this mixture into the remaining batter. Because genoise is leavened solely by air, place the batter into prepared pans and into the oven as quickly as possible.

STEP PROCESS

Making a Genoise Sponge Cake

The warm folding method for making genoise sponge cake allows the cake to reach full volume quicker than the cold foaming method.

Follow these steps to make a genoise sponge cake using the warm folding method:

1 Heat whole eggs and sugar over a double boiler, constantly stirring at all times, to 110°F (43°C).

2 Transfer the mixture to mixer and whip on high speed to full volume. Full volume can be determined by the 5-second track method. Run ½ inch (13 mm) of your finger through the batter and the track should last for 5 seconds.

3 Immediately add dry ingredients. Use your hand along with a plastic scraper to distribute the sifted ingredients.

4 Continue folding until all the ingredients are completely incorporated.

5 Place a small amount of batter in a small bowl. Quicky stir melted butter into this batter until completely incorporated.

6 Gently fold the butter mixture, back into the remaining batter and scale for baking.

A genoise sponge cake can also be made using the cold foaming method. In the ***cold foaming method*** the eggs and sugar are whipped without being heated. The eggs and sugar are placed into a bowl and whipped until they reach maximum volume. The sifted dry ingredients are then gently folded into the mixture followed by the inclusion of fat. The resulting foam will take longer to reach full volume than the one developed in the warm foaming method.

Chiffon Cake

Chiffon cakes are leavened by air but contain oil and/or egg yolks as well to produce a tender, moist crumb and a more stable product than the classic genoise. Chiffon cake is

Assembling a Torte Using a Ring

Joconde, the almond sponge cake used to make this Raspberry Bavarian torte, is named after the Mona Lisa.

To assemble a torte using a ring, follow these steps:

1 Spread decorating paste onto a non stick baking sheet. Pass a decorating comb over the paste to make strips. Freeze.

2 Spread the joconde batter over the decorating paste.

3 Bake on an inverted sheet pan according to formula.

4 To make strips of joconde, first mark the cake so that the strips will fit into the ring as desired.

5 Use a straightedge to cut the strips.

6 Line the ring with an acetate strip and then fit the joconde strip into the ring to form the sides of the torte. Trim as needed.

7 Cut two layers of genoise sponge to fit the ring. Place one on the bottom and top with half the filling.

8 Add the second layer of genoise sponge.

9 Top with remaining filling. Chill to set the filling.

10 Pour mirror glaze onto the top of the torte.

11 Swirl to even off the mirror glaze.

12 Pipe cream rosettes and garnish with raspberries.

STEP PROCESS

Preparing Angel Food Cake

The whipping method is used to make angel food cake. No fat is used and whipped egg whites provide the only leavening. A tube pan helps the center of the cake rise and dry evenly.

Follow these steps to make angel food cake:

1 Whip egg whites and cream of tarter until foamy.

2 Add sugar gradually as the whites increase in volume. Whip until soft to medium moist peaks.

3 Remove the egg whites from the mixer and fold in the sifted, dry ingredients.

4 Scale batter into an ungreased tube pan and bake.

---

typically used for jellyroll sponge cakes. The addition of oil to the formula produces a cake that is pliable and can be easily rolled. Chiffon cakes are made with a *two-step foaming method of mixing* whereby the yolks and whites are whipped separately. The dry ingredients are first sifted with a portion of the sugar. Next, the yolks are whipped and the oil is drizzled in once the yolks have doubled in volume. The remaining portion of the sugar and the egg whites are used to make a French meringue. **(See Chapter 20: Meringues and Soufflés, for more information on making French meringues.)** French meringue should be made while the yolks are whipping. Once the yolks have reached maximum volume, the sifted dry ingredients are folded in. The dry ingredients are then incorporated, and the French meringue is folded in. To preserve the air in the French meringue it must be folded into the rest of the batter in stages. Begin by adding about one-third of the whites to the batter to lighten the batter and to make it easier to fold in the remaining meringue with a minimal loss of volume.

Angel Food Cake

Angel food cake is made from sugar, egg whites, and flour. It is leavened solely by the air whipped into the meringue. When making an angel food cake proper preparation of cake pans and oven temperature are vital so that a minimal amount of air and leavening power is released. Because angel food cake contains no fat it is drier than either a genoise or a chiffon sponge cake. The cake batter begins with a French meringue. To make the meringue, whip the egg whites until foamy. Add the sugar in a slow and constant stream. Remove the whites from the mixer when they have reached soft to medium, peaks. Gently fold in the dry ingredients by hand. It is important that this step is done quickly to avoid excessive loss of air. Place the batter in an ungreased tube pan and bake. An ungreased pan is important to the success of an angel food cake because the cake will not rise if a greased pan is used. A tube pan is traditionally used for angel food cake to help the center of the cake rise and dry evenly.

Baking Cakes

Proper methods of preparation are only one step in making a successful cake. The scaling of the cake batter, preparation of the cake pans, baking, cooling, and storing of the cakes also contribute to the quality of the finished product.

Scaling Batter

Proper scaling of a cake batter is important to ensure consistent results and a minimal loss of product. Cake batter should always be scaled before panning. Scaling can be done using either a traditional baker's balance beam scale or a modern digital scale. For best results, scale the batter directly into the prepared cake pan.

Preparation of Cake Pans

Cake pans should be prepared before the cake batter is mixed to ensure a minimal loss of aeration. The proper preparation of cake pans depends upon the type of cake being made. Most cake pans are usually greased and lined (usually just the bottom) with parchment paper. Angel food cake pans are an exception and are usually baked in an ungreased tube pan.

Panning

In most cases, cake pans should be filled from one-half to two-thirds full. Once the batter is in the cake pan, the top is evened out with an offset spatula. Be careful to not overwork the batter or its air cells will collapse.

Baking Cakes

The oven should be preheated before the cake batter is mixed. If a cake has to wait for an oven to reach the proper temperature, it will lose some, or all, of its leavening and will not rise as it should. The oven must also be correctly calibrated. An oven temperature that is too high or too low will produce an inferior product. When placing cake pans in the oven, the pans should not touch one another so that the air is able to flow freely around each cake pan to help the cake to rise evenly. The oven door must also be kept closed while cakes are baking. A cake that is disturbed before it has set will fall.

Testing Doneness

There are various ways to test a cake for doneness. A cake may pull away slightly from the sides of the pan when done, or the center of the cake may spring back when lightly pressed. A cake tester, or thin skewer that comes out clean when inserted into the center of the cake indicates "doneness." Cakes should never be removed from the oven just because they "look" done. Color alone cannot be relied upon to determine doneness, especially with cakes that are dark in color, such as chocolate.

Cooling and Storing Cakes

For best results allow cakes to cool for at least 10–15 minutes before removing them from their pans. Once the cakes are completely cooled they can be turned out on cardboard rounds. If a cake is wrapped while still warm, condensation will develop on the top of the cake and make it soggy. To maintain moistness, cakes should be wrapped as soon as they are cool. Wrap cakes tightly with plastic wrap, label, date, and refrigerate. When properly wrapped, undecorated cakes can keep in the freezer for up to a month.

High-Altitude Baking

Cake making is greatly affected by high-altitude baking. Liquids boil at a lower temperature at high altitudes and also evaporate more readily. Consequently, if a cake loses too much liquid while baking, it will have a tough texture and little flavor. Ingredients, oven temperatures, and pan preparation must all be adjusted to counter the effect of high altitude on cake making. Some general guidelines to making successful cakes at high altitudes include:

- Reducing the amount of leaveners. Baking soda and baking powder expand more at high altitudes.
- Creaming or whipping batters for shorter periods of time. Less incorporation of air is needed at high altitudes.
- Increasing the amount of eggs and flour to provide more structure.
- Decreasing tenderizers that soften a cake's structure.
- Adding more liquids to counteract the increased evaporation of liquids.
- Coating pans with a heavier layer of grease because layer cakes are more apt to stick to cake pans at high altitudes.

See Figure 26-1 for high-fat cake formula adjustment at high altitudes.

Figure 26-1

High-Fat Cake Formula Adjustment at High Altitudes

Ingredient	Approximate Adjustment 2,500 feet (762 m)
Baking powder	−20%
Eggs	+2½%
Flour	—
Fat	—
Sugar	−3%
Liquid	+9%

Ingredient	5,000 feet (1,524 m)
Baking powder	−40%
Eggs	+10%
Flour	+4%
Fat	—
Sugar	−6%
Liquid	+15%

Ingredient	7,500 feet (2,286 m)
Baking powder	−60%
Eggs	+15%
Flour	+10%
Fat	−10%
Sugar	−10%
Liquid	+22%

STEP PROCESS

Making Buttercream Decorations and Borders

The following step process demonstrates the techniques for making a 9-petal rose and rosettes as well as illustrates several borders and common designs.

To make a 9-petal buttercream rose, rosettes, and other decorations, follow these steps:

1A Using a rose nail, make a base cone and an inside petal.

1B Starting at the base of the cone, pipe the next three petals.

1C Add the remaining 5 petals.

1D Scissors can be used to snip the rose from the nail and place it on the cake.

2A To make a rosette, apply pressure and pipe down about ½ inch (13 mm) to the 6 o'clock position. Continue applying pressure, reversing direction, and start making a circle toward the 12 o'clock position. Allow the buttercream to curl from right to left as you begin the spiral.

2B Allow the spiral to come around to close the circle. Finish by pressing in slightly with the tip and stopping pressure. Pull away quickly toward the 5 o'clock position.

2C One way to practice making borders and designs is to place a piece of parchment paper on a wooden workbench and use the lines as a guide.

389

STEP PROCESS

Assembling and Decorating a Wedding Cake

Wedding cakes come in countless shapes and designs and utilize many different ingredients. This basic version incorporates many of the techniques that must be mastered before moving on to more complex wedding cakes.

Follow these steps to make a wedding cake:

1 Use a serrated knife to level the top of the cake.

2 Carefully cut the cake into even layers.

3 Brush the layers with simple syrup to moisten.

4 Pipe a ring of icing around two layers and pipe filling into the rings and spread.

5 Stack the layers with the uniced and unfilled layer on top and apply a skim coat.

6 Apply a final coat of icing.

7 Insert cut straws into the cake to use as dowel supports for the top layer.

8 Snip the straws even to the top of the first layer.

9 Repeat steps 1–6 to assemble the top layer and add the top layer to the bottom layer.

10 Pipe drapes using a rose tip.

11 Pipe borders using a star tip.

12 Add roses to the cake.

13 Pipe leaves and other decorations to complete the cake.

14 A finished wedding cake.

STEP PROCESS

Piping Chocolate Decorations and Words

Small pastry bags or cones made from parchment paper work best for these techniques. If the chocolate begins to harden, warm briefly in a microwave oven. These techniques work for piping filigree designs on parchment paper or when writing directly on a cake.

To use chocolate for decorations and writing, follow these steps:

1 Fill a small pastry bag with chocolate. Remove a small point from the tip to make an opening. Begin piping with the bag elevated over the parchment paper. Use your fingers to guide and steady the bag.

2 Allow the chocolate to fall from the bag as you continue piping. Do not allow the tip to touch the parchment paper or the surface you are decorating.

3 Use the same technique when piping borders and scrolls.

4 Chocolate filigree decorations.

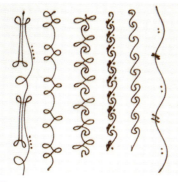

5 Chocolate borders or scrolls.

6 Chocolate script.

CHAPTER 27
Chocolate

The versatility of chocolate has made it an extremely important product that is used in desserts, candies, showpieces, and in a variety of ethnic cuisines as well. Professionals or novices working with chocolate must have a good understanding of how this complex food should be handled and used to obtain optimum results. Focus, determination, practice, and passion are key to success with chocolate.

KEY TERMS

conching	tempering	sugar bloom	nougat
cacao	precrystallized	couverture	hygroscopic
fermentation	polymorphic	praline	density
nibs	seed	ganache	invertase
cocoa liquor	fat bloom	gianduja	enrobing
cocoa butter	overcrystallize	marzipan	compound chocolate
lecithin			

An Introduction to Chocolate

Research studies have shown that the consumption of chocolate can aid in alleviating pain and improving moods through the release of the stimulants dopamine and serotonin, which are naturally found in chocolate. Cocoa contains PEA (phenyl ethylamine), which can elevate the mood, raise the heart rate, and boost energy. Dark chocolate is also rich in polyphenols and flavonoids that act as antioxidants that affect the lipoprotein (LDL) components in the "bad cholesterol" and reduce the risk of heart disease and certain types of cancer. The daily consumption of 46 grams of dark chocolate containing 70% cocoa solids is now said to help expand blood vessels to reduce the risk of heart attacks, strokes, and circulatory diseases. Numerous studies also laud the medicinal and mood altering effects of chocolate.

History

The story of chocolate begins in South America, where valued cocoa beans have been grown for centuries. These beans were later introduced in Mexico by the Mayans who prepared bitter drinks made of chocolate and water flavored with vanilla and various chilies. These drinks, known as Xocolatl, were also consumed by the Aztecs. Although Columbus is said to have introduced cacao beans in 1504, the Spanish conquistador Hernando Cortés is credited with popularizing chocolate.

The first sweet chocolate that was made by adding honey, cinnamon, and cane sugar was introduced by nuns in Oaxaca, Mexico. Spanish monks created a similar sweet delicacy circa 1590 that included honey and vanilla. By the mid-1600s the use of chocolate had spread throughout most of Western Europe and chocolate houses were becoming important social meeting places, particularly in London. The Latin botanical name *Theobroma cacao*, literally meaning "cocoa, food of the gods," was coined by Linnaeus in 1737.

Chocolate was rarely enjoyed as anything but a drink until 1828, when Conrad van Houten, a Dutchman searching for a way to make the chocolate drink less oily and filling, developed a process that removed most of the cocoa butter from the cocoa bean. In the process, van Houten "invented" cocoa powder. The excess cocoa butter that was extracted from the beans was added back to the original cocoa paste to greatly improve the texture. This single accomplishment made the development of chocolate candy possible. In 1828, van Houten also introduced an alkalized cocoa powder that made it more soluble in liquid. Today, chocolate can be alkalized, but it is rarely found on the market.

The first chocolate developed specifically for eating enjoyment was introduced by the English chocolate manufacturer Fry and Sons in 1847. Within 30 years from that time, the popularity of chocolate candy nearly rose to the status it holds today. The Swiss confectioner, Daniel Peter, developed milk chocolate in 1876. In 1879, Rudolphe Lindt introduced the technique of *conching* (slowly kneading chocolate to produce a smoother texture) named after the trough used to hold the chocolate, which was shaped like a conch shell. Swiss confectioner Jules Séchaud created the world's first molded and filled chocolates in 1913.

Cacao Growth

Cacao, which comes from the cacao tree, is a main component in chocolate. The cacao tree is an evergreen that grows to approximately 30 feet in height and flourishes in warm climates close to the equator. The name cacao is a corruption of the Aztec word *cocao*, meaning "bitter water." The cacao tree must have the right amount of moisture and heat for proper growth. Banana, lemon, or other trees are usually planted between cocoa trees to shield them from excessive wind and direct sunlight that can kill them.

Cacao is cultivated around the world in forests that lie within 20° north or 20° south of the equator, in an area known as the cocoa belt. Most of the world's cacao plantations are found in Western Africa, South and Central America, and in parts of South East Asia. Currently, over 40% of the world's cacao supply is grown in the Ivory Coast in Western Africa. Cacao trees also flourish in some of the islands of the West Indies, in Sri Lanka, the Philippines, and other South Pacific islands. Other leading cocoa-producing countries in the world include Ghana, Brazil, Nigeria, Cameroon, and Ecuador. Every year, people in the United States consume more than one third of all the cocoa grown.

The Cacao Bean

Figure 27-1 Chocolate comes from the bean pods of the cacao tree.

The cacao tree produces a tiny flower that may be white, pink, yellow, bright red, or two-toned, depending on the variety of the tree. The fruit of the cacao tree is called a pod, which when ripe, resembles a small red, orange, or gold football. The pods range in size from 6 to 10 inches (15-25 centimeters) in length, and 3 to 4 inches (8-10 centimeters) in diameter. They grow directly on the trunk of the tree, as well as on the older branches. Some pods are smooth, some are deeply ridged, and many are flecked with dark brown or black. *See Figure 27-1.*

The harvesting of the pods is conducted twice a year. When they are ripe, they are picked and split open to reveal almond-shaped cocoa beans that are nestled amid a whitish pulp. Each pod contains up to 50 cacao beans. Although there are multiple species of cacao beans, the varieties that are primarily cultivated are *Criollo, Forastero,* and *Trinitario.* These beans serve as the main ingredients in the manufacturing of chocolate. The beans are fermented, dried, and passed through many processes before they are transformed into the chocolate delicacies that are so widely enjoyed.

The Manufacturing of Chocolate

The creation of chocolate begins from the time the cacao beans are removed from their pods. The beans are fermented, dried, and then finally packed for transport to processing plants. At the plant, the cleaning, roasting, and grinding occur to prepare the product for mixing, refining, conching, and in some cases tempering.

Preparation of Cocoa Beans

Once the cacao beans have been removed from the pod they are placed in large piles or in wooden crates, covered with banana leaves or burlap, and left to ferment for up to 9 days. *Fermentation* enhances and develops the flavor of the beans, takes away some of the bitterness, and destroys their ability to germinate or sprout. The beans are then sun-dried to prevent spoilage, and to lower their moisture content. The beans are then packed into large bags and shipped to processing plants throughout the world.

At the processing plant, the beans are cleaned utilizing brushes, air, and gravity. They are brushed, blown, and shaken over screens to remove the dirt and foreign particles. Water is never used to clean cocoa beans. After cleaning, the beans are roasted in special ovens at up to 300°F (149°C) for 15 to 20 minutes. Roasting the beans enhances their flavor, darkens their color, dries them further, and loosens the shell for easy removal. The beans are then cracked, and the shell is removed through the use of air. The cracked bean without the shell is referred to as the *nib*.

The nibs, now in small pieces, are placed in a liquor mill and finely ground to produce a liquid substance resembling melted dark chocolate. This mass is called chocolate or *cocoa liquor*. The nibs contain up to 55% *cocoa butter* (the primary fat in cacao beans), and the heat from the friction of the mill causes the dry cocoa mass to liquefy and resolidify upon cooling. Chocolate liquor is very bitter and is sold in most grocery stores as unsweetened chocolate. It is primarily used in the manufacturing of sweet, eating chocolate or for making cocoa powder.

Mixing

To produce chocolate, the liquor is put in a mixer (mélangeur), and various ingredients are added to create the different types of chocolate: dark chocolate (sweet, semi-sweet, and bitter), milk chocolate, and white chocolate. The ingredients for each chocolate variety follow:

Figure 27-2 Chocolate, (dark, milk, or white) is a popular bakeshop and pastry shop ingredient.

Dark Chocolate liquor, cocoa butter, sugar, vanilla or vanillin, and lecithin.
Milk Chocolate liquor, cocoa butter, sugar, milk solids, vanilla or vanillin, and lecithin.
White Sugar, milk solids, cocoa butter, vanilla or vanillin, and lecithin. There is no chocolate liquor in white chocolate. *See Figure 27-2.*

The Addition of Soy Lecithin

Lecithin is used to emulsify the components of chocolate and to give melted chocolate a more fluid consistency. It is also used by manufacturers to preserve chocolate.

Refining

The chocolate is still harsh and gritty after mixing so it must be put into a refiner. It is sent through a roll press that crushes the particles and gives the chocolate a finer texture. The particles in chocolate are reduced to a micron size although they are not dissolved.

Conching

After the initial refining, the chocolate is conched. Conching further refines the chocolate by kneading, shaking, and pressing it for up to 72 hours. It affects the flavor and texture of the chocolate, reduces the particle size (good qualities are brought to 16–30 micron), and creates a smooth texture. It is important to remember that the particles are ground, not dissolved. The shaking, sometimes referred to as kneading, further emulsifies the ingredients, and creates a layer of cocoa butter that surrounds each of the particles to increase the smoothness of the chocolate and to allow for even melting.

Conching aerates the chocolate to help evaporate unwanted volatile acids and reduce the harshness of the flavor. Aeration also removes some of the moisture that is still present in the chocolate to help make the chocolate more fluid. Conching creates a chocolate that is more fluid, has better melting qualities, and is smoother in texture and flavor.

Tempering

Tempering is a heating, cooling, and reheating process that is used to stabilize chocolate. It creates an even suspension of solid particles and cocoa butter, and guarantees that the chocolate is perfectly emulsified. Chocolate manufacturing plants use specialized computerized machinery to perform this task. Once tempered, the chocolate is poured into molds, cooled in cooling tunnels and wrapped for shipment. Chocolate that is used for showpieces, molding, or dipping must be tempered to achieve an appropriate shine, smoothness, and firmness. Tempered chocolate is also known as *precrystallized* chocolate.

Chocolate is a suspension of sugar and cocoa (milk) particles in cocoa butter. The solids within the chocolate maintain their physical solid form when melted and during the tempering process while the cocoa butter begins to melt. At a temperature of 97°F (36°C) the cocoa butter is completely dissolved and turns into a liquid, transparent oil. The chocolate takes on a fluid state although the solid particles in the chocolate have not changed in physical form. As it cools down, the liquid cocoa butter begins to form crystals that cause it to firm up (cocoa butter crystals are always solid, never liquid.) Cocoa butter is comprised of a variety of fatty acids that have different melting and setting temperatures. Cocoa butter is said to be *polymorphic* (many forms) because it crystallizes (solidifies) in several ways, forming crystals of different sizes, shapes, and melting temperatures.

Crystals that are unstable will ruin the appearance and the texture of the chocolate, making it unappealing and unsuitable for sale. During the cooling procedure, a large number of proper "B" cocoa butter crystals are formed, in addition to some unwanted "A" crystals. The tempering procedure ensures that the cocoa butter in the chocolate will form the proper, stable "B" crystals as it cools and solidifies.

Well-tempered chocolate has the following properties:

- A good gloss (shine)
- No fat bloom (visible surface fat)
- A hard surface (consistency)
- Good shrinkage
- A clean snap when fractured
- A good shelf life (stable crystal formation = bloom resistant)

When the chocolate is heated to a maximum working temperature, most of the existing "B" crystals are stable at this temperature and will not melt. The unstable "A" crystals, which have a lower melting temperature, on the other hand, will dissolve and disappear. The chocolate now consists of a mixture of properly crystallized cocoa butter and liquid cocoa butter. The tiny invisible "B" crystals, known as "*seed*" are dispersed in the liquid chocolate to promote the growth of additional "B" crystals when conditions are favorable. At this stage the chocolate is fluid at a good working consistency, and is ready for use.

As the temperature of the tempered chocolate begins to drop, the melted cocoa butter begins to crystallize following the pattern of the already present "B" crystals. The melted cocoa butter bonds itself to the "B" crystals and begins to crystallize following the same pattern, and creating a strong bond. All the particles within the chocolate are now evenly dispersed preventing the cocoa butter from separating or rising to the surface. "B" crystals have a small uniform shape, that gives the chocolate a smooth texture and a shiny appearance while "A" crystals are bigger and uneven in shape and size and give the chocolate a coarse texture and an unappealing appearance.

Properly tempered melted chocolate sets faster than untempered chocolate due to the presence of properly formed cocoa butter crystals that cause the remainder of the melted cocoa butter to bond and crystallize more quickly. Melted tempered chocolate should never be overheated or it will lose its tempering. Overheating causes the "B" crystals in tempered chocolate to melt away as the chocolate cools and allows large amounts of unstable "A" crystals to slowly form. Chocolate that takes a longer period of time to set allows the cocoa butter to separate and rise to the surface, and cause a discoloration known as *fat bloom*.

After the tempering procedure is completed, the chocolate is heated to a maximum working temperature that should not be exceeded. The temperature, setting time, consistency, and texture of tempered chocolate must be checked before it is used. Working temperatures vary depending on the amount of excessive "B" crystals present.

Working Temperature Range of Tempered Chocolate

While working with tempered chocolate, the temperature should not drop more than 3 degrees below the maximum working temperature. If tempered chocolate is constantly used below working temperature it will begin to *overcrystallize,* thickening and remaining thick even at maximum tempered temperature due to an overabundance of cocoa butter crystals and insufficient melted cocoa butter.

To bring this chocolate back to a proper fluid working consistency, it must be reheated to a few degrees over the maximum tempered temperature. Heating a cooled chocolate considerably increases the chance of untempering the chocolate. To avoid this, never allow tempered chocolate to get too cold. Also, occasionally add some hot, untempered chocolate at a minimum 115°F (46°C) for milk and white chocolate, and at 122°F (50°C) for dark chocolate.

How to Fix Slightly Overheated Tempered Chocolate

A tempered chocolate that has been overheated no more than 4 degrees can be fixed without being completely retempered by either adding some tempered chocolate shavings (the finer, the better); cooling a small amount (20%) on a marble table to a soft paste; or adding some colder tempered chocolate.

Using Untempered Chocolate

If melted chocolate is used without being tempered, the cocoa butter will improperly crystallize, and form unstable "A" crystals as it cools. The chocolate will take too long to set and will prevent the cocoa butter from remaining emulsified with the other ingredients. Some of the cocoa butter will begin to separate and migrate to the surface, causing an excessive amount of fat to rest on the surface of the chocolate. Once it is fully set, this layer of cocoa butter will bloom. White/gray forms and shapes, ranging from a plain film to streaks, lines, dots, or circles will appear. The chocolate will also have a grainy, gritty texture, because of the formation of improper cocoa butter crystals and the lack of emulsification within the ingredients.

An untempered, or poorly tempered chocolate will not be stable during storage, and will display fat separation and a grainy texture. Chocolate pieces that have been stored in an untempered state for several weeks or months will also fail to melt properly when reused. Melted chocolate that is untempered will be warm, runny, and will take longer to set. Untempered chocolate is more likely to absorb moisture during storage, and to melt less quickly when exposed to heat.

Tempering Methods

There are a variety of methods that may be used to temper chocolate. These methods include: the table method, vaccination method, direct method, resting method, machine method, and cold-water bath method. Temperatures used in tempering methods vary and are based on specific manufacturer's recommendations.

TABLE METHOD Heat the dark chocolate to 122°F (50°C) and the milk or white chocolate to 115°F (46°C). At these temperatures the cocoa butter is sure to be completely melted. Never heat dark chocolate higher than 130°F (54°C), and milk or white chocolate higher than 120°F (49°C). Next, pour two-thirds of the chocolate onto a clean, sanitized marble slab. Scrape and spread the chocolate back and forth using an offset spatula and a spackle knife. Continue to move the chocolate around until its consistency begins to slightly thicken. It should then be at a temperature of 80°F (27°C) for dark chocolate, 78°F (26°C) for milk chocolate, and 76°F (24°C) for white chocolate. Quickly add the thick, cooler chocolate back into the remaining one-third of warm chocolate and stir immediately to prevent cool chocolate from setting. If necessary, slowly reheat the chocolate to 90°F (32°C) for dark chocolate, 88°F (31°C) for milk chocolate, or 86°F (30°C) for white chocolate.

Chocolate will not be tempered if:
- The chocolate was not heated enough before being poured onto the table.
- Not enough chocolate was poured onto the table.
- The chocolate on the table was not cooled enough.
- The chocolate was reheated at too high a temperature after the tempering procedure.

VACCINATION METHOD Heat the dark chocolate to 122°F (50°C), and the milk or white chocolate to 115°F (46°C). At these temperatures the cocoa butter is sure to be completely melted. Never heat dark chocolate higher than 130°F (54°C), and milk or white chocolate higher than 120°F (49°C). While constantly stirring, gradually mix in tempered chocolate shavings (or chocolate coins), slowly cooling the chocolate down to its tempered temperature of 90°F (32°C) for dark chocolate, 88°F (31°C) for milk chocolate, and 86°F (30°C) for white chocolate.

NOTE: All added shavings (or coins) must be completely melted when the chocolate has reached its working temperature!

Chocolate will not be tempered if:
- The chocolate was not heated enough before shavings (or coins) were added.
- The shavings used were made with untempered chocolate.
- Not enough shavings (or coins) were added.
- The chocolate was reheated at too high a temperature after the tempering procedure.

DIRECT METHOD (MICROWAVE METHOD) Place finely chopped tempered, set chocolate in a microwaveable bowl. Slowly melt the chocolate at short intervals. Stir well between the segments until the appropriate chocolate temperature is reached. Do not exceed the proper working temperature.

RESTING METHOD The resting method is used only for small quantities of chocolate. Heat the chocolate to 115°F (46°C), and then allow it to rest. Stir occasionally, until it cools and becomes pasty. Then reheat it to the proper working temperature. A large piece of chocolate can be added to the melted chocolate to speed the cooling process. This piece is removed when the chocolate begins to thicken.

MACHINE METHOD The machine method is also a direct tempering method. The tempering machine slowly melts the chocolate to the proper working temperature without exceeding it, to prevent untempering.

COLD-WATER BATH METHOD Chopped dark chocolate is heated to 115°F (46°C). The chocolate is cooled over a cold-water bath until it reaches a pasty consistency. It is then

STEP PROCESS

Tempering Chocolate

The following process demonstrates the use of the table method and the vaccination method for tempering chocolate.

To temper chocolate using the table method or the vaccination method, follow these steps:

1 Heat chocolate to the proper temperature and pour two-thirds of the chocolate onto a marble table.

2 Spread chocolate on the table using a back and forth motion.

3 Scrape and push the chocolate while spreading.

4 Move the chocolate around until it begins to thicken slightly.

5 Transfer chocolate from the table into the original batch of melted chocolate.

6 Quickly blend the chocolate together to a liquid state and the proper temperature is achieved.

7 To use the vaccination method, heat chocolate to the proper temperature. Drop pieces of chocolate into the melted chocolate and stir until blended and chocolate reaches the desired temperature.

reheated to the proper working temperature. Care should be taken to ensure that no water from the bath is splashed into the chocolate. This method may be used for all three types of chocolate, but works particularly well for white chocolate.

Why Chocolate Thickens (Seizes Up) When Melted

If melted chocolate becomes too cold, it will turn thick and pasty. To return it to its liquid state, chocolate should be slightly heated. Additionally, cold chocolate should not be stirred too much without reheating it or the chocolate will retain a large percentage of air bubbles and have a dull appearance.

Melted chocolate turns pasty, grainy, and loses some flavor if it is overheated beyond the maximum melting temperature. If it is just slightly overheated, it can be strained and used as a flavoring for sponges, creams, fillings, or in showpieces. Chocolate that has been overheated well beyond its maximum melting temperature cannot be used for pralines, fine fillings, or coatings. Melted chocolate that comes into contact with moisture becomes thick and pasty and cannot be used for coating. It may be used for ganaches, flavoring, sauces, and some fillings, as long as the moisture comes from a noncontaminated source.

If chocolate is mixed incorrectly or mixed when it is too cold, too much air will incorporate and bubbles will appear on the surface. The chocolate will also have a dull appearance when set. If air bubbles are present, the cool chocolate can be placed in a warm area until the air is released and the chocolate returns to a liquid consistency. If time will not allow the chocolate to sit for a while, it should be reheated just to 115°F to 122°F (46°C–50°C) and then retempered. A whipping motion should never be used when stirring melted chocolate.

Chocolate that is reused repeatedly without the addition of fresh chocolate becomes thick in consistency. A small amount of new, tempered chocolate can be added to the used batch each time it is worked.

Enemies of Chocolate

Moisture, excessive heat, and improper storage are enemies of chocolate that can cause gradual or rapid deterioration of the product.

Moisture

If moisture droplets fall into melted chocolate, it will thicken and become pasty and sluggish, and can be used only for fillings. Moisture surrounds the sugar particles in the chocolate, increasing the friction between particles and causing them to move more slowly, which in turn, increases the viscosity of the chocolate.

Set chocolate may become covered with a layer of sugar syrup, which eventually dries and forms a dried sugar crust on the surface known as *sugar bloom*. Sugar bloom looks similar to fat bloom, but unlike fat bloom, it has an even, whitish appearance and does not dissolve when the chocolate is remelted. Sugar bloom is always caused by moisture on set chocolate. It should be scraped off before the chocolate is remelted. To prevent sugar bloom, chocolate should never be stored in the refrigerator.

Excessive Heat

Chocolate is very sensitive to high temperatures and when overheated, becomes pasty, grainy, and loses its fine flavor. It also becomes grainy due to the lumping together of sugar crystals in the chocolate. If the chocolate is not too grainy, it can be strained through a cheesecloth and used for fillings of lower quality confections and/or sauces. If the chocolate has too much graininess, it should be discarded.

Chocolate Storage

Chocolate must be wrapped properly and protected from light, strong odors, and foreign particles such as dust. Aluminum foil is a good wrapping material that eliminates the penetration of light, which is known to cause chocolate to spoil more quickly. Chocolate should also be kept cool and dry, at temperatures between 60°F (16°C) and 65°F (18°C). The humidity in a room where chocolate is stored should not exceed 50%. Chocolate should never be stored in a refrigerator because moisture will collect on its surface and cause sugar bloom.

Chocolate may be frozen to increase its shelf life, but must be done carefully. It should be wrapped airtight and quickly frozen in a very cold freezer that keeps a constant temperature. The thawing of chocolate should be gradual to prevent condensation from developing and causing sugar bloom.

Buying Chocolate

Purchase chocolate that is wrapped in an opaque wrapper to prevent its exposure to sunlight and foreign particles. Its surface should be appealing and shiny. Chocolate should have a sharp, clean break and should not crumble in the hands. It should always be tasted before it is purchased to guarantee that it has an appealing aroma and smells like chocolate with a deep, rich flavor. It should not taste chalky, like cocoa powder. When tasted, the chocolate should melt evenly on the tongue and have a smooth texture with a soft flavor that is not harsh.

Melting or Reheating Chocolate

It is sometimes necessary to melt or reheat chocolate to an appropriate temperature to prepare it for use. A water bath, an oven, a warming cabinet, or a microwave may be used to do so.

Water Baths

Chop the chocolate into small pieces and place it in a dry bowl. Place the bowl over a water bath of simmering, but not boiling, water. Continue to stir the chocolate with a rubber spatula. Occasionally take the bowl off the water bath and keep stirring until the bowl cools a bit. Once it is cooled return it to the heat. Boiling water forces the steam out of the water bath. The steam then settles on the surface of the chocolate to thicken it.

Ovens

Ovens are excellent for reheating tempered chocolate. Place a bowl of chocolate in the oven for 20 seconds at a temperature of 250°F (121°C). Remove the bowl and stir the chocolate until the bowl has cooled a bit. Repeat the process until the chocolate has reached the proper temperature. If the bowl is too hot to be held with bare hands, it will cause the chocolate along the edge of the bowl to overheat and turn grainy.

Warming Cabinets

Large pieces of chocolate should be placed in a bowl or hotel pan, covered with aluminum foil, and placed into a 90°F–110°F (32°C–43°C) warming cabinet. The chocolate should be allowed to melt overnight. Stir the chocolate well before use.

Microwaves

When using a microwave to melt or reheat chocolate, chop the chocolate into small pieces and place in a microwave safe bowl. Using short time intervals heat the chocolate on high power. Stir the chocolate well between each interval. As the chocolate begins to melt, reduce the time segments accordingly. The last few intervals should only be a few seconds long.

Chocolate Equipment List

CHOCOLATE MOLDS made of plastics of varying densities and of metal. Molds are used in the production of pralines and showpieces. High-quality, sturdy molds are made of polycarbonate, a thick, more expensive durable plastic. Less expensive, flimsy vacuum-shaped molds are also available, but are primarily suitable for occasional use because they damage easily. Molds that are made of polished metal are less commonly used but also work well provided that they do not show any signs of corrosion.

CHOCOLATE STENCILS—used to create simple shapes such as circles. Stencils may be cut out of thin, soft rubber mats. Chocolate is spread over the open cavities of the mat, and when the mat is removed, a perfect chocolate duplicate of those shapes is reproduced.

CHOCOLATE TEMPERING MACHINES—serve to hold chocolate at its tempered state. The tempering process allows the melted cocoa butter in the chocolate to properly crystallize and to give the finished product the even, glossy shine traditionally seen on fine chocolates.

CHOCOLATE WARMER (EFFECTOR)—a warming unit used to safely and efficiently melt chocolate and to keep tempered chocolate at its correct working temperature.

DIGITAL THERMOMETERS—the safest and most accurate tools for checking the exact temperature of melted chocolate.

DIPPING FORKS—delicately fabricated tools used to dip (enrobe) pralines in chocolate. Most commonly used dipping forks consist of a handle and 2, 3, or 4 thin, metal wire prongs. Loop-shaped or round forks are used to dip oval or round candies.

HOT AIR GUNS—produce hot air and serve as an excellent heat source to reheat melted chocolate that has dropped below working temperature. Care should be taken not to hold the gun too close to the chocolate as it will burn the chocolate.

IMMERSION BLENDERS—have a fast moving, rotating blade that is used primarily to emulsify ganache and to speed up the melting of small chocolate pieces.

MARBLE SLABS—cold, polished stone surfaces that are best suited for tempering chocolate and for making chocolate decorations.

METAL BARS—made of steel, brass, or aluminum and used as guides to form even and level sheets of praline fillings. They are also employed to mold chocolate.

SPACKLE KNIVES—triangular-shaped metal scrapers that are 4 inches to 6 inches (10 centimeters to 15 centimeters) in size and used during the table tempering method of chocolate. They are an excellent all-around tool to scrape chocolate off a surface and may be purchased in any hardware store.

Spray Gun—can create a fine mist of chocolate on the surface of showpieces, pastries, and desserts. Two types of spray guns are available: the electric spray gun that is suitable for spraying simple, small surfaces, and the heavy-duty spray gun with a compressor that produces a finer mist and is mostly used for large chocolate showpieces.

Textured Surfaces—consist of any food safe plastic, paper, or rubber surface that chocolate can be poured onto to obtain a desired texture.

Texturing Combs—toothed scrapers used to give spread chocolate a linear design.

Transfer Sheets—colored designs printed on shiny plastic sheets. When chocolate has been spread on the design, it releases from the plastic and transfers to the chocolate.

Wood Grainers—painting tools used to give chocolate a wood grain appearance.

X-acto Knife—a small, very sharp, hobby knife used to cut spread chocolate into precise shapes.

Couverture

Couverture is a bulk chocolate that is used for making confections, pralines, and fine pastries. Couverture is a finer, higher-grade product than regular chocolate, and is therefore more expensive. It is used for dipping, enrobing, and molding chocolate products, and in the making of special fillings. Couverture is produced in the same way as chocolate, but includes a higher percentage of cocoa butter. The high percentage of cocoa butter provides greater fluidity when it is melted, and produces a better shine and a harder snap when it sets. Couverture is sold in dark, milk, and white forms.

Pralines and Other Fillings

Chocolate is a treat that becomes a delicacy when filled with praline and other such fillings. The types of fillings that can be created are endless, but as with all chocolate work a knowledge of ingredients and flavors, and their interactions, is imperative for success.

Pralines (Bonbon au Chocolat, Chocolate Candies)

The term ***praline*** is commonly used to describe a small, bite-sized, chocolate-based product. It is also used to refer to the flavor of a mixture composed of toasted hazelnuts or almonds and caramelized sugar. In the southwestern United States and Mexico, pralines are candies made from a sugar and pecan mixture. They were named in honor of the Duke of Plessis-Pralin, a field marshal in the French army under Louis XIII and Louis XIV, whose chef accidentally mixed toasted nuts with caramelized sugar to create this candy.

Types of Fillings

Countless combinations of fillings can be produced by mixing chocolate-based fillings, nut fillings, fruit fillings, sugar fillings, and liquid fillings. In Europe, after the filling is shaped, it is referred to as the center, or intérieur.

Chocolate-Based Fillings

Widely used fillings containing chocolate include ganaches, butter ganaches, and giandujas.

Ganaches The two main ingredients in ***ganache*** are chocolate and water-based liquids such as milk, cream, teas, or fruit juices. Other ingredients, such as fat, flavorings, liqueurs,

STEP PROCESS

Making Ganache

Ganache is a popular filling for pralines. It is used in the preparation of chocolate truffles, and it can also be used to make a pourable coating for other products such as cakes and tortes.

Follow these steps to make ganache:

1 Heat cream and glucose (if using) and pour the hot cream mixture over the chocolate. Let rest about 1 minute.

2 Stir chocolate and cream until both are blended together and fully melted.

3 An immersion blender can be used to emulsify the ganache.

4 Two examples of ganache. One properly emulsified (top) and one in which the ganache has separated (bottom).

5 To make a sheet of ganache to use as a filling for pralines, pour the liquid ganache between bars.

6 Spread and even off the ganache.

and distilled spirits, are optional. Additional sugars, including glucose, corn syrup, and granulated sugar, can also be added to improve the texture and shine of ganache. Pasteurized egg yolks are sometimes used to make a richer ganache, although precaution should be taken to avoid possible contamination of the ganache.

Ganache is prepared by bringing a chosen liquid to a boil. The liquid is removed from the heat and added to finely chopped chocolate. The mixture is then carefully stirred using a rubber spatula until the ingredients are well blended, the chocolate is completely melted, and the mixture is well emulsified and smooth. The sides of the bowl should be well scraped to ensure that all of the chocolate is incorporated. Any chocolate that is not properly blended will result in solid lumps in the set ganache.

Fats that are used in ganache should be added with the chocolate, or later incorporated. Care should be taken not to boil the mixture to avoid separation. When sugars are used, they should be brought to a boil with the liquid to prevent graininess in the final product. Flavorings, liqueurs, and distilled spirits are added after the ganache is completely smooth to prevent the evaporation of the alcohol and a loss of flavor. A cold liquid can also interfere with the melting of chocolate if added too soon. Pasteurized egg yolks, if used, are heated with the liquid until they are thickened, as in the preparation of crème anglaise.

Points to remember when making ganache for filling pralines:
- The boiling of the liquid not only raises the temperature high enough to melt the chocolate, but also prolongs the shelf life of the ganache by destroying some of the microorganisms present in the liquid.
- Using cream or milk that is too old will result in the separation of the fat(s).
- Overheating the ganache causes separation of the fat.
- Overwhipping or overmixing ganache once it cools will cause the fat to separate.
- If the fat content in the ganache is too high, the ganache will separate. Most ganaches should have a fat content of approximately 30%.
- If the amount of liquid added to a ganache is too low, the ganache will separate. This can be corrected by adding more liquid to it.
- Ganache has a very short shelf life due to its high content of water and/or dairy products.
- It should never be handled without gloves to avoid cross-contamination.
- Ganache will stay fresh for at least 2 weeks if properly handled. After this time, a flavor change may occur, diminishing the quality of the praline. The shelf life is determined by the composition of the ingredients, the handling, and the storage conditions.
- When used in pralines, ganache should be completely covered with chocolate to prevent it from drying out and spoiling. The ganache filling will spoil long before the chocolate.
- Ganache may include fat-based or water-based flavorings. Nuts, candied fruits, praline paste, crushed nougat, or dried fruit are also used in ganache. Cereals are not suitable for crunch effect because they quickly become soggy.

BUTTER GANACHE The two main ingredients in a butter ganache are butter and chocolate. Ingredients such as sugar, liqueurs, distilled spirits, and flavorings may also be added. To prepare butter ganache, room-temperature butter and sweetener are whipped to a light consistency, and flavorings are then added. The mixture is whipped again until incorporated and then folded into the tempered chocolate using a rubber spatula. Butter ganache should be used quickly to avoid setting.

Points to remember when making a butter ganache:
- Butter must be brought to room temperature (70°F/21°C) before whipping. Cold butter will not incorporate air well, and will take a long time to become light and fluffy.
- The butter should be whipped well before adding any flavorings. Butter that has a small amount of air incorporated will not emulsify properly with a liquid.
- The amount of air incorporated into the butter can vary depending on the firmness of the praline. If the filling is to be cut, less air should be incorporated. When the filling is piped, more air should be incorporated.
- Tempered chocolate should be used in the filling to ensure a smooth, homogenous product that is consistent in texture. If an untempered chocolate is used, the result may be a product that is not well emulsified, and may become grainy once set.
- Tempered chocolate must be used at the maximum working temperature to allow enough time for the chocolate to be folded into the butter without solidifying. If the chocolate is untempered or too warm, the filling will separate and give the filling a soft consistency with a grainy texture when set.
- Filling should be used immediately because it will firm up rapidly once it is made.
- A butter ganache should be quickly dipped completely in chocolate so that it does not dry out or absorb foreign odors that may change its taste.

STEP PROCESS

Making Chocolate Truffles

Truffles are members of the fungi family. They grow near the roots of oak or beech trees and are prized for their intense earthy flavor. Truffles come in white and black varieties and are extremely expensive. Chocolate truffles are made to resemble small fungi that grow in woodland areas.

To make chocolate truffles, follow these steps:

1 Briefly "massage" ganache to incorporate air.

2 Pipe ganache onto parchment paper. Chill.

3 Put a small amount of confectionery sugar onto your hands to facilitate rolling the truffles.

4 Roll into small balls using the powdered sugar to prevent sticking.

5 Precoat the ganache by applying a small amount of chocolate in the palm of your hand and roll to coat.

6 Dip the truffles into melted chocolate and place on a dipping screen.

7 Roll truffles on the dipping screen to create an interesting texture.

- The shelf life of a butter ganache is longer than that of a regular ganache. As with most fillings, the flavor of a butter ganache can change during storage. The finished praline should be consumed within one month.

GIANDUJA   *Gianduja* is a filling made of toasted nuts and sugar, ground to a very fine consistency, and mixed together with chocolate and/or cocoa butter. Gianduja can be made by mixing together dark, milk, or white chocolate and praline paste. Praline paste

407

STEP PROCESS

Preparing and Working with Butter Ganache

Butter ganache is a combination of whipped butter and flavorings folded into tempered chocolate. To finish the pralines, the butter ganache was piped onto "coins" made of tempered chocolate, dipped into additional tempered chocolate and finished to make both small mice and round pralines.

When preparing butter ganache and pralines, follow these steps:

1 Cream butter. Add flavorings if desired.

2 Add tempered chocolate to creamed butter.

3 Fold the tempered chocolate into the butter.

4 Use a stencil to make chocolate coins.

5 Pipe the ganache onto the coins.

6 To make mice, add sliced almonds for ears and dip into tempered chocolate. Another shaped praline can be made by topping with a second coin, dipping, and then using a transfer sheet to decorate the praline.

is a mixture of sugar and toasted hazelnuts (almonds also are used) that has been refined to a peanut butter consistency.

Gianduja can be prepared by heating dark chocolate to 122°F (50°C), and milk and white chocolate to 115°F (46°C), and then adding praline paste. The mixture is then tempered in the same manner in which chocolate is tempered. The cold-water bath method works particularly well for giandujas.

Another method for preparing gianduja consists in mixing the praline paste with tempered chocolate. Because it contains real chocolate, the handmade gianduja should always be tempered to ensure the proper setting of the cocoa butter in the filling. Using untempered gianduja will result in a filling that does not set well and is difficult to cut. The texture of the filling will also be less smooth than that of a tempered gianduja.

Gianduja can be purchased premade and used in a variety of ways as a filling. It also can be made with other nuts, such as peanuts, coconuts, or pistachios. Gianduja is waterless and has a much longer shelf life than ganache. It can keep well for several months.

STEP PROCESS

Preparing Gianduja Filling for Pralines

Traditionally, hazelnuts were used to make the nut butter for gianduja filling. Though praline paste is still made with hazelnuts, any nut butter can be used to make gianduja. For this preparation, creamy peanut butter was used.

To make gianduja filling for pralines, follow these steps:

1 Scale peanut butter onto plastic wrap. Fold the wrap into a bundle and cut a small opening. Squeeze the peanut butter into the melted chocolate.

2 Blend the peanut butter into the chocolate.

3 Gianduja should have a fluid consistency before it is cooled.

4 Chill using an ice-water bath.

5 Gianduja will stiffen after cooling.

6 Add Rice Krispies® to gianduja.

7 Spread gianduja between bars to make an even sheet.

8 Use a paring knife to cleanly separate the gianduja from the bars.

9 Precoat the gianduja by pouring tempered chocolate over it.

10 Spread the tempered chocolate to form an even coat. Turn over and coat the other side of the gianduja sheet.

Nut Fillings

The main ingredient in nut filling is plain or toasted nuts that are chopped, whole, or ground. Marzipan and nougat are both nut fillings.

Marzipan Fillings

Marzipan is a soft filling commonly used in European pralines. It is an almond dough composed of almond paste, fondant, confectioner's sugar, and glucose. Marzipan is flavored and sometimes softened with liqueurs or flavorings to obtain a range of consistencies when used as a praline filling. Avoid overmixing or overhandling marzipan because the almond oil will separate and cause the paste to become crumbly and difficult to work with. Marzipan may be purchased premade for quick and easy use.

Nougat Fillings

Another type of filling commonly used in the preparation of pralines is *nougat*. Nougat has a sugar base with nuts added. There are two types of nougat, brown and white. Both differ in their methods of preparation and the types of ingredients used. Brown nougat is also called croquant (French) or krokant (German). It is a mixture of caramelized sugar and lightly toasted nuts, classically almonds. Soft nougats used in pralines are made by mixing the nougat with other ingredients, such as cream, corn syrup or glucose, honey, or fruit juices. Candied fruits, chocolate, almond paste, praline paste, or other flavorings may also be added.

White nougat, also called Torrone (Italian) or Montélimar (French), is similar to what Americans call "divinity," but is a bit firmer. It is a mixture of a thick Italian meringue composed of honey, toasted nuts, flavorings, and sometimes candied fruits. The mixture is spread on edible wafer paper, allowed to set, then cut into rectangles, and sometimes dipped in tempered chocolate.

Due to a high sugar content, most nougat fillings are *hygroscopic* (able to absorb moisture from the atmosphere) and therefore sensitive to moisture. If the set fillings are exposed to the air for a long period of time, prior to dipping, they will become sticky, soft, and lose their shape. Nougats have a long shelf life and can keep for several months especially when fresh nuts are used.

Fruit Fillings

Fresh fruit should never be used in a filling because the water content is too high and causes spoilage. A high water content also makes the praline difficult to dip. The acid content in fresh fruit is also very high causing the chocolate to dissolve and the filling to leak out of the chocolate shell. Preserved fruits should be used in pralines. Types of fruits that may be used in pralines include:

- Fruits soaked in liqueurs or distilled spirits until the water within the fruit is actually replaced and completely saturated by the soaking liquid.
- Candied fruits that are completely saturated with a sugar syrup.
- Dried fruits that have been dried of their moisture. The dried fruits are usually cut and cooked or soaked in distilled spirits, liqueurs, or sugar syrup to ensure that they are not too tough.
- Fruits cooked in a high percentage of sugar such as preserves, jellies, and jam compounds. The high sugar content and the cooking process preserve the fruit.

Sugar Fillings

Sugar fillings are very inexpensive to use and have an extremely long shelf life. They are also easy to produce by machine. Sugar fillings are commonly used in the production of less expensive chocolate candies. These fillings consist of a high amount of different types of sugar in a concentrated cooked sugar solution. After cooking the solution, the mixture is cooled. The formation of small, fine crystals can be seen in fondant, cream fillings, or fudge.

Liquid Fillings

Liquid fillings are completely liquid sugar syrups that are entirely covered with chocolate. A good example of this type of filling is found in liquor cordials. Although there are numerous methods used to prepare liquid centers, only a couple of common methods are discussed in this chapter.

Methods for Incorporating Fillings

Fillings are incorporated in a number of ways that vary in complexity and style from the simply prepared shell method to the sophisticated starch method.

The Starch Method

The starch method is one of the oldest methods used to prepare fillings and also one of the most difficult to execute. Cornstarch is dried for several days in a low-temperature (140°F/60°C) oven and is then placed in a 3-inch-high frame. The starch is lightly packed down using a ruler until it is level with the sides of the frame, and a praline stamp mold is pressed into the starch to create cavities for the fillings.

A sugar syrup is then prepared for the filling. To determine the precise *density* (ratio of sugar to water) of a sugar syrup, a French instrument known as a Baumé measure is used. The initial syrup must have a 28° Baumé reading and after being flavored must read at 32° Baumé.

The starch molds are then filled with the warm sugar syrup using a sauce gun that can be controlled to prevent leakage of the syrup. At this point, a half-inch layer of starch is sifted over the filled molds so that they are completely covered. The molds are then allowed to rest for at least 24 hours. During this time, the sugar begins to crystallize around the starch, and form a crust around the filling, which will remain liquid in the center. Once a crusty layer has formed, the sugar shells are removed from the starch. The starch is then eliminated with a soft brush and the fillings are carefully dipped in thin, tempered chocolate.

The Prepared Shell Method

The most popular way to incorporate liquid fillings is to use already prepared chocolate shells. These shells are filled with a flavored sugar syrup and then sealed and dipped into tempered chocolate.

Chocolate Molds

Regular candy molds can be used for liquid fillings. The shells are filled with the syrup and then sealed by spraying a thin layer of chocolate over the liquid filling. Once set, additional chocolate is spread over the mold to give it a stronger shell.

Fruit and Fondant

If using a combination of liquor-soaked fruits and fondant as a filling, the fruit will dissolve the creamy fondant and turn it into a liquid after a few days. Cherry cordials are made using this method.

The Use of Invertase with Fondant

In commercial operations, invertase is used with creamy fondant to create liquid fillings. *Invertase* is an enzyme that expedites the breaking down of the sucrose within the fondant into a sugar syrup, after the filling has been surrounded by a shell of chocolate. Invertase is sometimes added to soften marzipan and other fillings.

Items That Should Not Be Used for Fillings in Pralines

- Fresh fruit. Spoilage is high and their acidity dissolves the chocolate.
- Products made with flour or starch. Pralines become petits fours when flour or starch is added to the mixture.
- Compound chocolate. Because of its inferior taste, quality, and consistency, most fillings made with compound chocolate will not have the proper consistency or flavor.
- Unpasteurized egg products. If eggs are used, they should be heated to the correct temperature and/or pasteurized because of their high spoilage factor and the threat of contamination. Egg products will also change the flavor of a filling after a short period of time and reduce the shelf life of a praline.

Working Conditions for the Dipping of Pralines

When preparing to dip pralines, careful consideration and control of the temperature of both the physical environment and the product are of the utmost importance.

Temperature to Consider During the Dipping

The process of dipping pralines is known as *enrobing*. The temperature of the filling, the room, and the chocolate are all very important factors when dipping pralines. These conditions greatly determine the final appearance, quality, and the shelf life of a finished product. The room temperature should be 68°F (20°C). The temperature of the filling should be the same as the room temperature, or a little higher, but no more than 74°F (23°C). The temperature of the filling should never be lower than room temperature and the chocolate should be tempered to the highest working temperature possible. If the filling is too warm when dipped, the chocolate around the filling will take too long to set and will allow adequate time for some of the cocoa butter to separate and rise to the surface and form fat bloom.

When the filling temperature is colder than the room temperature, condensation of water will collect around the filling and make the filling difficult to dip. Some of the moisture surrounding the filling will also end up in the dipping chocolate, gradually increasing the viscosity of the chocolate as more pralines are dipped. A small amount of water will also remain trapped between the filling and the chocolate shell, causing the filling to become grainy and to spoil prematurely. Due to the cold temperature of the filling, the chocolate will not run off as easily during the dipping and will create a chocolate layer around the filling that is thicker than desired. The praline will be too large, the chocolate will overpower the taste, and the food cost will rise due to the overuse of chocolate.

In addition, the cocoa butter will set unevenly from the inside outward, and will produce a dull appearance. To ensure a good shine on the set chocolate, the cocoa butter must set evenly, or must set from the outside inward. When the chocolate cools down to the temperature of the filling, which is colder than the room temperature, condensation will form on the chocolate and eventually cause sugar bloom. As the filling eventually warms up to the temperature of the room, it will expand and cause the chocolate shell to crack, exposing the filling to the air and causing it to spoil or dry out.

Room temperature is also important to consider when dipping chocolate. A room temperature that is too warm will increase the setting time of the chocolate around the filling, while a room temperature that is too cold for dipping will cause the chocolate to set prematurely. Constant reheating of the chocolate must be performed to minimize the occurrence of overcrystallization, which will make the dipping procedure difficult and inefficient.

STEP PROCESS

Making Pralines

To most pastry chefs, pralines are small, bite-sized, chocolate candies. The term also refers to various confections made with nuts and sugar. The following pralines are made with gianduja and milk chocolate.

Follow these steps to make milk chocolate and gianduja pralines:

1 After preparing gianduja, mark and square off the gianduja sheet.

2 Trim gianduja using a knife dipped in hot water.

3 Measure and cut into strips.

4 Cut gianduja strips into squares.

5 Dip squares into tempered chocolate to coat.

Chocolate temperature is the third factor and is an equally important consideration. Chocolate that is too warm will become untempered and will result in fat bloom. Chocolate that is too cold for dipping will cause the chocolate surrounding the filling to be too thick. The chocolate will also have a tendency to incorporate a lot of air when stirred and will appear dull when set.

Molded Chocolates

Molded chocolates are created using a mold designed specifically for pralines that are coated with chocolate, filled, and then sealed. These molds can be made from a variety of substances, including hard plastic, metal, silicone, and soft plastic. Hard plastic molds are the best to use, but are also the most expensive. Soft plastic molds are inexpensive, but do not last as long, and can be more difficult to handle. A key point to remember regardless of the type of mold used is that the interiors of the cavities must be shiny and free of scratches or defects to guarantee ease in unmolding the finished product. Molding chocolates is a time-efficient method of making pralines.

Procedures for Making Molded Pralines

1. The molds should be at room temperature. The cavity of the molds must be rubbed with cotton before use to remove impurities and excess cocoa butter. The tempered couverture chocolate should always be used at maximum working temperature for maximum fluidity. There should be no lumps in the chocolate.

2. A warmed ladle is used to fill the chocolate into each of the mold's cavities. If the molds are overfilled, the excess chocolate should be scraped off using an offset spatula. Molds that have very detailed designs within their cavities should be brushed with a thin layer of chocolate before they are filled to avoid air pockets.

3. Once filled, the molds should be tapped with a rubber mallet in two different directions to remove any air bubbles that may be trapped.

4. The tapped molds should then be inverted directly over the source of chocolate and allowed to drip, thus releasing any excess chocolate. If necessary, the molds should then be tapped again very lightly to speed up the flow of a more viscous chocolate. While the mold is still upside down, scrape any excess dripping chocolate using a spackle knife. If light can penetrate the layer of chocolate, it is too thin and steps 2 and 3 must be repeated.

5. The molds are then placed upside down, suspended between two metal bars and left to set until the wet, melted appearance of the chocolate has disappeared. The chocolate should still be soft.

6. When the wet, melted appearance of the chocolate is no longer visible, the excess chocolate should be scraped off with a spackle knife or a bench scraper. If the chocolate sets completely, it will crack and destroy the shell when scraped. If the chocolate is scraped too soon, all of the scraped chocolate will be pushed into the cavity of the mold and will remain inside.

7. At this point, the shells are ready to be filled. A filling that has a somewhat fluid consistency is piped in to fill the shells evenly. A filling that is too stiff, will produce a dome shape when piped into the chocolate shell and will leave an uneven space to be sealed off with chocolate. The temperature of the filling should be between 80°F (27°C) and 85°F (29°C) so as not to affect the chocolate. The filling will shrink a bit as it cools. A 1/16-inch space should be left between the top of the mold and the filling. This space is necessary for the final covering and for sealing the filling with chocolate. The filling should be allowed to set a bit until it is no longer sticky to the touch.

8. The remaining space is then covered with tempered couverture at maximum working temperature. The chocolate should be gently poured over the filling using a spatula and spread quickly and evenly. Any excess chocolate should be scraped off using a spatula, while the chocolate is still melted.

9. The chocolate is then left to set completely. It can be put in the refrigerator for a few minutes if necessary. When the chocolate has released itself from the mold the appearance of a gap between the mold and the chocolate will be visible. The finished chocolates may now be removed by carefully twisting the molds, inverting them, and gently tapping them against the table.

Reasons Pralines Do Not Release From A Mold

- The mold may be damaged or scratched.
- The chocolate was not tempered.
- The filling was too warm and melted the chocolate.
- The layer of chocolate on the mold was too thin.
- The chocolate had not completely set and released itself.

Points to Remember

If a mold gets too cold before the unmolding, allow it to warm to room temperature before unmolding to prevent condensation from forming on the pralines. After using a mold, wipe it with cotton to remove fingerprints and small chocolate fragments. If they are to be stored, molds should be washed with a very mild soap and warm water, and then dried completely.

Molding Chocolate Pralines

Molded chocolates are some of the most popular "candies" eaten around the world.

To make molded chocolate pralines, follow these steps:

1 Clean the mold with soft cotton balls.

2 Pour chocolate into the mold.

3 Tap mold with a mallet to release air bubbles.

4 Turn the mold over to allow the chocolate to drip out.

5 Scrape excess chocolate from the mold.

6 Drain upside down until the chocolate begins to set.

7 Turn the mold over and scrape to clean the shells.

8 Fill the molds. It is best to use a filling with a fluid consistency at a temperature between 80°F (27°C) and 85°F (29°C).

9 Pour chocolate onto the mold and spread.

10 Use a spatula to scrape off the excess chocolate.

11 Remove the chocolate pralines from the mold.

Chocolate that is too cold and thick should not be used for molding as it will produce a chocolate shell that is too thick. When air bubbles are present in the chocolate they may cause holes in the shell as well. In addition, the base of the praline will fail to adhere to the shell and the filling will leak or dry out. Ultimately the praline may not easily release from the mold and may have less shine.

Storage and Shelf Life of Pralines

The storage of pralines is basically the same as for all chocolates, although the shelf life will vary depending upon the filling. Moisture, excessive heat, and storage will all determine the extent to which pralines retain their original quality.

Chocolate Showpieces and Decorations

There are a variety of methods that exist to transform chocolate into showpieces and decorations. Consideration must be given to both the type of chocolate that will be used and the procedure to be employed. Tempered chocolate, molded chocolate, piping chocolate, modeling chocolate, compound chocolate, gianduja, and colored cocoa butter are frequently used for showpieces.

Care must also be taken in the handling of chocolate used in this type of work. Fingerprints are a hazard that can ruin the final appearance of a showpiece. It is for this reason that pastry chefs frequently wear rubber gloves to protect the integrity of the product. Body temperature easily melts chocolate, so pieces should not be held or touched for long periods of time because they will melt, or display fat bloom. As with any pastry work, chocolate showpieces are more successful when they are created in a clean and organized environment.

Tempered Chocolate

Tempered chocolate can be spread, cut, and then "glued" with piping chocolate, or molded. The following procedure is used for spreading chocolate:

1. For maximum fluidity, use tempered couverture that is at maximum working temperature and free of lumps.
2. Pour the couverture onto plastic, foil, parchment paper, or butcher paper that is placed on a smooth, movable surface (e.g., a level board).
3. Spread the chocolate evenly and quickly, using an offset spatula.
4. Knock on the surface that is used to remove air bubbles.
5. Run a yardstick back and forth underneath the paper, plastic, or foil to even out the surface of the chocolate, and to help remove air bubbles.
6. Slide the paper or plastic onto another smooth surface so that the chocolate is able to set more evenly. The original pouring surface will retain some of the warmth from the chocolate.
7. Once the wet, melted appearance of the chocolate has disappeared, the chocolate is ready for cutting.
8. When cutting chocolate, use an x-acto knife with a very sharp blade, or other tools, such as a slicing knife or cookie cutter.
9. As the chocolate continues to set, release it from the paper so that it does not warp. Setting chocolate shrinks and pulls the edges of the spreading surface, causing chocolate to warp. To prevent warping, the chocolate must not be allowed to shrink freely. Release it by running a spatula between the paper and the chocolate. If possible, evenly weigh down the entire surface.

    NOTE: Chocolate that is spread onto a plastic surface must not be released until fully set. This is done by flipping the spread chocolate piece over and slowly peeling off the plastic surface to ensure a maximum shine on the chocolate.

STEP PROCESS

Using a Transfer Sheet

Transfer sheets are acetate sheets or strips with designs made of cocoa butter and coloring. Chocolate is spread onto the acetate and the designs and colors transfer onto the chocolate when the sheet is removed. When making pralines, small transfer sheets can be placed on top of the chocolate and then removed.

To use a transfer sheet, follow these steps:

1 Spread chocolate onto the transfer sheet over a piece of parchment paper.

2 When the chocolate begins to set, lift the transfer sheet away from the chocolate.

3 The strip of chocolate can be left flat, wrapped around an object to make a particular shape or base, or wrapped around a cake for a dramatic effect.

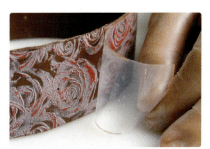

4 Remove the acetate and the design will transfer onto the chocolate.

5 Transfer sheets can be cut and pressed onto chocolate as well.

---

10. Allow the cut pieces to set completely on a very flat surface to give them their full strength and to make them less sensitive to fingerprints.
11. Glue the pieces together with either tempered chocolate or piping chocolate. Always use piping chocolate for large, vertically glued pieces.

For chocolate pieces that are not flat but have a bent or curved shape, cut the chocolate slightly sooner while it is still pliable and not fully set. Bend the pieces to the desired shape with the parchment paper or plastic still attached. Remove the plastic or parchment paper once the chocolate is fully set. A piece of plastic or transfer sheet can be cut to the desired shape and placed on a piece of parchment paper, and the chocolate can be spread over the sheet. Peel the plastic piece off the parchment paper, and place the chocolate on a flat surface until it sets slightly. Just before it is set, wrap or shape a chocolate strip around an appropriate mold. Once the chocolate has set completely, the plastic can be peeled away.

Molded Chocolate

Large or small molds can be implemented into a chocolate showpiece. The method used is the same as that used for filling a praline mold. Commonly used molds are plastic or metal (such as bowls and cups). Any item that has a smooth, polished surface and has a

STEP PROCESS

Molding Chocolate

Though molds are made from various materials, hard plastic molds are usually the best to use though they are also the most expensive. Any food safe item that has a smooth, polished surface and has a suitable shape can be used for molding chocolate.

To mold chocolate, follow these steps:

1 Prepare a clean mold by wiping it with soft cotton balls. Brush the mold with tempered chocolate.

2 Snap the mold together and pour chocolate into the mold.

3 Invert the mold and drain the chocolate.

4 Pour some chocolate onto parchment paper and place the mold on top to form a base.

5 Unmold the chocolate when set.

suitable shape can be used for molding chocolate showpieces, although glass or ceramics will not work because chocolate adheres to them.

To give the molded item a thicker, stronger shell, repeat the molding procedure several times until the desired thickness is achieved. A mold can also be filled with chocolate and then let to set until a layer of chocolate along the mold has formed. The mold is then inverted and the excess melted chocolate is drained.

Piping Chocolate

Piping chocolate is a combination of tempered chocolate and a small amount of water or distilled spirits. The addition of liquid thickens the chocolate and makes it more suitable and easier to pipe. To prepare piping chocolate, bring the tempered chocolate to its maximum working temperature, add a few drops of liquid and stir until the correct consistency is reached by adding more drops, if necessary. Check the temperature again, and make sure that the chocolate is kept at maximum working temperature. Piping chocolate can be used immediately and should be treated as tempered chocolate. If piping chocolate is not tempered or becomes untempered during use, it will continue to spread out during the piping and result in fat bloom. It will also produce a very weak structure when set, and will lose details. Piping chocolate for showpieces is used in three different ways:

1. It is piped onto plastic or paper, using a template as a guide, and then released before it is fully set to prevent warping. After it is set, the pieces are glued together with tempered chocolate or more piping chocolate.
2. The piping chocolate can be piped directly onto prepared pieces for decoration.
3. Piping chocolate is also used to glue chocolate pieces together.

Modeling Chocolate (Chocolate Paste)

Modeling chocolate is a combination of chocolate and glucose. It is sometimes referred to as chocolate paste or plastic chocolate. Because of its glucose content, it is much more flexible and pliable than tempered chocolate. The ratio of ingredients needed vary according to the type of chocolate that is used and the desired texture. The basic formula for modeling chocolate is:

Dark 2 parts chocolate to 1 part glucose.
Milk 2½ parts chocolate to 1 part glucose.
White 3 parts chocolate to 1 part glucose.

For faster drying modeling chocolate, 10% of the glucose can be replaced with a simple syrup.

To prepare modeling chocolate, temper the chocolate and bring it to maximum temperature. Heat the glucose to the same temperature as the chocolate. Combine the chocolate and the glucose using a rubber spatula and spread the mixture 1 inch thick onto parchment paper. Cover it tightly with plastic wrap so that no moisture collects on the surface and the chocolate does not dry out. Let the chocolate rest in the refrigerator overnight. After refrigeration, the modeling chocolate is ready to use.

When using modeling chocolate, work with a small amount at a time to prevent it from setting and drying out. Knead the chocolate paste in both hands and on the marble table until it has reached the consistency of a plastic, pliable paste. Take care that the heat of the hands does not warm the paste too much to cause it to become sticky or cause fat separation. After kneading, the modeling chocolate can be rolled out to a desired width and thickness using sifted cocoa powder for dark chocolate or confectionery sugar for white chocolate. Rub the surface with the palm of the hand to create a nice shine and to remove some of the cocoa powder or confectionery sugar. The pieces can now be cut and shaped as desired. Let the pieces set and dry, then glue them together with tempered chocolate. Modeling chocolate can also be used in the same manner as marzipan to model figures such as roses or animals.

Remember these points when working with modeling chocolate:

- If the modeling chocolate is overheated or overmixed, the fat will separate, and the mixture will become crumbly.
- If the modeling chocolate is undermixed, the paste will have a grainy texture and will appear dull.
- If the mixture separates, dark or milk modeling chocolate can be fixed by adding corn syrup or simple syrup and cocoa powder or confectionery sugar.
- The strength of the modeling chocolate comes from both the setting and the drying of the chocolate. It sometimes takes a week or more for modeling chocolate to dry to a sufficient strength to hold its shape.
- For large showpieces, increase the amount of chocolate in the formula. For very small items, add a bit more glucose or simple sugar.

Confectioner's Coating (Compound Chocolate)

Compound chocolate may be used in the same way as tempered chocolate, but because it has very little or no cocoa butter, it sets faster and does not allow as much time for cutting as does tempered chocolate. Compound chocolate is available at a relatively low cost, but cannot be used to make strong piping chocolate. White compound, in contrast, may sometimes be used to make modeling chocolate.

When working with compound chocolate it is important to remember that although it sets faster than real chocolate, it is not always as strong. Compound chocolate also shrinks as it sets, so it should be released in the same manner as real chocolate. Compound chocolates, like real chocolates, vary from manufacturer to manufacturer so it is important to use the manufacturer's temperature guidelines. Most compound chocolates should not be heated over 105°F (41°C). Failure to follow prescribed guidelines may result in an overheated chocolate with a dull appearance or fat bloom.

Chocolate Carving

Creating a chocolate carving is one of the most difficult of all procedures in making a showpiece. Tempered chocolate, chocolate coating, and gianduja are usually used as a base. Most tempered chocolate is too hard for carving, so shortening or oil must be added to the tempered chocolate. Chocolate used for coating is softer and does not need to be tempered. It can be used in the same way as tempered chocolate without worrying about fingerprint damage or fat bloom. Gianduja is very soft and may be easily carved.

When carving chocolate, pour the base into a shape similar to the desired final product. Use carving tools of the appropriate size, begin by carving a rough outline of the piece, and then carve the details.

Cocoa Painting

When painting with cocoa, a mixture of melted cocoa butter and sifted cocoa is used for the "paint." Shortening can be used in place of cocoa butter if cocoa butter is not available. Imitation cocoa painting can be done with brown food coloring that is mixed with distilled spirits. The three "canvas" bases that can be used are marzipan, pastillage, and rolled fondant. Marzipan should be dried before it is painted.

When cocoa painting, outline the painting with a pencil, and then paint directly on the "canvas." Because everything is done with only one pigment color, there should be preparation of different shades of cocoa powder mixture. As more cocoa powder is added to the cocoa butter, the shade will become darker. Use light shades first, then move to the darker shades. Cocoa painting is never used alone in a showpiece. It can be used on a torte, in a frame, or as part of another showpiece.

Postproduction

After the showpiece or individual pieces are made, they may be left as they are, or sprayed or brushed over with a mixture of tempered chocolate couverture and extra tempered cocoa butter to achieve a finer finish. The mixture should contain at least 50% cocoa butter. The ratio of tempered couverture to the tempered cocoa butter can vary according to the grain and surface texture desired. Generally, for a fairly coarse appearance, the ratio is two parts tempered couverture to one part cocoa butter. For a finer texture, the mixture is a one-to-one ratio of couverture to cocoa butter. For a satin, velvet-type finish, allow the piece to cool in the freezer for approximately 30 minutes, then remove and spray the piece while it is still cold.

The tempered chocolate mixture can be sprayed over the showpiece using an electric spray gun or a spray gun attached to a compressor to cover fingerprints, small scratches, and imperfections. The chocolate must be strained through a nylon stocking before it is sprayed. The spray gun should be warm and the room and the chocolate must be at the proper working temperatures. Showpieces can also be brushed with the same mixture, using different brushes or paint rollers.

Coloring Chocolate

Liquid chocolate is colored using oil-based food colorings or colored cocoa butter. A dull set chocolate piece is achieved by airbrushing it with a water-based food color. A shiny chocolate piece is sprayed or airbrushed and has tempered cocoa butter applied over it.

Storage

When completed, chocolate showpieces are stored in the same manner as other chocolate products. The showpiece is kept in a cool, dry place, and if not on display, away from light, and well wrapped to prevent it from discoloring and getting dusty.

STEP PROCESS

Making a Chocolate Box

Chocolate can be transformed into stunning showpieces and centerpieces. The steps to creating a showpiece must be well conceived before beginning to work on it.

To make a chocolate box, follow these steps:

1 Spread tempered chocolate horizontally onto parchment paper.

2 Spread vertically and then horizontally again until you have a smooth, thin sheet of chocolate.

3 For a marbled effect, swirl light and dark chocolate together and pour and spread the chocolate in a similar manner.

4 A textured mat can also be used.

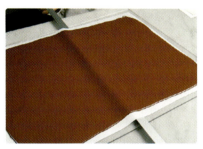

5 Spread the sheet to the desired thickness and then loosen the sheet with a metal ruler to prevent warping.

6 Use a ruler and knife to square off the sheet and then mark the chocolate sheet with a template.

7 Using the ruler as a guide, use a knife dipped in hot water to begin cutting the chocolate to conform to the template.

8 Trim away the outside edges.

9 Continue cutting chocolate using the template and ruler as a guide.

10 Pipe chocolate filigree ornaments.

11 Assemble the chocolate box using paper tabs to transfer and handle the cut chocolate pieces.

12 Use tempered chocolate to "glue" the pieces together.

13 Mark the lid of the box to help place the ornaments.

14 Pipe the chocolate "glue" onto the marks.

15 Add the ornaments.

16 Pipe chocolate along the ornaments to hold them together.

17 Attach the ornaments to complete the chocolate box.

18 A finished chocolate box.

CHAPTER 28
Pastillage, Sugar Artistry & Marzipan

Artistic creations with pastillage, sugar, and marzipan rely on the mastery of a variety of techniques and an in-depth knowledge of the handling of the necessary mediums. Intense concentration, dexterity, creativity, discipline, and practice are all skills required to produce the showpiece that is ultimately referred to as the "showstopper" or the "pièce de résistance."

KEY TERMS

showpiece/centerpiece
pastillage
flow icing
fondant
rolled fondant

crystallization
caramelization
titanium dioxide
calcium carbonate

pulled sugar
blown sugar
invert
Isomalt

silica gel
quick lime
calcium chloride
marzipan

An Introduction to Showpieces/Centerpieces

A *showpiece/centerpiece* is an object designed to attract attention, whether displayed on a table, on a platter, or used as a competition piece in and of itself. It brings harmony and beauty to its location and is pleasing to the eye.

There are numerous considerations to keep in mind when producing showpieces/centerpieces. Theme, color, contrast, flow, balance, symmetry, shape, size, and medium are all very important. Most pastry chefs hone their abilities for many years before becoming truly proficient in creating showpieces and constantly practice to develop the skills they needed to create beautiful works of art. Progressive pastry chefs also indulge in traditional art classes, airbrushing, pottery, glass blowing, and architecture courses to gain fresh and innovative perspectives. They visit art galleries and museums for exposure to a multitude of classical and modern forms of expression not typically found in the pastry field. To gain "an edge," a pastry chef must be highly resourceful and observant of nature and the surrounding environment. Steps for creating a showpiece must be well conceived before beginning a project.

A focus on the theme and the main subject of the showpiece is the first consideration. Deciding on the materials (mediums) that will be used is the next step. Have a basic knowledge of the materials (mediums) which can be used, and choose those with which you are most comfortable. Examples of mediums used in pastry arts include: pastillage, icings, fondant, rolled fondant, poured sugar, and marzipan. Although pastry chefs use mediums (materials) of a nonpermanent nature, rather than permanent materials such as acrylics, clay, and glass, they are still able to render a variety of creative and artistic expressions.

A basic drawing of the showpiece to provide an illustration of the finished pieces is always necessary. Consider the structural, visual, and color balance of all the components. Draw templates of all the parts needed for each element of the showpiece. A simply made cardboard or foam board model is an excellent guide to obtain the proper composition.

Lastly, consider the most appropriate environment in which the piece should be created. The work area must be a calm, quiet environment with ample working space. Adequate time for finishing the showpiece without rushing should also be considered. Working fast and efficiently may take many hours of practice.

Pastillage

Pastillage comes from the French word *pastille*, which is a small, usually peppermint-flavored sugar candy. Pastillage, the mixture used to make pastille, consists of confectionery sugar, cornstarch, cream of tartar, cold water, and gelatin. Confectionery sugar is the main ingredient that gives body to the mixture. Cornstarch is used to make pastillage stronger when it dries and also serves to expedite the drying time. Cream of tartar is an acid that helps to make the mixture smoother, and serves to slow the granulation of the sugar to prevent graininess. It also bleaches the mixture and makes the pastillage whiter. Cold water hydrates the gelatin and causes the gelatin granules to swell to facilitate melting. A small amount of glucose or shortening may also be added to slow down the drying of the pastillage.

Figure 28-1 Pastillage tools.

When preparing pastillage, do not mix it at a high speed or for a long period of time because too much air will become incorporated and will create air pockets. Add hot water to adjust the consistency, if necessary. For smaller showpieces use softer pastillage, and for larger pieces, select one that is stiffer. It is best to use pastillage immediately. Pastillage that is too cold will be rubbery and difficult to work with. Cold pastillage pieces will also shrink and lose their shape when cut. Pastillage must be softened and warmed before use. It can be stored several weeks in a refrigerator if well wrapped before it is shaped, but it must first be resoftened before it is used. Because pastillage is white, the work area must be extremely clean. *See Figure 28-1.*

Procedure for Rolling, Cutting, Sanding, and Gluing Pastillage

The following steps are necessary in creating a successful pastillage centerpiece:

- Lightly dust a very clean surface with sifted cornstarch and/or confectionery sugar. Use small amounts of product to prevent the pastillage from absorbing the sugar or starch that will cause it to dry prematurely.
- Soften the pastillage to body temperature before rolling, by kneading it or placing it in a microwave for a few seconds.
- Roll the pastillage to an even, desired thickness (the thinner the better).
- Cut the rolled-out pastillage into pieces using a mat knife, x-acto knife, or scalpel. Guide the cuts by using templates. Straight cuts can also be made with a thin, nonserrated slicing knife. After each cut, the blade should be dipped in pure grain alcohol to prevent the pastillage from sticking to the cutting device.
- Cut and shape quickly to prevent the pastillage from drying.
- Transfer the cut pieces to a flat surface. Styrofoam is an excellent surface for drying pastillage. Other materials such as glass, Plexiglass, cork, wood, and plastic sheet pans are also acceptable. Do not use paper because it will absorb the moisture in the pastillage and cause wrinkling.
- Dry the pieces at room temperature, carefully inverting the pieces periodically so that they dry evenly and do not warp. As the pastillage dries, it will whiten and cause undried sections to appear off-white or somewhat yellowish.
- When the pieces are completely dry, cautiously sand the surface and rough edges of each piece with a fine grit sandpaper.
- Brush the sanded pieces carefully with a clean, dry brush to remove any sugar dust that might prevent the finished pieces from gluing together properly.
- Glue the pieces together with royal icing or with pastillage that has been melted. If the icing is too stiff, it will dry too quickly, and there will not be enough time to glue the pieces. Icing that is too soft will run and fail to provide enough structure to hold the pieces together.

STEP PROCESS

Making a Pastillage Showpiece

Because pastillage is white, make sure the surface is very clean before you begin to work. The work surface is dusted lightly with a small amount of confectionery sugar or cornstarch to avoid sticking. If too much sugar or cornstarch is used, the pastillage may dry prematurely and become difficult to work with.

To make a pastillage grand piano showpiece, follow these steps:

1 Soften the pastillage to body temperature before rolling. "Wedging" helps to smooth the pastillage and make it easier to work with.

2 Roll pastillage to desired thickness.

3 Use a template to cut shapes.

4 Small shapes and intricate cuts are best cut with a small, sharp knife.

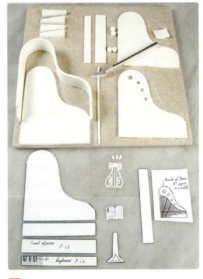

5 Once the shapes are cut from templates, allow the pastillage to dry.

6 Carefully sand the rough edges of each piece, using fine grit sandpaper. Use a clean, dry brush to remove any sugar dust.

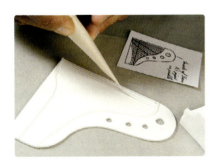

7 Royal icing is used to pipe decorations and to "glue" pieces together.

8 Glue all the pieces together to complete the showpiece.

The Coloring of Pastillage

Pastillage dough can be colored using water-based food colorings. The dried pieces may be painted or airbrushed. Because of its smooth surface, pastillage is a great medium for coloring with an airbrush or for painting with a brush.

Other Uses of Pastillage

Pastillage can also be used to make decorations for special-occasion cakes. These ornaments are commonly rolled out thinly and cut into crescents, hearts, and other shapes that adorn the celebration cake. To create three-dimensional pieces, the pastillage is rolled out, pressed into the mold, and allowed to dry undisturbed. Pastillage can also be heated to a liquid with a royal icing–like consistency, and poured into silicon molds. Keep in mind that pastillage dries extremely fast, so it is important to work quickly and with small quantities.

Storage of Pastillage Showpieces

If stored properly, a pastillage showpiece can have a shelf life of several years. After the finished pieces have dried, wrap them well and store them covered at room temperature in a dry area to protect them from light and to prevent discoloration. Store the finished pastillage showpiece in a place with low humidity to prevent the pieces from softening.

Icings

Icing is a sweet, cooked or uncooked mixture that may be employed for a variety of purposes. Sometimes referred to as frosting, icing is used to cover cakes and decorate chocolates, to create ornaments and decorations for showpieces, or as glue for assembling showpieces.

Royal Icing

Royal icing is a mixture of pasteurized egg whites, confectionery sugar, and cream of tartar. It is used in much the same way as piping chocolate. It frequently serves as "glue" for assembling pastillage showpieces, and is also used to make piped ornaments, using a template as a guideline. When the pieces are dry, they are glued together to create a showpiece/ornament. Royal icing should be piped onto plastic or foil that has been treated with a small amount of shortening, or onto waxed paper. Do not pipe royal icing on parchment paper because the paper will curl due to the moisture in the icing. Royal icing may also be piped as a decoration directly on prepared pieces such as chocolate and sugar to highlight the work.

When preparing royal icing be careful not to overmix because it will become stiff and airy, and difficult to pipe. Overmixing also weakens the structure of the icing once it is dry. If the icing is undermixed, it will be runny and off-white in color. A very small amount of blue food coloring may be added to cover the yellowish tint of the icing and to allow it to appear white.

The egg whites used for the icing should be used at room temperature of 72°F (22°C) to ensure that the viscosity of the finished icing does not change during the piping. If the icing is left out for 2 or more hours, it should be lightly rewhipped before it is used to give it more body.

Storage of Royal Icing

While working with the icing, place it in a small stainless steel or plastic container, and cover the container with a damp towel and plastic wrap. Make sure that the moist towel does not come in contact with the icing as it may add unwanted moisture and prevent a sugar crust from forming on the surface, making it grainy and difficult to pipe. When storing the icing, place it in a plastic bag and seal the bag, pushing out any air that remains in the bag. The finished ornaments made with royal icing are best stored in the same manner as are pastillage showpieces.

Flow Icing (Run-out Icing)

Flow icing is royal icing with extra egg whites added to bring the icing to a flowing consistency. Flow icing should be made the day before it is used to allow adequate time for the air to escape, so that the piece is not covered with unattractive air bubbles. Flow icing is used to create ornaments and decorations for showpieces and to decorate cakes and small chocolates.

Procedure for Flow Icing

When preparing flow icing, work on plastic sheets or wax paper. Do not use parchment paper because it will absorb too much moisture and cause the pieces to wrinkle. Place the paper template of the design to be duplicated under the plastic. Pipe an outline of each differently colored segment of the design with royal icing, and allow the outline to dry completely. Fill the outlined sections with colored or white flow icing. Allow each section to dry briefly before continuing to pipe the next section. If the different sections are piped too quickly the icing will run together and cause the colors to bleed into one another. Let the pieces with the most food coloring dry completely before filling the adjacent section. Once the pieces are dry, glue them together with royal icing for use as components of a cake or showpiece.

Fondant

Fondant is a cream of tartar mixture containing gelatin, glucose, glycerin, and shortening. Gelatin is a binding ingredient, while glucose serves as a softening agent that allows the rolled fondant to stay soft for an extended period of time so that it is easier to cut. Glycerin and shortening make the mixture smoother, more pliable, and slightly shiny, and are important ingredients that prevent the fondant from cracking when it is rolled out.

Rolled Fondant

Rolled fondant is used to cover both edible and dummy cakes. It creates an extremely smooth surface that allows for the application of very fine decorations or brushwork. Rolled fondant, also known as Australian fondant, may be used as a modeling paste for small showpieces. Premade rolled fondant of excellent quality can be purchased to shorten the production time of elaborate specialty cakes. Fondant can be flavored, colored, or painted when dry.

Procedure for Covering a Cake with Rolled Fondant

A dummy cake may be masked with a thin layer of piping gel or apricot glaze to provide stickiness. Edible cakes, on the other hand, should be covered with a stiff icing or with jam or jelly that has a flavor that complements the flavor of the cake. Dense cakes such as pound cake are best suited to hold the weight of rolled fondant.

- When covering a cake with rolled fondant, first soften the fondant to body temperature by kneading it lightly or placing it in a microwave for a few seconds.

- Dust the table periodically with a 1:1 ratio of sifted cornstarch and confectionery sugar. Excessive dusting will dry out the dough.
- Roll to a ¼-inch (.6 cm) thickness, moving the rolled fondant constantly to prevent sticking, without turning the surface over.
- Rub the surface with the palm of the hand, and gently lift the fondant off the table.
- Lay the rolled fondant over the cake from one side to the other, slowly working across the cake.
- Gently rub the top surface of the cake from the center outwards, pushing out all of the air pockets trapped between the cake and the fondant.
- Using the palm of both hands, rub the fondant on the side of the cake, working from the top edge down toward the table. Slowly remove any folds and creases that have formed.
- Cut the excess, leaving approximately ½ inch (1 cm) of fondant attached to the cake.
- Lift the cake with one hand and use the other hand to rub the sides so that the excess fondant hangs vertically below the cake. Use a sharp paring knife to cut the excess fondant horizontally at the base level of the cake.
- Smooth the surface, using the palm of the hand and a plastic scraper. Work on a turntable for best results.

Sugar Artistry

All sugar candies were at one time made by hand. Large batches of sugar were boiled, colored, flavored, pulled, and shaped to form individual sweets. In time, candy makers began to experiment with more elaborate ways of shaping sugar, and creations such as candy canes and ribbon candy were born. The art of glassblowing actually served to inspire some confectioners to blow sugar in a similar manner, creating the art of sugar artistry.

Sugar artistry is divided into three categories: poured sugar, pulled sugar, and blown sugar. All of these forms of sugar work share a common cooked sugar syrup that is made of the same basic ingredients.

Sugar Equipment

Before attempting to master the art of sugar work, it is of the utmost importance to have a good understanding of the equipment used, the main ingredients needed, the cooking procedure of the sugar solution, and the behavior of the sugar and water in the syrup while it is boiling.

Some of the equipment necessary for the preparation of sugar work includes: airbrushes, cooking pots, metal bars, paintbrushes, sugar pumps, and stamps.

AIRBRUSH—used to apply various shades of colors or detailed brushwork onto the surface of the set sugar. A more realistic rendering of the artwork can be achieved using an airbrush.

COOKING POT—should be made of stainless steel or copper and is used to cook the sugar syrup. It should be of high quality and preferably only used for sugar artistry.

METAL BARS—can be arranged to form frames for a variety of geometrical shapes into which cooked sugar can be poured.

PAINTBRUSHES—used for detailed brushwork on the surface of set sugar. Quality natural brushes such as camel hair brushes can be purchased at a moderate price.

RUBBER MOLDS—most often made of silicon, these molds provide quick and efficient reproduction of detailed sugar pieces. They can be used to give the sugar a desired shape, and also to texture the surface of the sugar. These types of molds are quite expensive and should only be purchased if repeated use is intended.

Sugar Pump—a necessity for blowing sugar. This pump, made of rubber, is used to inflate warm, pliable sugar pieces with air to produce bigger, fuller, and three-dimensional shapes.

Texturing Molds (stamps)—usually made of silicon, plastic, or metal. Soft sugar is pressed onto molds to obtain a desired surface texture, such as, the veins on a leaf or the petals of a flower.

Ingredients for Poured Sugar

A granulated sugar of a high quality that is well refined, has a low mineral content, and contains no impurities should be used. Sugar that is poorly refined and is not clean will turn the solution off-white (yellow) and possibly crystallize. The water must also be very clean and have a low mineral content. Impurities and minerals can turn the sugar off-white (yellow) and may crystallize it. Use distilled water if tap water is not suitable. It is important to make sure that the sugar is properly dissolved, because undissolved sugar crystals that remain in the solution will cause the dissolved sugar to recrystallize during the cooking process.

To help reduce the chance of early *crystallization*, glucose is then added to the solution. During crystallization, the liquid sugar syrup forms large, visible crystals and turns grainy, dull, and eventually solid and unusable. Crystallization may occur during the cooking of the sugar, while working with it, or after it has cooled and set.

Mise en Place for Cooking Sugar

Before cooking sugar syrup for a showpiece, many items must be gathered, organized, and prepared. These items include pots, thermometers, spoons, spatulas, a pastry brush, a mesh skimmer, and a tea strainer.

Sugar cooks best in a copper or stainless steel pot. Copper conducts heat faster and more evenly than other metals, and guarantees fast and even cooking. Copper does oxidize and must be cleaned with soapy water, vinegar, or lemon juice, and rinsed to prevent oxidation that can discolor the sugar. A stainless steel pot used on an induction burner provides an even heat source and is another good choice for cooking sugar.

Because the temperature of the sugar syrup is so important, a reliable thermometer is invaluable. Mercury thermometers are the most accurate instruments to use when cooking sugar syrups, but because of their potential hazards, good-quality digital thermometers are recommended. The thermometer probe should never touch the sides and the bottom of the pot. Alcohol (blue liquid) thermometers are not recommended because they fail to provide accurate readings above 300°F (149°C). A clean, metal spoon or rubber spatula should be used to stir the sugar at the beginning of the cooking stage, just before the sugar begins to boil.

A container filled with clean distilled water and a clean pastry brush with natural bristles for brushing down the sides of the pan should also be kept nearby. The brush should have natural bristles because synthetic bristles can melt or dislodge. A clean, fine-mesh skimmer to skim off any impurities that rise to the surface before the boiling point is reached is another important tool.

Have a large bowl that is at least twice the size of the sugar pot filled with cold water near the stove to shock the sugar when it has reached the proper temperature. A 1-to 2-quart container filled with hot water is needed to store the thermometer while it is not being used in the sugar syrup. To skim impurities from the surface of the sugar syrup during cooking a small tea strainer is invaluable. The tea strainer should be kept in a small container that is filled with hot distilled water.

Poured, pulled, and blown sugar solutions are hygroscopic and will absorb moisture from the air that causes the sugar to dissolve. A cooked sugar solution that is kept

Cooking Procedure for a Sugar Syrup

- Clean the pot with an acid (lemon or vinegar), salt, and hot water to remove oxidation and other impurities that might discolor or crystallize the sugar. Stainless steel pots should be cleaned with soapy water and well rinsed. Do not dry pots using a cloth towel because fibers that shed can promote the crystallization of the sugar syrup.
- Put water in the pot and then add the sugar to avoid compression of the sugar. Stir with a clean spoon or rubber spatula.
- Using a low flame, stir the solution to dissolve the sugar until it begins to boil. Sugar that is not stirred with the water before coming to a boil will sink to the bottom of the pot, caramelize, and turn the batch of sugar yellow or cause it to crystallize.
- When a sugar solution crystallizes, it forms large, visible crystals that turn the syrup grainy, dull, and eventually solid. Crystallization may occur during the cooking process, while the sugar is worked, or after it has set and is cooled. Crystallized sugar is not sensitive to moisture.
- Before the sugar comes to a boil, use a fine-mesh skimmer or tea strainer to remove the white scum that forms on the surface to prevent the sugar from turning more yellow and increasing its chance of crystallizing.
- Just after the sugar has boiled, add the glucose. Skim again before the sugar comes to a second boil to remove the remaining impurities. At this point, if not used immediately, the sugar syrup can be stored in an airtight container overnight.
- To obtain an accurate reading of a cooked sugar solution, suspend a thermometer in the center of the pot. When using a digital thermometer never allow it to touch the sides or the bottom of the pot because it will register an inaccurate reading.
- Increase the heat to the highest setting possible, making sure that the flames do not shoot along the side of the pot if a gas range is used. Cook the sugar as quickly as possible on high heat. Do not cook too much sugar at one time, and do not interrupt the cooking process. The faster the sugar reaches its final temperature, the whiter it will be. If the sugar is cooked too slowly, it will turn yellowish and will have a greater chance of crystallizing.
- Occasionally brush down the sides of the pot and the thermometer with a clean brush that has been dipped in water. Sugar that collects along the sides of the pot may burn, discolor, or crystallize the batch. Sugar on the sides of the pot may also be removed by covering the pot with plastic wrap or with a lid to allow the trapped steam to wash down the crystals. This method allows steam to collect and slows down the cooking of the sugar causing the particles to fall back into the syrup.
- When the sugar is nearing the required temperature (approximately 10° lower than the desired temperature), stir it lightly with the thermometer to more evenly distribute the heat and to provide a more accurate reading.
- Cook the sugar until it reaches its desired temperature between 305°F /152°C (poured sugar), and 320°F/160°C (pulled and blown sugar). Immediately remove the thermometer, and place it in hot water. Cooked to a higher temperature, sugar will be firmer and less sensitive to moisture, but more yellowish (off-white) in appearance. When cooked to a lower temperature the sugar will appear whiter, but softer in texture and more sensitive to moisture. Sugar and other substances that have the capacity to absorb moisture from the atmosphere are considered to be hygroscopic.

Sugar that is left to cook until it reaches 335°F (168°C) will caramelize. Caramelization *occurs when the sugar begins to break down and create a rich brown color and bittersweet taste. If heated at an even higher temperature, the sugar will begin to carbonize and turn black.*

- To prevent the temperature of the sugar from rising any further, shock the sugar by dipping the bottom of the pot in a bowl of cold water for a few seconds. Sugar that has not been shocked will turn yellowish because of the carryover cooking (the residual internal heat of the product that continues to cook it slightly). Reheat the sugar to maximum fluidity without bringing it to a full boil. Sugar that is too cold will be thick and difficult to pour. Bubbles may also appear and remain on the surface of the set sugar because of the thick viscosity.
- The sugar is now ready to be colored and poured. Use the cooked sugar as soon as possible so that it does not turn yellowish and crystallize.

Figure 28-2

Skimming helps remove the white scum that forms on the surface before the sugar comes to a boil.

Brushing the sides of the pot with water helps prevent sugar from collecting on the sides of the pot.

Dipping the bottom of the pot in cold water prevents the temperature of the cooked sugar from rising.

warm for a long period of time or reheated too many times will be more hygroscopic once set. *See Figure 28-2.*

Boiling Point and Sugar Concentration

When sugar is dissolved in water, it raises the boiling point of the solution above that of water (boiling water is 212°F/100°C). The magnitude of the change depends on the amount of sugar dissolved in the water: the greater the amount of sugar, the higher the boiling point. The boiling point of a sugar solution is an indirect indication of the amount (concentration) of dissolved sugar it contains. A sugar solution that is cooked to a high temperature contains a greater quantity of sugar and very little water; one that is cooked to a lower temperature contains more water.

When cooking a solution of water and sugar, the water gradually evaporates, while the sugar remains. The syrup becomes more and more concentrated (with sugar) as the water boils off, causing the temperature of the solution to rise. The greater the amount of water a cooked sugar solution contains, the softer it is when cooled, and the more sensitive it is to moisture. Conversely, a solution that is cooked to a high temperature contains very little or no moisture and is firmer when cooled and less sensitive to moisture.

Coloring the Sugar

Sugar may be colored at about 20°F (-7°C) (before the desired temperature is reached), to allow the color to cook itself into the mixture. Color may also be added after the sugar is shocked and before it is poured. Sugar that is used for pulling or blowing can be colored after it is poured to allow for the production of multiple colors from one batch. The addition of acids such as tartaric acid or lemon juice provides the sugar with additional pliability. When using cream of tartar, it must be added at the beginning of the cooking process to ensure that it is properly dissolved. Cane sugar that is pulled or blown requires the addition of an acid for more pliability.

Use paste, liquid, or powdered food colors to color sugar. Paste or powdered food colors are more concentrated and will not change the consistency of the cooked sugar. Powdered food colors must be diluted in water or alcohol prior to their addition to the solution. Keep the colors clean and as new as possible to prevent the sugar from crystallizing.

To make a black sugar, start with a clear, cooked solution that does not add white color (white and black produce grey). Most colors (secondary colors) can be made from the three primary colors (red, blue, and yellow).

Use the following chart when mixing colors:

Red + Yellow =	Blue + Yellow =	Red + Blue =	Yellow + Red + Blue =
Orange	Green	Purple	Brown/Black (depending on amount used)

Four additional colors (secondary) can be created by combining the primary colors for a total of seven pigments. When mixing shades, add a small amount of color until the desired shade or color is obtained.

Whitening Agents That Make Cooked Sugar Opaque

Titanium dioxide is one of the most common whitening agents used in the food industry. It can be purchased in a variety of forms such as white-white or milchweiss in liquid form, and added to the sugar solution during or after the cooking procedure. If not fully incorporated, it gives the sugar a very attractive, marbleized effect. No more than four drops of liquid should be added per pound of cooked sugar syrup because the glycerin (a moistening agent found in plants) contained within the agent may cause the sugar pieces to soften and bend.

Calcium carbonate is one of the oldest types of whitening agents used. It is of an organic nature (naturally occurring) and is found in chalk, limestone, and marble. Calcium carbonate is commonly used in most toothpastes, anti-acid medications, and in plasters. This chalk-like powder forms a paste when mixed with a small amount of water. It transforms clear sugar into an opaque sugar. The paste is added to the sugar solution when the mixture reaches a temperature of 260°F (127°C). Use approximately ½ ounce of calcium carbonate for each pound of sugar.

Appropriate Surfaces for Poured Sugar

Appropriate bases for poured sugar include silicon paper, marble slabs, aluminum foil, polyvinylchloride (vinyl), and silicon rubber. Silicon paper is parchment paper that has been treated with silicon. Sugar will not stick to this surface, so it does not need oiling. Marble slabs should be well polished, free of cracks or holes, and slightly oiled. If the sugar is poured directly onto the marble, it should be released from the marble using a lightly oiled scraper, while the sugar is still slightly warm. When sugar is allowed to cool completely on marble a vacuum is created between the sugar and the marble causing the sugar to stick. Aluminum foil that is oiled and then wrinkled may be used to achieve a wrinkled effect.

Polyvinylchloride (vinyl) is another surface that accommodates sugar well. It is available in two forms: a soft and plastic-like form that is commonly used in refrigerators and freezers to keep the cold air in; or a hard variety, similar to that used in PVC pipes. These versatile bases can be used to bend and shape sugar pieces as well. The sugar is bent around the pipe while pliable and warm, and then peeled from the PVC when it has cooled down.

Silicon rubber mats with simple or elaborate textures are yet another choice of pouring base that does not require oiling. Silicon rubber also allows the texturing of both sides of the sugar so that the sugar can be sandwiched between two mats. It is one of the best surfaces upon which to pour sugar and is also the most expensive.

Molds for Poured Sugar

Molds are used for poured sugar to retain a specific shape and uniform thickness. Common mediums for molding sugar include plastiline clay (an artist sculpting clay, usually grey in color), steel bars, metal bands, aluminum foil, cake rings, flan rings, cookie cutters, soft PVC, silicon, and food safe rubber mats such as white neoprene. Materials made of petroleum by-products, which are toxic and can release chemicals or vapors, should not be used.

Plastiline/clean clay (#4) is artificial clay that does not dry out or spoil and can be reused many times. It is carefully softened in the microwave (in order not to melt it), rolled out ¼- to ⅓-inch (.6 cm to 1 cm) thick with the aid of two wooden rulers that serve as a guide, and lightly dusted with cornstarch to keep it from sticking to the rolling surface. It is then cut into desired shapes with cutters, an x-acto knife, or a slicing knife. Plastiline needs to be lined with thin strips of lightly oiled aluminum to keep the hot sugar from sticking to the clay.

Lightly oiled stainless steel bars can be used to form geometrical shapes with straight sides. They are also used in conjunction with lightly oiled cake rings or other materials to create a variety of shapes. Metal bands and heavy-duty aluminum foil strips that are lightly oiled can be easily bent to any desired shape and used to create simple, curved forms.

Lightly oiled cake rings, flan rings (used to make thin European tarts), cookie cutters, and any other enclosed molds are good for forming simple shapes as well. When these forms are used, the sugar must be released from the mold while it is still slightly warm. If the sugar is allowed to cool completely within the mold, the mold will adhere tightly to the sugar causing the sugar to crack.

Soft PVC, silicon, and other types of food safe rubber mats are also practical for making molds. Cut the desired shape with the aid of an x-acto knife or a utility knife, making sure that the blade is kept vertically positioned at all times. Three-dimensional silicon molds of varying sizes, small to very large, can be purchased premade or hand constructed using liquid silicon. Using these types of molds allows for a perfect duplication of an object in sugar. Molds made of silicon are expensive yet very durable.

Gluing Poured Sugar Pieces

The proper gluing of all the components used in a showpiece is extremely important to support the shelf life of the showpiece. One improperly glued or misplaced piece can cause the collapse and disintegration of an entire showpiece.

Poured sugar has to be completely set before it is glued. Sugar that is too warm will not attain its full structure and will bend when glued. Solid sugar pieces can be dipped into cooked, clear sugar, or sugar that is the same color as the piece being glued. The pieces should then be quickly attached one to another. For sticking small pieces together, heat the back of the piece over a clean flame until the surface starts to bubble and slightly melt. Use alcohol, gas flames, or a gas torch to produce a clean flame without soot residue. The more concentrated the flame, the better the control.

Pieces may also be glued together by placing a set piece of sugar onto a piece that is still hot and soft. The pieces must be completely dry and clean for gluing. Pieces that are sticky from the moisture or have a lot of oil on them do not adhere properly. Stickiness can be removed by scraping the sticky area carefully with a small knife. Oil should be wiped away and the area should then be scraped in the same way. Pieces that are cold will crack when exposed to a sudden, intense heat change produced during the gluing. A piece that has become too cold may be slowly warmed to approximately 90°F to 100°F (32°C to 38°C) with the use of a heat lamp, heat gun, or hair dryer. It may also be placed in a warm oven or cabinet for a few minutes.

Pulled Sugar

Pulled sugar is produced by pouring a cooked sugar solution onto a non-sticking baking mat or lightly oiled marble surface and occasionally folding and flipping it until it becomes a cooled, pliable mass with a consistent temperature throughout. Pulling introduces air into the sugar to give it a satiny and opaque appearance. It is necessary to understand that sugar can only be pulled if it has the correct pliable consistency, which is determined by its temperature. If the sugar is too cold, it will be too hard to pull. If it is too hot, it will be soft and will not trap and hold the air needed. Sugar must be kept under a heating unit, such as a heat lamp, to keep it at the right working consistency.

STEP PROCESS

Making a Sugar Showpiece

Once the underwater theme was selected for this showpiece, several mediums such as poured sugar, pulled sugar, and piped sugar were chosen to provide visual interest and enhance the overall appearance. Most of this step process involves working with poured sugar and assemblage. For additional information on making the components of this showpiece and the mediums used, see information in this text and follow other step processes.

When making sugar showpieces, follow these procedures:

1 Make sugar syrup and color it as needed.

2 Place the molds over foil and pour the sugar into the molds.

3 When the sugar has set, remove the mold to release the sugar.

4 Continue pouring sugar to make the components of the showpiece.

5 Score sugar as needed to complete the design.

6 An airbrush can be used to further enhance the design and color.

7 Pour sugar syrup onto molded pieces as "glue" and attach components. Be sure all pieces are fully set before gluing them together.

8 Complete all the components of the showpiece and attach them with warm syrup to complete the design.

9 The finished showpiece.

Figure 28-3 Using a sugar pump to blow sugar.

Once the appropriate consistency is achieved, the sugar is placed on a piece of soft PVC canvas, silk, or copper screen that has been stretched over a wooden frame. The choice of these materials is obvious due to their nonstick quality. The elevated surface also keeps the sugar from cooling on the cold table to ensure a more uniform temperature.

Blown Sugar

Blown sugar is made by blowing air into a piece of pliable pulled sugar, similar to glassblowing, until the desired shape and size are obtained. A sugar pump is commonly used to inject air into the pliable paste, although pastry chefs in the past have used straws and copper pipes to blow the sugar. When the desired form is obtained, the blown sugar must be cooled in front of a fan to keep it from collapsing. This technique requires a lot of practice. *See Figure 28-3.*

Cooking Procedure for Pulled and Blown Sugar

The cooking procedure for the sugar solution used in pulled sugar and blown sugar is the same as that provided for poured sugar with the addition of an acid.

The Function of Ingredients Used in Pulled and Blown Sugar

Glucose is added to pulled sugar to lower its risk of crystallization and to make it more pliable. It is important to remember that the amount of acid found in glucose varies from brand to brand and that glucose becomes more acidic as it ages, so the amount of additional acid needed in a formula varies.

Acid is used to reduce the chance of crystallization and to make the sugar more pliable. The acid, in combination with heat and time, actually *inverts* (breaks down) the sugar to make it softer and more yellowish in appearance. The longer the sugar is exposed to the acid, the more intense the reaction. A minimal amount of inversion also occurs naturally in boiled sugar solutions that do not contain acid or enzymes.

Tartaric acid, citric acid, malic acid, cream of tartar, lemon juice, and vinegar are all commonly used in sugar. Each type of acid has a different pH level and reacts with the sugar in a slightly different manner. Test batches of sugar should be made to determine the exact amount of acid needed.

The amount of acid added to a batch of sugar influences the pulling temperature and the consistency of the mass. If too much acid is added, the sugar feels softer and cooler to the touch when pulled. The pulled pieces also lose their shape shortly after they have set. Although this type of sugar will not crystallize as readily, it will be more hygroscopic and have a very short shelf life. If less acid is added, the sugar will feel hot to the touch, will be less pliable, and prone to crack when shaped. Sugar with less acid crystallizes more readily, but is stronger and holds its shape better when set.

Cooking Temperature for Pulled and Blown Sugar

The final cooking temperature of a sugar solution is of the utmost importance because it determines the working temperature (feel temperature) of the sugar while it is being pulled, and its consistency once it has cooled and hardened. If sugar is cooked to a lower temperature, more moisture is retained and the product is softer and more pliable for pulling and pouring. The sugar is also cooler to the touch and easier to work with, but has a shorter shelf life as it is more hygroscopic. Additionally, the pieces may bend or lose form when cooled.

A sugar solution cooked to a higher temperature will feel hotter to the touch. It will be stiffer to work with, and will firm up very fast while being shaped. The set sugar will have a stronger structure, resulting in a longer shelf life.

Cooking Time for Pulled and Blown Sugar

Pulled sugar formulas vary from larger to smaller batches if cooked on a different heat source or even in a different pot. The total cooking time of a sugar solution (the time it takes until it has reached its final temperature) also has a significant effect on the feel, the consistency, and the structure of the pulled sugar. A sugar solution cooked on a commercial heat source or on an induction burner will cook much more quickly than one that is cooked on a small, domestic, household-size stove.

Preparation for Pulled and Blown Sugar

Once the sugar has been cooked, it is poured onto a silicon mat or on a very clean, lightly oiled marble tabletop. If different colors are to be made from one batch, the sugar can be poured onto separate silicon mats and each section can be tinted with food color as desired. If only one color is needed, the sugar is tinted in the pot at the end of the cooking process.

Once the edges of the poured sugar have slightly cooled and are still pliable (not hard), they are quickly flipped toward the center using the fingertips or a lightly oiled spackle knife. Foodhandlers' gloves should be worn at all times when working with hot sugar to prevent burns.

The procedure is repeated until the sugar is firm enough to be handled by hand. The sugar is now folded with the cold side (the side that was against the marble) to the inside. It is folded several times until the sugar stops spreading. At this stage the sugar should be somewhat firm yet pliable. Holding the sugar in both hands, stretch it and then quickly fold it back together. Repeat the stretching-folding technique until the sugar is opaque and silky. Fold and form the sugar into a ball and place it under a heating unit (heat lamp). The sugar is now ready to be shaped or cut into smaller pieces and stored in a dry place for later use.

The Pulling of Sugar

Sugar should be placed on a piece of PVC, canvas, silk, 100% polyester, or a copper screen that has been stretched over a wooden frame, or placed on a dipping screen to keep it soft and pliable, and to prevent it from sticking to the work surface.

Sugar must be kept under a heating unit to keep it pliable as it is being pulled. One or more infrared heating light bulbs, inserted into a flexible, heat-resistant light stand, provides a good heating unit. Specially made sugar heaters can be purchased, but their cost is significantly higher. There should be a minimal amount of air flowing over the sugar as a constant draft will cool down the sugar and make it too stiff to pull. It is best to shield the sides and the top around the heating unit and working area with heat-resistant panels. Plexiglas is often used to allow spectators to view the process as it is performed.

Adjacent to the heat lamps, surfaces should be covered with nonstick materials such as parchment paper, aluminum foil, plastic wrap, or silicon mats. Very few tools are required for pulled sugar work: a good pair of scissors, a small paring knife, a few leaf stamps (usually made from silicon), and a small alcohol burner or gas torch are usually sufficient.

At this point, the sugar is ready to be shaped into various shapes and forms. Expertise in pulling sugar is developed through observation and constant practice. It is best to begin with simple shapes such as flowers and leaves to master the basic skills. Refined skills are necessary for elaborate showpiece work. Once the pulled sugar pieces are set, they are glued together by melting a small amount of set sugar over a clean flame and then quickly assembled.

Points to Remember

- Do not stretch pulled sugar more than necessary as it will result in a very white, dull, and grainy product, referred to as "sugar dying."
- Do not let the sugar get too hot or begin to bubble under the heat lamp or in the microwave oven, because this will cause the sugar to crystallize more readily and may also cause dangerous burns.
- Wear high-quality vinyl gloves (or other durable, heat-resistant substitutes) at all times to work with sugar. Moisture on the hands will cause sugar to become sticky and difficult to shape and may eventually cause the sugar to crystallize. Bare hands also leave fingerprints on the sugar.
- If the humidity in the air is too high, the pulled sugar pieces will become sticky, impossible to assemble, and will eventually crystallize. The moisture content in the air should not exceed 50%.
- Do not put small hard pieces of sugar back into the soft sugar as this may cause the sugar to crystallize.
- Unshaped or leftover sugar pieces can be cooled, wrapped in plastic, and stored in airtight containers in a dry area. When ready to use the stored pieces, slowly reheat them under a heat lamp or in a microwave.
- All sugar pieces will eventually crystallize if used for an extensive amount of time. Some formulas will last a bit longer than others, but in time, all will become grainy and unusable.
- Sugar may also be poured into lightly oiled pie tins rather than on a marble table. Allow the sugar to cool, remove it from the tins, wrap it in plastic, or seal in an airtight plastic bag, and store in a dry area. When ready to use, slowly reheat the sugar carefully under a heat lamp or in a microwave. Then stretch the sugar to make it opaque and silky.

Items Made from Pulled Sugar

Once the sugar is prepared and pulled it can be shaped into a variety of forms such as flowers, leaves, and ribbons.

STEP PROCESS

Pulling Sugar

Pulled sugar is produced by pouring a cooked sugar solution onto a nonstick baking sheet or lightly oiled surface. It is folded and flipped until pliable and worked into a variety of shapes.

To pull sugar, follow these steps:

1 Pour sugar solution onto a nonstick baking sheet.

2 Flip the sugar inward allowing the cooler edges to cool the warmer center.

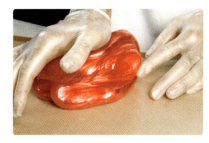
3 Fold sugar together several times until the sugar no longer spreads.

4 Roll sugar into a log.

5 Fold and stretch the sugar.

6 Continue folding and stretching until the sugar is opaque and silky.

7 Sugar may be cut into pieces as desired.

Flowers and Leaves To create flowers it is best to first learn two basic petal forms: round and pointed. The mastery of these two forms will eventually allow for the creation of rather sophisticated floral structures. Use sugar that is similarly colored to that of natural flowers; red or white for roses, violet for violets. When forming petals for roses, pull the sugar and form the petals using the thumb and index finger. To glue the petals together, hold each petal briefly over a clean flame and attach it at the base, taking care not to close the flower. Adjust the shape of a petal by carefully and briefly warming it over a flame. It

STEP PROCESS

Making Pulled Sugar Roses

Follow these steps when pulling sugar for roses:

1 Pull off a round piece of sugar.

2 Shape the piece in the palm of your hand.

3 Soften the edges over a clean flame.

4 Bend down the edges of the rose petal for a realistic look.

5 Repeat these steps, pulling and completing all the petals.

6 Heat the base of the petal over a clean flame to slightly melt the sugar and attach the petals.

7 Finish attaching petals to complete the rose.

8 Attach the leaves in the same manner used to attach the petals.

9 The finished rose.

is always a good idea to have a natural flower as a model. Each petal on a flower must be made separately and then joined to the others. Use green sugar for the stems, and as soon as the sugar is pulled, gently push a stem wire through the sugar to cover it before it emerges at the other end.

To attach the flowers to the stems, wrap a small amount of soft sugar around the end of the stems and touch them to the heater or briefly heat them over a flame. Make many

Crystal Sugar (Rock Candy)

The process to create crystal sugar is a lengthy one that may take up to two weeks to merely produce ¼ inch (.6 cm) of sugar crystals. First combine three parts high-quality sugar with one part distilled water and bring to a boil. Pour the mixture into a clean plastic container and allow it to cool to room temperature, undisturbed. If the plastic container is not completely clean or if it is scratched, the sugar will prematurely crystallize on the sides of the container, rather than around the intended item.

When the mixture is cooled, gently place an object coated with granulated sugar into the sugar syrup. Wrap the entire container in plastic wrap, and set it aside to allow large sugar crystals to slowly form around the granulated sugar. Allow the crystals to grow until the sugar syrup is no longer saturated. When the proper crystallization has been achieved, remove the piece from the syrup, rinse it off under cold water, and let it fully dry.

Spun Sugar

Spun sugar is frequently used in pastry shops as part of a showpiece, or as a decoration for classic buffet items, such as croquembouche. It resembles cotton candy but has slightly thicker sugar strands. Spun sugar is produced from recently cooked sugar that sits in a pan until it reaches the consistency of corn syrup. A brush or a wire whisk that has been cut off at the end is used to form spikes. Dip the brush or whisk into the sugar, and lift the brush or whisk away from the bowl using a quick back and forth motion. "Spin" the sugar among three yardsticks that have been lightly oiled and set to extend from the sides of a table. Repeat the motion several times until enough sugar has been spun. Gather all the sugar threads with your hands and roll them together to form a nest shape that resembles cotton candy. Remember to wear fitted, nonlatex gloves when doing so. Spun sugar is extremely sensitive to moisture and will melt very easily. It must be stored in a very dry environment with less than 50% humidity. Do not use spun sugar for showpieces that will be kept for a long period of time because of its susceptibility to melt with the slightest of humidity changes.

Sand Casting

To duplicate free-form pieces that have the appearance and texture of sand, the method known as sand casting is used. Fill a deep hotel pan with white or colored granulated sugar and with gloved hands make channels into the sugar, using a zigzag or spiral motion. Pour thick, cool sugar into the cavities and cover the liquid sugar with more granulated sugar, up to about 1 inch thick. As the sugar cools and thickens, twist the pliable sugar into a desired abstract shape. Keep the shape embedded inside the granulated sugar until it has fully hardened to ensure that it retains its shape while cooling.

Starch Molds

Starch molds may also be used to create seashells or other small objects. Fill a 1- or 2-inch-high (2.5 cm- or 5 cm-high) frame pan (such as a sheet pan or hotel pan) with sifted cornstarch. Level the surface of the starch using a long ruler or stick and press the model (for example a plastic seashell) into the starch. A perfect duplicate of the desired shape should now be embossed in the starch. Fill the cavity with liquid sugar until cool. Remove the sugar piece from the starch, brush off the excess starch and if necessary, rinse the pieces under warm water. Let the sugar fully dry before placing it onto a showpiece.

Finishing Touches

Dry sugar pieces can be painted or shaded with food color with an airbrush, atomizer, or a paintbrush. They can also be sprayed or brushed with an edible food lacquer to make the sugar somewhat more resistant to moisture. Sugar pieces will still require a dry storage condition. Royal icing, pastillage, marzipan, chocolate, and other decorative mediums can also be used with sugar pieces to enhance their appearance.

Storage

A sugar showpiece can keep for several months if properly stored, and can be destroyed in a few days when storage conditions are inappropriate. Because sugar is extremely hygroscopic, it attracts the moisture in the atmosphere. Moisture begins to dissolve set sugar, make it sticky, and eventually cause it to crystallize. Humidity within the environment should not exceed 50%, and the sugar pieces should be wrapped airtight or in an airtight container. Silica gel, quick lime, or calcium chloride may be added to keep the air in the storage container dry.

Silica gel is a chemical substance that absorbs moisture. The gel crystals are blue when they are activated, and turn pink or white when no longer active. They can be redried in an oven or microwave to be reactivated.

Quick lime is a natural, processed rock that is arid and absorbs moisture. It is used in the making of mortar, plaster, cement, and ceramics. A piece of quick lime may be placed next to the showpiece and replaced when it softens and turns to powder.

Calcium chloride, which is often used as a road deicer, is also very effective in reducing humidity. Like quick lime, it can be used only once and will turn to a liquid state when saturated with moisture. Calcium chloride may be placed in a perforated container with a second, watertight container underneath to capture the liquid.

Marzipan

Marzipan is a pliable dough made of almonds and sugars. It originated in the Orient as early as A.D. 800. Early recipes of marzipan consisted of a simple mixture of sugar, almonds, and rose water. Today, marzipan is graded by the quality and type of ingredients used. High-quality marzipan is white and is made of the finest almonds and sugar. A lower-quality marzipan is yellowish or light brown and is composed of a lower-quality grade of almonds and unrefined sugar. Almond content varies depending on the use of the marzipan. A modeling marzipan contains more sugar and fewer almonds, while a marzipan used in baking and candy making has a higher ratio of almonds and less sugar.

Marzipan may be molded or shaped into fruits, vegetables, animal figures, flowers, or any other variety of items. It may be colored with paste or liquid coloring. Finished marzipan pieces may be airbrushed with color and sprayed with a thin layer of cocoa butter or edible food lacquer. Like rolled fondant, marzipan can be used to cover and decorate cakes and tortes. *See Figure 28-4.*

Figure 28-4 An assortment of whimsical and realistic pieces can be made from six basic shapes.

Storage and Handling of Marzipan

Because of its susceptibility to fermentation, gloves should be worn when handling marzipan. The ingredients used to make it should be as fresh as possible, and the area for working with or rolling it out should be very clean. Sifted confectioner's sugar can be used to prevent marzipan from sticking to table surfaces during roll-out. Great care must be taken not to overwork or overmix marzipan, or the oil will separate and cause it to crumble and crack. Marzipan pieces should be stored covered in a cool, dry area away from direct light. *See Figure 28-5.*

Figure 28-5

A small ball of marzipan can be transformed into a beautiful apricot with a simple shaping tool, an airbrush, and a clove.

A textured plastic sheet can be used to create an effect that duplicates the rind of citrus fruit.

A garlic press can be used to make marzipan hair or grass.

STEP PROCESS

Making a Marzipan Alligator

Because marzipan is a favorite confection among young children, chefs traditionally create an array of whimsical animals and creatures. A basic cylindrical shape was used as a base for this marzipan alligator.

Follow these steps to make a marzipan alligator:

1 Roll marzipan into a small ball.

2 Roll the ball into a cylinder and taper one end for a tail.

3 Make an indentation in the other end of the cylinder for the head.

4 Use a scissors to cut the mouth.

5 Make small snips to create a scale pattern.

6 Bend the alligator for a more interesting and realistic look.

7 Use a round-edged tool to create the eyes.

8 Roll out smaller cylinders for legs and mark them to create feet.

9 Use an airbrush to paint the alligator with food coloring.

10 Finish the alligator by using black coloring for the eyes and snout.

Making Marzipan Roses

Marzipan is a perfect medium for making flowers and leaves to adorn special-occasion cakes. The closer the colors mimic those found in nature, the more realistic the flowers will look.

To make a marzipan rose, follow these steps:

1 Color the marzipan as desired.

2 Form the colored marzipan into a cylinder and cut off pieces for petals.

3 Cover the pieces with plastic wrap and use a rolling pin to flatten the pieces.

4 Use your fingers to thin the edges of each piece.

5 Use your fingers and the palm of your hand to shape each petal.

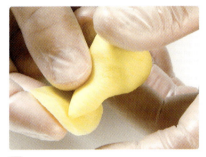

6 Make a conical-shaped base and wrap the first petal around the base.

7 Pinch the bottom of the petal and attach it to the base.

8 Continue attaching petals in a manner so that the petals appear to be overlapping.

9 Use your fingers or a paring knife to bend back and open up the outside petals.

10 Use scissors to snip the rose from the base.

11 To make leaves, roll out marzipan between plastic wrap. Cut out leaf shapes and re-cover with wrap. A knife can be used to make the vein patterns.

Appendix

Common Measurement Conversions

U.S. Standard	Metric
Weight	
0.035 ounce	1 gram
1 ounce	28.35 grams
16 ounces (or 1 pound)	454 grams
2.2 pounds	1 kilogram
Volume (Liquid)	
0.034 ounce	1 milliliter
1 ounce	29.57 milliliters
8 ounces (or 1 cup)	237 milliliters
16 ounces (or 1 pint)	474 milliliters (or .47 liter)
2 pints (or 1 quart)	946 milliliters (or .95 liter)
33.8 ounces (or 1.06 quarts)	1,000 milliliters (or 1 liter)
4 quarts (or 1 gallon)	3.79 liters
Volume (Dry)	
1 pint	.55 liter
0.91 quart	1 liter
2 pints (or 1 quart)	1.1 liters
8 quarts (or 1 peck)	8.81 liters
4 pecks (or 1 bushel)	35.24 liters
Length	
0.39 inches	1 centimeter
1 inch	2.54 centimeters
39.4 inches	1 meter

COMMON CONVERSION FACTORS

Weight	*To convert:*	*Multiply by:*
	ounces to grams	28.35
	grams to ounces	.03527
	kilograms to pounds	2.2046

Volume	*To convert:*	*Multiply by:*
	quarts to liters	.946
	pints to liters	.473
	quarts to milliliters	946
	milliliters to ounces	.0338
	liters to quarts	1.05625
	liters to pints	2.1125
	liters to ounces	33.8

Length	*To convert:*	*Multiply by:*
	inches to millimeters	25.4
	inches to centimeters	2.54
	millimeters to inches	.03937
	centimeters to inches	.3937
	meters to inches	39.3701

Common Temperature Conversions

Degrees Fahrenheit	Degrees Celsius
32°F	0°C
41°F	5°C
140°F	60°C
150°F	66°C
160°F	71°C
166°F	74°C
180°F	82°C
212°F	100°C
300°F	149°C
325°F	163°C
350°F	177°C
375°F	191°C
400°F	204°C
425°F	218°C
450°F	232°C
475°F	246°C
500°F	260°C

Temperature Conversion Factors

To convert Fahrenheit to Celsius:

Subtract 32, multiply by 5, and then divide by 9.

To convert Celsius to Fahrenheit:

Multiply by 9, divide by 5, and then add 32.

Bibliography

BOOKS

Amendola, Joseph. *The Baker's Manual*. 4th ed. Hoboken, NJ: John Wiley & Sons. Inc., 2004.

Amendola, Joseph, and Donald Lunberg. *Understanding Baking*. 2nd ed. New York: John Wiley & Sons, Inc., 1992.

Baker, William. *The Modern Pâtissier*. London: Northwood Publications Ltd., 1978.

Brenner, Joël Glenn. *The Emperors of Chocolate: Inside the Secret World of Hershey & Mars*. New York: Broadway Books, 2000.

Calvel, R., J. MacGuire, and R. Wirtz. *The Taste of Bread*. Gaithersburg, MD: Aspen, 2001.

Charpentier, H., and B. Sparkes. *Life a la Henri—Being the Memories of Henri Charpentier*. New York: The Modern Library, 2001.

Coe, Michael and Sophie. *The True History of Chocolate*. London: Thames & Hudson Ltd., 1996.

Connelly, Paul, and Malcolm Pittman. *Practical Bakery*. London: Hodder & Stoughton, 1997.

Duchene, Laurent, and Bridget Jones. *Le Cordon Bleu Dessert Techniques*. New York: William Morrow & Company, Inc., 1999.

Escoffier, A., D. Hervé, and J. M. Pouradier. *Special Decorative Breads*. New York: Van Nostrand Reinhold, 1987.

Figoni, Paula. *How Baking Works: Exploring the Fundamentals of Baking Science*. Hoboken, NJ: John Wiley & Sons, Inc., 2004.

Flandrin, Jean-Louis, and Massimo Montanari, ed. *Food: A Culinary History*. New York: Penguin Books, 1999.

France, W. J. *The Student's Manual of Breadmaking and Flour Confectionery*. London: Routledge, 1966.

Friberg, Bo. *The Advanced Professional Pastry Chef*. Hoboken, NJ: John Wiley & Sons, Inc., 2003.

Friberg, Bo. *The Professional Pastry Chef*. 3rd ed. New York: John Wiley & Sons, Inc., 1996.

Friberg, Bo. *The Professional Pastry Chef*. 4th ed. New York: John Wiley & Sons, Inc., 2002.

Gisslen, Wayne. *Professional Baking*. 3rd ed. Hoboken, NJ: John Wiley & Sons, Inc., 2001.

Gisslen, Wayne. *Professional Baking*. 4th ed. Hoboken, NJ: John Wiley & Sons, Inc., 2004.

Glezer, M. *Artisan Baking Across America*. New York: Artisan, 2000.

Greenspan, D. *Baking With Julia*. New York: William Morrow & Company, Inc., 1996.

Hamelman, J. *Bread. A Baker's Book of Techniques and Recipes*. New York: John Wiley & Sons, Inc., 2004.

Haroutunian, A. *Pâtisserie of the Eastern Mediterranean*. New York: McGraw-Hill, 1989.

Herbst, Sharon Tyler. *The New Food Lover's Companion: Comprehensive Definitions of Nearly 6,000 Food, Drink, and Culinary Terms*. 3rd ed. Hauppauge, NY: Barron's Educational Series, 2001.

Johnson & Wales University. *Baking & Pastry Fundamentals*. 6th ed. Dubuque, IA: Kendall/Hunt Publishing Co., 2000.

Johnson & Wales University. *Culinary Fundamentals*. Providence, RI: Johnson & Wales 2003.

King Arthur Flour Co., Inc. *The King Arthur Flour Baker's Companion*. Woodstock, VT: The Countryman Press, 2003.

Labensky, Sarah R., and Alan M. Hause. *On Baking: A Textbook of Baking and Pastry Fundamentals*. Upper Saddle River, NJ: Prentice Hall, Inc., 2005.

Larousse Gastronomique. New York: Clarkson Potter/Publishers, 2001.

MacLauchlan, Andrew. *The Making of a Pastry Chef: Recipes and Inspiration from America's Best Pastry Chefs*. New York: John Wiley & Sons, Inc., 1999.

Matz, Samuel A. *Technology of the Materials of Baking*. McAllen, TX: Pan-Tech International, Inc., 1989.

Mazda, M. *In a Persian Kitchen: Favorite Recipes from the Near East*. Rutland, VT: Charles E. Tuttle Co., 1990.

McGee, H. (1984). *On Food and Cooking. The Science and Lore of the Kitchen*. New York: Charles Scribner's Sons, 1984.

McVety, Paul J., Susan Desmond Marshall, and Bradley J. Ware. *The Menu and the Cycle of Cost Control*. 3rd ed. Dubuque, IA: Kendall/Hunt Publishing Co., 2005.

Milton, Giles. *Nathaniel's Nutmeg: Or the True and Incredible Adventures of the Spice Trader Who Changed the Course of History*. London: Penguin Books, 1999.

National Restaurant Association. *ServSafe Coursebook*. N.p.: Chicago, National Restaurant Association Education, 2004.

Nicolello, I., and R. Foote. *Complete Confectionary Techniques*. London: Hodder & Stoughton, 1994.

Reinhard, Peter. *The Bread Baker's Apprentice*. Berkeley, CA: Ten Speed Press, 2001.

Rogers, Ford. *Nuts: A Cookbook*. Edison: Chartwell Books, 1993.

Sizer, Frances Sienkiewicz, and Eleanor Noss Whitney. *Nutrition Concepts and Controversies*. 8th ed. Belmont, CA: Wadsworth/Thomson Learning, 2000.

Sokol, Gail. *About Professional Baking*. Clifton Park, NY: Thomson Delmar Learning, 2005.

Stewart, M. *The Martha Stewart Cookbook*. New York: Clarkson Potter, 1995.

Sultan, William J. *Practical Baking*. 5th ed. New York: Van Nostrand Reinhold, 1990.

Symons, Michael. *A History of Cooks and Cooking*. Chicago: University of Illinois Press, 1998.

The Culinary Institute of America. *Baking and Pastry: Mastering the Art and Craft*. Hoboken, NJ: John Wiley & Sons, Inc., 2004.

Toussaint-Samat, Maguelonne. *History of Food*. Oxford: Blackwell Publishers, 1998.

Trager, James. *The Food Chronology: A Food Lover's Compendium of Events and Anecdotes, from Prehistory to the Present*. New York: Henry Holt & Company, 1995.

Wing, D., and Scott, A. *The Bread Builders. Hearth Loaves and Masonry Ovens*. White River Junction, VT: Chelsea Green Publishing Company, 1999.

PERIODICALS

Bishop, Jack. "Shaping Foolproof Meringues," *Cook's Illustrated* (May & June 1995): 12–13.

Bruce, Erika. "Lemon Cheesecake," *Cook's Illustrated* (May & June 2003): 22–23.

Figoni, Paula. "Mastering the Three Types of Meringue," *Chef* (March 2002): 31–33.

Freeland, Gene. "Perfect Crème Brûlée," *Cook's Illustrated* (March & April 1995): 6–7.

Kimball, Christopher. "Mastering the Chocolate Souffle," *Cook's Illustrated* (September & October 1996): 18–19.

Kimball, Christopher. "The Perfect Chilled Lemon Souffle," *Cook's Illustrated* (July & August 1999): 20–21.

Labensky, Sarah R. "Classic and Contemporary Crème Brûlée," *The National Culinary Review* (November 2001): 37–39.

Matuszewski, Barbara Bell. "Cheesecake," *Chef Magazine* (August 1999): 58–59.

Morris, Jamie and Marie Piraino. "Classic Crème Caramel," *Cook's Illustrated* (September & October 1998): 24–25.

Pence, C. and J. "Discovering Semifreddos," *Pastry Art & Design's Frozen Desserts* (January 2007): 56–60.

Yanagihara, Dawn. "Perfecting Crème Brûlée," *Cook's Illustrated* (November & December 2001): 22–23.

Yanagihara, Dawn. "Perfecting New York Style Cheesecake," *Cook's Illustrated* (March & April 2002): 22–23.

Yankellow, J. "Lamination: Layers Beyond Imagination," *What's Rising*. San Francisco Baking Institute Newsletter (spring 2005): 1–5, 8.

Yankellow, J. "Yeast: Facts Every Baker Should Know," *What's Rising*. San Francisco Baking Institute Newsletter (fall 2003): 1, 3–5, 6.

WEB SOURCES

A.C.H. Food Companies, Inc. (2006). *History of Fleischmann's Yeast*. Retrieved July 2006 from http://www.breadworld.com/sciencehistory/history.asp

"A Chocolate Timeline." *Diner's Digest*. Retrieved June 2006 from http://www.cuisinenet.com/glossary/chocolate.html

"A Timeline of Spice History. . ." *ASTA Spice History Timeline*. Retrieved June 2006 from http://www.astaspice.org/history/timeline.htm

Bellis, Mary. "The History of Cheesecake." *About.com*. Retrieved June 2006 from http://inventors.about.com/library/inventors/blcheesecake.htm

Brennan's Recipes. (n.d.) Retrieved February 2007 from http://www.brennansneworleans.com/recipes.html

Celiac Sprue Association. Retrieved June 2006 from http://www.csaceliacs.org/celiac_treatment.php#treated

Charlotte Celiac Connection. Retrieved June 2006 from http://www.charlotte-celiac-connection.org

Cherkasky, Shirley "The Birthday Cake: Its Evolution from a Rite of the Elite to the Right of Everyone." Retrieved June 2006 from http://www.chowdc.org/Papers/Cherkasky2000.html

Crystallization. Retrieved February 9, 2007 from http://food.oregaonstate.edu/learn/crys.html

Dipping, Caroline. "Fine Fluff." Retrieved June 2006 from http://www.signonsandiego.com

Fallon, S., and Enig, M. (1999). *Be Kind to Your Grains and Your Grains Will Be Kind To You*. Retrieved July 2006 from http://www.westonaprice.org/foodfeatures/be_kind.html

"Flan." *Gourmet Sleuth*. Retrieved June 2006 from http://www.gourmetsleuth.com/flam.htm

Jaworski, Stephanie. "New York Style Cheesecakes." Retrieved June 2006 from http://joyofcakes.com/cheesecakes.html

Loren, Karl "The History of Oleo Margarine: The Cause of Many of Today's Health Problems." Retrieved June 2006 from http://www.karlloren.com/Diabetes/p46.htm

Medline Plus (National Institute of Health and U.S. National Library of Medicine). Retrieved June 2006 from http://www.nlm.nih.gov/medlineplus/celiacdisease.html

Monticello, The Home of Thomas Jefferson, A Day in the Life of Thomas Jefferson. (n.d.) Retrieved March 2007 from http://www.monticello.org/jefferson/dayinlife/dining/at.html

North American Miller's Association. Retrieved July 2006 from http://www.namamillers.org/kids.html

Stradley, Linda. "History of Cheesecake." *What's Cooking America*. Retrieved June 2006 from http://whatscookingamerica.net/History/Cakes/Cheesecake.htm

Strassman, Patty. "The Influence of Spice Trade on the Age of Discovery." Retrieved June 2006 from http://muweb.millersville.edu/~columbus/papers/strass-1.html

"Subpart M—United States Standards for Wheat, Terms Defined." Retrieved August 2006 from http://archive.gipsa.usda.gov/reference-library/standards/810wheat.pdf

The Bread Bakers Guild of America Online. The Bread Bakers Guild of America. Retrieved June 2006 from http://www.bbga.org/

The Food Allergy and Anaphylaxis Network. Retrieved June 2006 from http://www.foodallergy.org

The History of Chutney. (n.d.) Retrieved February 2007 from http://www.virginiachutney.com/history.php

"The History of Kraft Foods." *About.com.* Retrieved June 2006 from http://inventors.about.com/od/foodrelatedinventions/a/kraft_foods.htm

"The History of Margarine." *Margarine & Spreads.* Retrieved June 2006 from http://www.margarine.org.uk/pg_his1.htm

"The History of Sara Lee." Retrieved June 2006 from http://saralee.com/saraleebrand/history.aspx

"The World's Healthiest Food." The George Matejan Foundation. Retrieved June 2006 from http://www.whfoods.com/foodstoc.php

Trowbridge, Peggy. "Canned Milk History: Evaporated and Sweetened Condensed Milk." *Your Guide to Home Cooking.* Retrieved June 2006 from http://homecooking.about.com/od/milkproducts/a/canmilkhostory_p.htm

U.S. FDA/Center for Food Safety and Applied Nutrition. Retrieved June 2006 from http://www.cfsan.fda.gov

Whole Grains Council. Retrieved June 2006 from http://www.wholegrainscouncil.org

Glossary

A

abrasion Scrape or minor cut.

acetic acid Organic acid most commonly produced during fermentation; imparts a sour flavor.

achenes Minute black seeds that cover strawberries.

additive Substances placed in foods to improve characteristics such as flavor, texture, and appearance, or to extend shelf life.

aerobic Need oxygen to survive.

albumen Protein found in egg whites.

American buttercream Made with a combination of vegetable shortening, powdered sugar, and flavoring.

American-style ice cream (a.k.a. Philadelphia-style) Ice cream that does not contain eggs; need not be cooked prior to freezing.

amino acid Building blocks of proteins.

amylase Enzyme that breaks down sugars in yeasted doughs.

anaerobic Ability to survive without oxygen.

angel food cake Made from sugar, egg whites, and flour; leavened solely by the air whipped into the meringue.

AP weight The weight of the product when purchased.

arabica Highest-quality coffee.

ash content Mineral content of milled flour; good indicator of the extraction rate.

avulsion Partially or completely removed portion of the skin.

B

baba, or baba au rhum Yeasted raisin cake that has been soaked in a sweet rum syrup.

bacteria Tiny, single-celled microorganisms; leading cause of foodborne illness.

bagged /pressed cookies Shaped soon after the batter is made; macaroon and spritz cookies are examples.

baisure de pain/kissing crust Undesirable soft, pale sidewalls that result from loaves touching.

baker's percentage Includes the percentage of each ingredient in relation to the weight of the flour in the final baked product.

baking Dry cooking technique that uses dry, hot air to cook food in a closed environment.

balance scale Two-platform scale used in bakeshops and pastry shops.

bar cookies Baked in a bar shape; biscotti and zwieback are examples.

batonnet Long, rectangular cut that is similar to julienne.

batter Produced by mixing dry and liquid ingredients; can either be thin or thick.

Baumé Instrument used to determine the density of a sugar syrup.

Baumé scale Hydrometer; device used to measure sugar density in a sugar syrup or other solution.

Bavarian cream Consists of a mixture of crème anglaise, gelatin, and whipped cream.

bench tolerance Indicates the amount of time dough or batter can retain reliable leavening action between the mixing and baking steps.

betty Consists of spiced fruit that is baked with a granular topping made of buttered bread crumbs or stale cake crumbs, combined with brown sugar.

beverage cost Total dollar amount spent to purchase all the ingredients needed to produce a beverage item.

biga Firm mixture of water, flour, and yeast, giving bread a closed crumb structure.

biological hazards Disease-causing microorganisms, such as bacteria, viruses, parasites, and fungi.

black tea Strongly flavored tea with an amber or coppery brown color; varieties include Earl Grey and English breakfast.

blanching Moist cooking technique used to partially cook food before using another cooking method to finish the cooking process.

blending Incorporating ingredients until they are evenly combined.

blending method Method used in quick bread formulas that call for oil.

blending/mixing method Most of the ingredients, including at least half of the liquid, are placed in the mixing bowl at the same time.

blind baking Baking an unfilled pie crust.

blitz or Scottish method Most commonly used in high-volume bakeshops because it reduces procedure time and quickly creates dough; easier and faster than block method.

block or French method Most common method of rolling fat into dough; characterized by using blocks of fat.

bloom   Dusty white coating on grapes indicating that they have been recently harvested.

bloomed   Gelatin that has been hydrated in a proper ratio of cold liquid.

blown sugar   Made by blowing air into a piece of pliable pulled sugar; similar to glassblowing.

blueprint   Drawing that shows what a platter will look like with food displayed.

boiling   Moist cooking technique that transfers heat to food through convection, using a greater amount of liquid and with greater agitation than poaching or simmering.

bolster   Metal point on the knife where the blade and handle meet; also called a shank.

bookfolds   Used in conjunction with singlefolds to complete the laminating process.

bran   The outer protective covering of a grain kernel.

brandy   Distillate of wine and fruits used as flavoring for desserts.

bread pudding   Dessert consisting of a custard base and a bread or bread-like product.

broiling   Dry cooking technique that uses radiant heat from above to cook food.

brûlée   Caramelized sugar on the top of custard.

brunoise   A very fine dice cut.

buttercream   Frostings that have a butter base that allows them to become aerated as they are creamed.

C

cacao   Main component in chocolate; comes from the cacao tree.

calcium carbonate   Oldest type of whitening agent; naturally found in chalk, limestone, and marble.

calcium chloride   Used to reduce humidity.

calorie   Food energy measured in units of heat.

candling   Process of holding an egg up to a light to check the shell for cracks and tiny holes.

cannoli   Cylinders of sweet, crispy deep-fried pastry traditionally filled with a sweetened ricotta cheese.

caramel sauce   Made from melted, caramelized sugar that is thinned to the proper consistency with water or other liquids and then made richer with the addition of butter or cream.

caramelization   Occurs when the sugar begins to break down and create a rich brown color and bittersweet taste.

carbohydrate   Compound made up of sugar units; classified as simple or complex.

carotenoid pigments   Responsible for desirable flavor development and creamy coloration.

carryover cooking   Continued cooking that occurs even after food is removed from the heat source.

celiac disease   Disease in the intestinal tract brought about by the consumption of gluten; caused by an intolerance rather than an allergy.

certified milk   Milk that has met strict sanitary conditions.

chalaza   Twisted white cord that holds the yolk of an egg in place within the shell.

Champagne   The wine that comes from a defined region in the north of France.

Chantilly cream   Whipped cream with the addition of sugar and vanilla.

charlotte royale   Molded dome dessert that utilizes Bavarian cream filling.

charlotte russe   Dessert constructed with ladyfingers and Bavarian cream filling.

cheesecake   Member of the custard family consisting of a dairy base or soft, bland cheese, bound by eggs, and baked in a water bath.

chemical hazard   Substances such as cleaning supplies and pesticides that may cause food contamination.

chemical leaveners   Produce carbon dioxide in the presence of water and/or heat.

chewy-textured cookies   Cookies that are relatively soft in texture and high in moisture; have a high ratio of eggs, or other liquids, to dry ingredients and usually contain less fat than crisp cookies.

chiffon cake   Leavened by air but contain oil and/or egg yolks to produce a tender, moist crumb and a more stable product than the classic genoise.

chiffonade   Cut used to create fine ribbons or strips.

chocolate sauce   Most commonly made by incorporating melted chocolate and/or cocoa powder in sugar syrup.

cholesterol   Waxy substance in the body cells of all animals, found only in foods of animal origin; necessary substance in the body but in excess may create problems that can lead to cardiovascular disease.

churn-frozen dessert   Dessert that is churned continuously during the freezing process.

clarified butter   Pure butterfat; also called drawn butter.

cleaning   Removing all visible dirt and grime.

cobblers   Deeper versions of fruit pies and traditionally baked in square pans.

cocoa butter   The primary fat in cacao beans.

cocoa liquor   Liquid substance resembling melted dark chocolate.

cold foaming method   Eggs and sugar are whipped without being heated.

complementary protein   Combination of amino acids derived from a variety of foods.

complete protein Proteins that contain all essential amino acids.

complex carbohydrate Composed of starch and fiber units that help maintain the digestive system's function.

compote Stewed fruit dessert prepared by simmering fruit in syrup until it begins to lose its shape and fall apart.

compound chocolate May be used in the same way as tempered chocolate, but because it has very little or no cocoa butter, it sets faster and does not allow as much time for cutting as does tempered chocolate.

conching Slowly kneading chocolate to produce a smoother texture.

conduction Direct exchange of heat by physical contact with another item.

contamination Occurs when harmful substances such as microorganisms adulterate food, making it unfit for consumption.

convection cooking Spreading of heat while cooking through air or water; can be natural or mechanical.

conversion factor Derived by dividing the desired yield by the existing standard yield.

cost of sales Total amount spent to purchase all food and beverage products needed to produce total sales.

couche Linen material used for baking bread.

coulis Dessert sauce made from puréed fresh, canned, or frozen fruit.

coupe French name for a sundae.

couverture Bulk chocolate that is used for making confections, pralines, and fine pastries.

cream cheese icing Consists of cream cheese, butter, and powdered sugar; commonly used on carrot cakes.

creaming Beating soft fat or butter with sugar and gradually adding the remaining ingredients.

creaming method Method which produces a cake-like texture by mixing solid fats with crystalline sugars and then adding dry ingredients and any liquid ingredients.

crème anglaise Classic vanilla custard sauce made from egg yolks, milk or cream, and sugar.

crème brûlée Thicker custard made with heavy cream, sugar, and egg yolks cooked over a water bath.

crème caramel Caramel infused custard.

Crème Chiboust Pastry cream that includes Italian meringue.

crème fraîche Heavy, cultured cream that resembles a thinner, richer version of sour cream.

crème parisienne Chocolate-flavored Chantilly cream.

crêpe Thin, French-style pancakes prepared from a very thin, unleavened egg-rich batter.

crisps Topping containing white sugar.

crisp-textured cookies Cookies with a crunchy texture; have a relatively low moisture content and are usually high in fat and sugar.

critical control points (CCPs) Last point in the flow of food where a potential hazard can be prevented.

cross-contamination Most commonly caused by people, rodents, and insects; involves the transference of microorganisms or other harmful bacteria from one source to another through physical contact.

crumb crusts Made from finely ground cookies or graham crackers.

crumbles Topping containing brown sugar and oatmeal.

crunch Component that offers textural contrast to the main plated dessert item; may be in the form of a decorative cookie or as a container for the main item.

crystallization Liquid sugar syrup forms large visible crystals and turns grainy, dull, and eventually solid and unusable.

curds Coagulated milk solids.

custard Classification of products that are dairy based, egg thickened, and cooked in a water bath at a low temperature.

D

daily production report Form that helps control and manage costs and shows how much product was prepared, how much was sold, and how much was not used.

daily values Amount of nutrients a person needs every day on the basis of a 2000-calorie diet.

death phase Period in time which more bacteria are dying than growing; a population decline.

décor Small, piped decorations and filigrees; made with royal icing.

deep-frying Dry cooking technique that cooks food by completely submerging it in hot fat.

density Ratio of sugar to water.

dessert syrups Flavored simple syrups that are used to add moisture, sweetness, and flavor to a variety of desserts.

détrempe Mixture of water, flour, butter, and salt.

diastase Enzyme that converts starches into sugar; essential in yeast breads.

diplomat cream Consists of pastry cream and whipped cream.

disaccharides Two or more monosaccharides linked together; include sucrose, lactose, and maltose.

doughs Produced by mixing dry and liquid ingredients; have a low moisture content and a thick, pliable consistency.

dropped /deposit cookies Made from a relatively soft dough that is simply dropped or deposited onto a sheet pan; chocolate chip and peanut butter cookies are examples.

drupe Stone fruit.

E

economies of scale Price breaks for purchasing in large quantities.

edible portion Cost of a usable portion of the product, incorporating loss from trimming, shrinking, and packaging.

edible yield Usable portion of the food product left after it has been trimmed.

edible yield percentage Percentage of a food product that is left after trim and/or shrinkage.

emulsified shortening Shortening that has been specially formulated to absorb large quantities of liquids and sugars.

endosperm Main part of the seed from which flour is made; innermost portion.

enrobing Covering baked goods with a glaze. Process of dipping pralines.

ethylene gas Colorless, odorless gas given off by fruit as it ripens.

evaporated milk Canned concentrated milk.

extensibility Ability of the dough to stretch and not pull back; easier to shape and is necessary for a long, thin loaf of bread; contributes to larger volume in the baked loaf.

extraction rate Percentage of flour produced from a given weight of a wheat kernel.

F

facultative Ability to survive with or without oxygen.

falling number Indicator or enzymatic activity.

fat bloom Discoloration of chocolate.

fermentation Chemical reaction in a yeast-raised dough brought about by carbon dioxide gas. Enhances and develops the flavor of cacao beans, removes some of the bitterness, and destroys their ability to germinate or sprout.

first aid Assisting an injured person until professional medical help can be provided.

First In, First Out (FIFO) Method of rotation of goods whereby older items are moved to the front and used before newer ones.

flambéed desserts Ignited with rum, cognac, or a fruit liqueur immediately before serving.

flan A Spanish baked custard coated with caramel.

flat icing Combination of powdered sugar and water; sometimes referred to as water or simple icing.

flow icing Royal icing with extra egg whites added to bring the icing to a flowing consistency.

flow of food Process by which food items move through a foodservice operation from receiving to reheating.

fluting Decorative method of finishing a pie crust by making folds or pleats in the dough at regular intervals around the edge of the pie.

folding Gently mixing a lighter ingredient into a heavier one. Strengthens dough; expels some of the accumulated carbon dioxide; stimulates fermentation.

fondant Cooled sugar syrup; best-known of all of the icings. Cream of tartar mixture containing gelatin, glucose, glycerin, and shortening.

food cost Total dollar amount spent to purchase the food and beverage products needed to prepare menu items intended for sale.

formula Standardized recipe used in baking.

formula conversion Changing a formula to produce a new amount or yield.

formula cost Total cost of measurable ingredients added to the Q factor of immeasurable ingredients.

formula cost per unit Cost of one formula unit of the ingredient.

fourfold Double turn.

fraisage Method of mixing dough by hand.

fraise des bois Small alpine strawberry.

French buttercream Consists of a pâte à bombe and butter.

French-style cheesecake Similar to the New York-style cheesecake; eggs must be separated.

French/common meringue Meringue that is made without any application of heat and requires the use of pasteurized whites if it is to be used as a component of a product that is not cooked or baked.

French-style ice cream Custard-based ice cream made with milk, cream, sugar, egg yolks, and flavoring.

frozen soufflé/soufflé glacé Frozen dessert made to visually resemble a baked soufflé.

frozen yogurt Lowfat alternative to ice cream; lowfat yogurt is substituted for the milk, cream, and eggs.

fruit chutney Sweet and spicy cooked fruit-based condiment with roots in the exotic cuisine of India.

fruit salad Simple combination of complementary fruits, decoratively cut, and often lightly sugared to encourage maceration.

fruit salsa Used to accompany desserts and entrées; fruit is finely diced or julienned and combined with herbs, spices, and other flavorings.

fruit sauce Base of puréed fresh, frozen, or canned fruit, thickened with a thickening agent such as cornstarch, arrowroot or instant starch, and briefly cooked.

fudge icing Made by cooking sugar syrup and adding it to dry ingredients; usually consisting of a mixture of sugar and cocoa powder.

functional food Foods that provide health benefits beyond their basic nutrients.

fungus Microorganisms found in plants, animals, soil, water, and the air; molds and yeast are two forms of fungi.

G

ganache Filling made of chocolate and water-based liquids such as milk, cream, teas, or fruit juices.

Gâteau St. Honoré Dessert that is filled with a praline-flavored Chiboust.

gelato Italian custard-based ice cream, similar to French-style; contains a higher percentage of egg yolks.

genetically modified organism (GMO) Foods that have undergone gene modification in a laboratory to enhance specific traits.

genoise sponge cake Leavened solely by the air whipped into eggs; traditionally mixed using the warm foaming method.

germ Embryo that produces a new plant; gives wheat its nutty taste.

German buttercream Made with a combination of pastry cream, butter, and flavorings; considerably richer than any meringue-based buttercream.

gianduja Filling made of toasted nuts and sugar, ground to a very fine consistency, and mixed together with chocolate and/or cocoa butter.

glazes Sweetened and flavored liquids that are used to cover baked goods.

gliadin Component that provides dough with elasticity.

glucose Single sugar that makes up half of sucrose (table sugar).

gluten Protein found in wheat flour; affects the texture of baked goods.

gluten window Thin membrane used to determine if dough is properly developed.

glutenin Component that gives dough strength and structure and the ability to retain the gases during leavening and baking.

granita Coarsely textured Italian ice made from a sweetened, flavored water base and still-frozen rather than churned.

green tea Unfermented, yellowish-green tea with a slightly bitter flavor.

grigne The area on the surface of the crust that was exposed by the score marks; should be hard and well colored.

grilling Dry cooking technique that uses radiant heat from below to cook food on an open grid.

grunts Similar to cobblers in that their biscuit dough is dropped on top of the fruit; steamed on the stove top rather than baked to make the dropped biscuit dough resemble dumplings.

H

hand tools Any handheld tool used in the preparation, cooking, baking, or service of food.

hard wheat Higher in protein than soft wheat and is more appropriate for breads and other yeast items that rely on a gluten matrix to trap carbon dioxide; coarse to the feel, does not pack when squeezed, and absorbs more than soft wheat.

hardening The time during which packed ice cream remains in a subzero hardening cabinet.

Hazard Analysis Critical Control Point (HACCP) System used to monitor the flow of food.

herbs Fragrant leaves, stems, buds, or flowers from aromatic perennial or annual plants.

high-ratio cakes Have a high ratio of liquid and sugar to flour to produce a cake that is more moist and tender than a pound cake.

high-density lipoprotein (HDL) Carry lipids to and from the body's cells; good cholesterol.

high-ratio shortening Vegetable shortening with mono- and diglycerides added to improve emulsifying properties.

homogenization Process of breaking down fat globules by forcing warm milk through a very fine nozzle.

hydration Percentage of water needed in a formula.

hydrogenation Process that alters the fatty acid chains in polyunsaturated vegetable oil to extend the freshness of oil, raise its smoking point, and change it from a liquid to a solid state.

hygroscopic Able to absorb moisture from the atmosphere.

I

ice cream Smooth frozen dessert made by simultaneously churning and freezing a base mixture of milk, cream, sugar, egg yolks (optional), and flavorings.

ice cream bombe A molded frozen dessert that classically consists of a parfait filling encased in concentric layers of ice cream in a dome-shaped mold, with an optional sponge cake base.

icebox/refrigerator cookies Made from a dough that has been made in advance and then chilled; most common in high-production bakeshops.

ices Frozen desserts made from sugar-syrup bases with the addition of fruit purée, fruit juice, and sometimes an alcoholic beverage.

icings Simplest member of the frosting family; fairly dense and usually used on products such as brownies, cupcakes, or Danish.

incomplete protein Proteins missing one or more essential amino acid.

induction cooking Magnetic energy is used to rapidly heat cookware.

infection Occurs when pathogens grow in the intestines of an individual who has consumed contaminated food.

insoluble fiber Fiber that absorbs water.

intoxication Results when the pathogen produces toxins that cannot be seen, smelled, or tasted.

invert Breaking down sugar to make it softer and more yellowish in appearance.

invertase Enzyme that expedites the breaking down of the sucrose within the fondant into a sugar syrup; sometimes added to soften marzipan and other fillings.

invoice List of the products that are being delivered, including amount, description, weight or count per package, unit price, and extension, as well as terms of payment and total cost to the operation, including taxes and transportation costs.

invoice cost per unit Cost of an ingredient in the specified unit in which it is purchased.

irradiation Application of ionizing radiation to foods to kill insects, bacteria, and fungi, or to slow the ripening or sprouting process.

Isomalt Only sugar substitute derived exclusively from sucrose; primarily from beet sugar.

Italian buttercream Consists of an Italian meringue with butter.

Italian meringue Strongest type of meringue; most time consuming to prepare.

J

Japonaise A meringue variation consisting of a meringue base into which ground nuts are folded; used for cake layers in tortes.

julienne A matchstick-shaped cut.

K

kataifi Shredded phyllo dough.

kneading Folding and pressing a dough with the hands to develop gluten.

L

labor cost Cost of paying employees wages, salaries, and benefits.

laceration Deep cut or tear in the skin.

lactic acid Organic acid most commonly produced during fermentation; imparts a mild flavor.

lag phase Period of time when bacteria adjust to their new environment.

laminated dough Dough with many layers of dough and fat, adding to the final product's flavor, tenderness, and flakiness.

lattice crust Crust of interwoven strips of dough evenly placed across the top of a pie.

layer cakes Primarily chemically leavened; examples are pound cakes and high-ratio cakes.

lean dough Made with the four basic bread ingredients; contain little or no added sugars or fats.

leavening agent A substance that causes a baked good to rise by adding carbon dioxide, steam, or air into the mix.

lecithin Natural emulsifier found in egg yolks. Used to emulsify the components of chocolate and to give melted chocolate a more fluid consistency.

levain or sourdough starter A flour and water culture populated with natural or wild yeast.

lipid Organic compound that is insoluble in water but can be used by the body.

liqueurs A sweetened spirit used in baking for flavoring; can vary widely in alcohol content.

log phase Period of accelerated bacteria growth.

long-flake pie dough Used only as a top crust; fat is cut into nickel-sized pieces.

low-density lipoprotein (LDL) Transports excess cholesterol to tissues where it slowly builds up as a thick, hard deposit; bad cholesterol.

M

macédoine A square dice cut.

manufactured yeast or baker's yeast Produced in a lab specifically for use in breadbaking.

marzipan Soft filling commonly used in European pralines. Pliable dough made of almonds and sugar.

mass Amount of matter, or weight of an object.

mealy pie dough Used only as a bottom crust; fat is cut into small, pea-sized pieces.

meringue Delicate, fluffy mixture of stiffly whipped egg whites plus sugar; three basic types include French/common, Swiss, and Italian.

mincing Process of cutting food into very fine pieces.

minerals Naturally occurring substances that the body requires in very small amounts; classified as major minerals and trace minerals.

mirror glaze Consists of fruit purée and gelatin; usually applied to molded cakes or to a dessert as a top layer and are primarily used with cream-based cakes.

mold Grows in any environment and is commonly found growing on bread or cheese.

molded cookies Cookies that are molded into shapes before baking; crescent cookies are an example.

monosaccharides   Composed of a single sugar carbohydrate unit; include glucose, fructose, and galactose.

monounsaturated fat   Fat that is liquid at room temperature.

mousse   Cream that is generally stabilized with gelatin.

N

naturally leavened or sourdough   Breads leavened with wild yeast.

New York-style/deli-style cheesecake   Baked on top of a graham cracker or cookie crust and served with a fruit topping; consists of cream cheese, eggs, and sugar.

nibs   Cracked cocoa bean without the shell.

no-bake-style cheesecake   Unbaked cheesecake that is set with gelatin and lightened with the addition of whipped cream.

nougat   Filling that has a sugar base with nuts added; commonly classified as brown and white.

nutrient   Chemical compounds that make up food and perform one or more functions: supply energy, build and repair tissue, or regulate body processes; includes macronutrients and micronutrients.

O

oolong tea   Flavored tea with the characteristics of both black and green teas; varieties include black dragon and pouchong.

open market   Common system of buying that requires a bakeshop or pastry shop to secure price quotes for identical items from several sources of supply.

operating expenses   The combination of labor and overhead costs.

oven spring   Dramatic increase in volume that occurs during the first minutes of baking due to the rapid production of carbon dioxide as fermentation rate spikes in the heat of the oven, and the expansion of water into steam.

overcrystallize   Results when tempered chocolate is constantly used below working temperature.

overhead cost   All expenses an organization incurs except, food, labor, and profit.

overproofed   Describes dough that does not spring back quickly when touched.

overrun   The percent by which an ice cream base increases in volume as air is incorporated into it.

P

pandowdy   Deep-dish, double crust fruit pie that has had its crust broken and mixed with the filling for the final minutes of baking.

panna cotta   A fruit mousse that consists of a sweetened and flavored dairy base set by the addition of gelatin.

parasite   Tiny organisms that must live in or on a host to survive.

parfait glacés   Desserts frozen in glasses or molded in terrines.

parstock   The amount of a product kept on hand from one delivery date to the next.

pasteurization   Processing of milk and other liquids to prevent spoilage.

pastillage   Consists of confectionery sugar, cornstarch, cream of tartar, cold water, and gelatin; used in sugar showpieces.

pastry cream   Crème anglaise that includes whole eggs as well as egg yolks and thickened with a starch.

pâte à bombe   A mixture that is used as a base for many desserts; it is made by cooking sugar and water to the softball stage, then poured into egg yolks and whipped to a ribbon stage.

pâte à choux   Basic pastry dough; finished product is light and tender with a hollow interior and a shiny outer shell.

pâte fermenté   Easy way to use a preferment without having to mix a separate preparation.

pathogen   Undetectable by sight or smell, these disease-causing bacteria are responsible for many serious foodborne illnesses.

paysanne   A triangular-shaped cut.

pentosans   Gas-trapping substance found in rye.

periodic ordering method   Establishes how much product will be used for a given period of time.

perpetual inventory   Inventory system used to maintain constant records of expensive and highly perishable inventory items.

petits fours   Small, bite-sized pastries.

petits fours demi-secs   Petits fours that are only slightly dry.

petits fours glacés   Small, glazed cakes.

petits fours secs   Small, delicate, bite-sized cookies; tuile and meringue are examples.

petits fours variés   Miscellaneous assortment of bite-sized baked goods.

phyllo dough   Paper-thin pastry dough.

physical hazard   Foreign particles such as glass chips, toothpicks, dirt, and hair that contaminate food.

phytochemicals   Nonnutritive chemicals made by plants that may help guard against disease.

phytonutrients   Antioxidants which appear to be active agents in fighting heart disease, cancers, and other age-related illnesses.

pièce montée   Centerpiece on a platter or buffet.

plate cost   Amount of a menu item, including accoutrements.

poaching   Moist cooking technique that cooks food gently in a small amount of flavorful liquid.

polymorphic   Having many forms.

polysaccharides   Compound, such as starch, that is made up of many long strands of sugar units linked together.

polyunsaturated fat   Fats that are liquid at room temperature.

poolish   Mixture of equal parts of flour and water with a small amount of yeast; used to break down some of the starches in flour, converting them to sugar.

portion cost   The cost of one portion of the standardized formula.

potentially hazardous food   Foods typically high in protein and moisture and low in pH levels that present a greater risk for harboring microorganisms that cause foodborne illnesses.

pots de crème   French dessert consisting of custard served in small containers; "pot of cream."

pound cakes   Leavened by both the air cells that are incorporated through the creaming method as well as by the addition of baking powder or baking soda.

praline   Commonly used to describe a small, bite-sized, chocolate-based product.

precrystallized   Tempered chocolate.

preferments   Made by combining a portion of flour, liquid, yeast, and sometimes salt.

preliminary selling price   The lowest suggested selling price to be listed on the dessert menu.

preserve-based sauce   Fruit sauce made by thinning fruit preserves with simple syrup.

product specification   Describes the quality and quantity of each standardized formula ingredient.

protease enzymes   An enzyme found in exotic fruits that breaks down gelatin.

protein   One of a class of complex compounds that are an essential part of living matter.

pulled sugar   Produced by pouring a cooked sugar solution onto a silpat or lightly oiled marble surface and occasionally folding and flipping it until it becomes a cooled, pliable mass with a consistent temperature throughout.

puncture wound   Deep hole in the skin.

purchase order   Form that communicates to the purveyor the items to be purchased.

putrefactive   Undesirable bacteria that cause food to spoil and produce off-flavors, odors, slimy surfaces, and discoloration.

Q

Q factor   Price charged to recover the cost of all the ingredients that are too small to calculate.

quick breads   Chemically leavened by carbon dioxide gas that is produced by a chemical reaction; distinguished by their mixing methods, which minimize gluten development.

quick lime   Natural, processed rock that is arid and absorbs moisture.

R

radiation   Energy (heat) transmitted through space by the propagation of a wave through any form of matter; can be in the form of infrared or microwave cooking.

receiving   The process by which the product is accepted from the purveyor.

recovery time   Time it takes for the fat in a deep-fat fryer to return to the required cooking temperature after the food has been submerged.

red wheat   Contains a bran pigment that makes it darker than white wheat; makes up the majority of wheat grown in the United States.

reduction   Sauce made from a sweet liquid that is thickened by cooking it uncovered, over low heat.

refined white flour   Contains only the endosperm of the wheat berry.

requisition   Internal invoice that allows management to track the release of inventory from storage.

retarding   Extending dough fermentation.

ribbon   Describes egg yolks that have thickened and turned a light lemon color.

ripening   Stage in the cheese-making process in which bacteria or molds begin to work on the fresh curd.

rivet   Flat piece of metal that holds the tang in a knife's handle.

roasting   Dry cooking technique that uses dry, hot air to cook food in a closed environment.

robusta   Hardier type of coffee with a heavy, earthy aroma.

rolled cookies   Cookies that are rolled with a pin or a sheeter; sugar cookies and shortbread are examples.

rolled fondant   Used to cover both edible and dummy cakes.

roller milling   Milling process that shears grain open with corrugated rollers that allow the germ and bran to separate from the endosperm.

royal icing   Made of powdered sugar, egg whites, and cream of tartar; used solely for decoration.

rubbing   Solid fat is rubbed or cut into dry ingredients to create a mixture of flour and large fat flakes.

rubbing method or biscuit method   Method which involves rubbing or cutting fats into flour to create a mixture of flour and large fat flakes.

russe   Saucepan with a long handle and straight sides.

S

sabayon   A variation of pâte à bombe containing wine or liqueurs.   Classic French dessert sauce of egg yolks, sugar, and white wine.

sanitizing   Reduces the number of pathogenic microorganisms to a safe level with the use of chemicals and/or moist heat.

saturated fat Fats that are solid at room temperature.

sautéing Dry cooking technique in which food is quickly tossed in a sauté pan in a small amount of hot fat.

sautoir Shallow sauté pan with straight sides.

savarin An adaptation of the Baba that does not contain raisins and is baked in a ring mold.

scald Preboil; reached when tiny bubbles form around the side of the pan.

scaling Term used in weighing bakeshop and pastry shop ingredients.

scoring Provides a predictable spot for a loaf to burst.

seed Tiny crystals dispersed in liquid chocolate during tempering.

semifreddo Means "half cold" in Italian; refers to a dessert that consists of partially to completely frozen ice cream, mousse, or custard, piped or packed in a mold, and sometimes layered with cake.

serum solids Solids found in milk and cream other than the fat.

sheet cookies Cookies that are made in a sheet pan and then cut into individual pieces for serving; brownies and pecan diamonds are examples.

sherbet Fruit-based ice made from sorbet syrup and a base containing milk or cream and/or egg whites.

short dough Mixture of sugar, eggs, butter, and pastry flour.

shortening Fat used in the bakeshop or pastry shop because of ability to shorten gluten strands.

short-flake pie dough Used as either top or bottom crust; fat is cut into hazelnut-sized pieces.

showpiece/centerpiece Object designed to attract attention; can be displayed on a table, platter, or used as a competition piece.

shrinkage The percentage of food lost during cooking.

sifting Transference of dry ingredients through a wire mesh in order to aerate, separate, and remove lumps from a fine mixture.

silica gel Chemical substance that absorbs moisture.

simmering Moist cooking technique that cooks food by means of mild convection in a flavorful liquid.

simple carbohydrate Sugars found in milk, honey, fruit, maple syrup, refined table sugar, and unrefined sugar help retain moisture.

simple syrup Concentrated solution of sugar and water, usually in a one-to-one ratio.

single-source buying Linking of a bakeshop or pastry shop with one purveyor for most of the products that will be bought.

smallwares Small, nonmechanical foodservice equipment that are used for food preparation and baking and pastry tasks in commercial bakeshops and pastry shops.

soft wheat Lower in protein and is more frequently used in cakes, cookies, and pastry items; soft and silky, packs when squeezed, and absorbs less water than hard wheat.

soft/cake-textured cookies Cookies that have a texture resembling that of a dense cake; relatively low in fat but high in moisture.

soluble fiber Fiber which dissolves in water.

sorbet Ice made with a sugar syrup and a base infused with fruit, wine, or fresh herbs.

soufflé Light dessert that is leavened with meringue.

soufflé glacé Frozen dessert made to resemble a lofty baked soufflé.

spelt Nutritious, flavorful subspecies of wheat; does not support extended fermentation because it is low in gluten.

spices Berries, fruits, flowers, bark, seeds, and roots of plants or trees.

spirit Highly alcoholic beverage produced by both fermentation and distillation.

sponge Preferment containing flour, water, and yeast; typically employed in enriched or sweet-dough products for the strength they provide.

sponge or foam cakes Rely on the air whipped into the eggs or egg whites for leavening; examples are genoise, chiffon, and angel food.

spoom Light, frothy variation of sorbet made by folding an uncooked Italian meringue into the sorbet base during the final stages of freezing.

spring wheat Planted in the spring, grows and matures over summer, harvested in autumn; generally contains more protein than winter wheat but is usually of a lower quality.

standard portion Amount of food that is served for each order, usually stated in ounces or count.

standard yield Consistent amount of product produced by a standardized formula.

standardized formula Written procedure customized to meet the needs of a bakeshop or pastry shop.

stationary growth phase Period in which bacteria grow and die at the same rate.

steaming Moist cooking technique that cooks food in a closed environment via the use of steam.

still-frozen dessert Frozen products that are not churned during the freezing process.

stirred crème brûlée Poured into a chocolate or a cookie cup and garnished with fresh berries, or used as a filling for tartlettes.

stirring Blending ingredients gently by hand until they are equally combined.

stone grinding Milling process that retains some of the germ and bran from the endosperm.

streusel topping Crumbled mixture of flour, fat, sugar, and often spices and nuts.

strudel dough Dough that has been stretched paper-thin and sprinkled with butter for added flavor and flakiness.

sugar bloom Dried sugar crust on the surface of chocolate.

sundae A scoop (or scoops) of ice cream served in a glass or dish, with sauce and topped with fruits or nuts spooned on top or layered with the ice cream.

supremes Citrus segments.

sweetened condensed milk Whole milk that has had 60% of its water removed and a large amount of sugar added.

Swiss buttercream Consists of a Swiss meringue with the addition of butter.

Swiss meringue Stable meringue that is used to make macaroon cookies.

T

tang Portion of the knife's blade that extends into the handle.

tare weight Counterweight of a balance scale.

temperature danger zone (TDZ) Temperature ranging between 41°F–135°F (5°C–57°C) in which bacteria may thrive.

tempering Heating, cooling, and reheating process that is used to stabilize chocolate.

theme Central concept or motif of a buffet.

threefold Single turn.

titanium dioxide One of the most common whitening agents used in the food industry.

tomato concassée Peeled, seeded, and diced or chopped tomato.

toxin-mediated infection Occurs when a person eats food contaminated with pathogens.

trans-fatty acid Result of hydrogenated and hardened polyunsaturated oils.

triglyceride Class of chemicals that include fats and oils.

trim Percentage of food lost during the preparation of a product.

trim loss By-product material trimmed from the food item.

trueing Process in which steel is used to keep the blade of a knife straight and smooth; it does not sharpen the knife.

tunneling The formation of holes within a baked good.

turn The step of rolling out and folding dough to create the alternating layers of dough and butter.

two-stage mixing method Sometimes referred to as the blending method because the liquid ingredients are added in two stages.

two-step foaming method of mixing Yolks and whites are whipped separately.

U

ultra high temperature (UHT) processing Combination of ultrapasteurization with special packaging.

ultrapasteurization Process of subjecting dairy products to much higher temperatures for shorter periods of time, killing nearly all bacteria in order to extend the product's shelf life.

underproofed Describes dough that does not spring back when touched.

V

virus The smallest known form of life that is responsible for the majority of foodborne illnesses.

vitamins Organic compounds that work in conjunction with enzymes and hormones to regulate bodily functions and to maintain health.

volume Amount of space an ingredient takes up.

W

wafer cookies Light, crisp cookies that have a characteristic waffle pattern.

warm foaming method Eggs and sugar are heated over a double boiler and the mixture is constantly stirred to prevent the eggs from coagulating.

water activity (a_W) Moisture needed for bacteria to survive.

water bath Used to bake delicate dishes such as custards and mousses; involves placing custard in a ramekin that is placed in a hotel pan filled with water.

weeping Separating of liquid in soft pie fillings.

weight Heaviness of ingredients.

wet slow baking Method of baking where the product is placed in a water bath in the oven and baked slowly at a low temperature.

whetstone Stone used to sharpen knives.

whey The part of milk that does not coagulate.

whipped cream Heavy cream that has been whipped to a thickened foam.

whipping Performed by beating ingredients for the sole purpose of aerating.

whipping method Eggs are whipped and air cells are formed and incorporated into the mix.

white wheat Characterized by its mellow, nutty taste.

whole-grain flour Contains all three parts of the wheat berry.

wild yeast Occurs naturally in the environment; captured by the baker and maintained by regular feeding and care.

winter wheat Planted in autumn, germinates soon after planting then lies dormant in the field during the cold months; sprouts resume growing in spring and are ready for harvest in the early summer; used in artisan hearth breads because of its high-quality gluten and superior fermentation tolerance.

withered Drying process for tea leaves.

X

xanthan gum Gum that is produced as the result of fermentation.

Y

yeast Most often associated with bread and the baking process and is considered to be beneficial but can also cause spoilage when present in other foods.

yeast A living single-celled fungus that multiplies when fed simple sugars in a warm, moist environment.

yield test Procedure used to determine the edible yield percentage.

Z

zones Buffet area.

Index

A

"A" crystals in chocolate, 397, 398
À la minute plating, 197
Abrasions, 33
Acetic acid, 214
Achenes, 143
Acid in pulled sugar, 438–439
Active dry yeast, 215
Additives, 54
    in ice cream, 327
    in wheat flour, 213
Aerobic bacteria, 22
Agar-agar, 106
Aging ice cream, 328
Aging wheat flour, 97
Air as leavening agent, 103
Airbrush for sugar work, 431
Albumen, 130
Alcohol in fermentation, 214
All-purpose flour, 99
All-purpose shortening, 99–100
Allspice, 162
Almonds, 165
Amaretto, 172
American buttercream, 368
American Heart Association, 34
American Red Cross (ARC), 33, 34
American-style ice cream, 324
Amino acids, 42
Amylase, 214
Anaerobic, 22
Angel food cake, 386
Anise seeds, 163
Apple Brown Betty, 342
Apple corer/slicer, 75
Apple fritters, deep-frying, 186

Apples, 145–146
    sautéing, 184
Apricots, 143
Armagnac, 173
Arrowroot, 105
Artificial sweeteners, 50
Artisan cheeses, 125–126
Artisan hearth breads, 228
As-Purchased (AP) weight, 62
As-purchased quantity, 63, 64
Ash content of flour, 212
Aspartame, 50
Australian fondant, 430
Autolyse, 219
Avulsion, 34

B

"B" crystals in chocolate, 397, 398
Baba, 276
Baba au rhum, 276
Back injuries, 31–32
Bacteria, 17, 18
    growth requirements of, 21
Bad cholesterol, 44
Bagged cookies, 359
Bags, pastry, 76
Baguette screens, 81
Baguettes, 228
Baisure de pain, 225
Baked Alaska, 330
    making, 331
Baker's cheese, 125
    in cheesecake, 291
Baker's margarine, 128
    as roll-in fat, 249
Baker's Math System, 93
Baker's percentage, 58, 93–94

Baker's yeast, 104, 214, 215
Bakery ovens, 73–74
Baking, 182
    carryover, 114
Baking bread, 225–227
Baking chocolate, 109
Baking laminated dough, 254
Baking pies and tarts, 269, 271
Baking powder, 104, 237
    in cookies, 357
Baking process, stages of, 113–114
Baking puff pastry, 248
Baking quick breads, 243
Baking soda, 104, 237
    in cookies, 357
Baking temperatures, adjustments in formula conversion, 61
Baklava, 277
Balance scale, 78, 92
Banana Flambé, making, 345
Bananas, 147–148
Bananas Foster, 344
Banquet style plating, 196
Bar cookies, 359
Base of ice cream, 328
Basil, 161
Batonnet cut, 86
Batter, 110
    pourable, 280–281
Baumé hygrometer, 79
Baumé measure, 411
Baumé scale, 334, 352
Bavarian cream, 299–301
    making, 300
Bavarois, 299–301
    making, 300
Beignets, 274
Ben & Jerry's ice cream introduced, 12

Bench rest for bread dough, 224

Bench scraper, 75

Bench tolerance, 237

Beneficial bacteria, 17

Berries, 141–143

Betty, 342

Beverage cost, 58

Biga preferment, 219

Biological contamination hazards, 16, 17
    types of, 189

Biscuit method, 239
    in pie dough, 259

Bitter chocolate, 109

Black breads, 231

Black tea, 177

Blackberries, 142

Blade, knife, 82

Blanching, 191

Blast chiller, 29

Bleaching wheat flour, 97

Blending, 110

Blending method, 241, 382, 383
    for cookies, 359
    of mixing, 241

Blind baking, 269, 271

Blitz method in puff pastry, 247

Block method in puff pastry, 247

Blooming gelatin, 267, 299

Blown sugar, 438–443
    cooking procedure for, 438
    cooking time and temperature for, 439
    function of ingredients in, 438–439

Blowtorch, 75

Blueberries, 142

Blueprints, platter, 204

Boiling, 190–191

Boiling point and sugar concentration, 434

Bolster, knife, 83

Bookfolds, 252

Borden, Gail, 8

Bourbon, 175

Bowl scraper, 76

Boysenberries, 142

Bran, 95, 211

Brandy, 172–173
    grading of, 172–173

Brazil nuts, 165

Bread
    basic ingredients of, 210–217
    cooling and storage, 227–228
    enriched, 232–233
    types of, 228–233

Bread flour, 98

Bread making
    history of, 4
    incorporating additional ingredients, 221–222
    steps of, 219–228

Bread ovens, 226–227
    unloading, 227

Bread pudding, 289

Bread spiral mixer, 72

Brick ovens, 226

Brie cheese, 124

Brioche, 233

Brioche tins, 81

Broiling, 188

Broken tea leaves, 177

Bromate, 213

Brown sugar, 102

Browning in baking, 113–114

Brûlée, 288

Brunoise cut, 86

Brushes, pastry, 76

Bubble sugar, 445

Buffet design, 199–201

Buffet planning, 198–206

Bulk fermentation, 223

Burns, 31
    first aid for, 33

Butter, 100, 126–127
    clarifying, 128
    as roll-in fat, 249
    types of, 127
    USDA grades of, 127

Butter ganache, 406–407
    preparing, 408

Buttercream, making decorations and borders, 389

Buttercreams, 368
    flavoring, 372
    types of, 368

Buttermilk, 121

Buying chocolate, 402

C

Cacao, 395

Cacao bean, 395
    fermentation of, 396

Cacao pods, early use of, 5

Cadbury, John, 7, 8

Cake flour, 98

Cake pans, preparing, 387

Cake-textured cookies, 356

Cake yeast, 215

Cakes, 380–391
    baking, 387–388
    cooling and storing, 388
    covering with rolled fondant, 430–431
    glazing, 377

Calcium carbonate, 435

Calcium chloride, 448

Calimyrna fig, 152

Calories, 38

Camembert cheese, 124

Candling, 132

Candy thermometer, 79

Canned fruit, 155

Cannoli, 278

Cantaloupes, 139–140

Carambola, 153

Caramel sauce, 349–350
    making traditional, 351

Caramelization, 226

Caramelized cages, making, 203

Caramelized sugar decorations, making, 204

Carbohydrates, 38, 102

Carbon dioxide in fermentation, 214

Cardamom, 163

Cardiopulmonary resuscitation (CPR), 34

Carotenoid pigments, 221

Carryover baking, 114

Carryover cooking, 182

Carving chocolate, 420

Casaba melon, 140

Casein, 118

Cashews, 165

Celiac disease, 51

Centerpieces
    buffet, 203
    introduction to, 426

Certified milk, 119

Chalaza, 130

Challah, 233

Champagne, 174

Champagne grapes, 147

Chantilly cream, 298

Charlotte royale, 301
    making, 301

Charlotte russe, 301–302
    making, 301

Chase, Oliver, 8

Checkerboard cookies, making, 361

Cheddar cheese, 123

Cheese
    handling of, 126
    receiving and storing, 126

Cheesecake, 290–295
    finishing, 295
    types of, 291–292
    unmolding, 295

Cheesecake family, defining, 284

Cheeses, 122–126

Chef's knife, 83

Chemical contamination hazards, 16, 24–25

Chemical leaveners for quick breads, 237

Cherimoyas, 151–152

Cherries, 143–144

Cherries Jubilee, 344

Chervil, 160

Chestnuts, 165

Chèvre cheese in cheesecake, 291

Chèvre frais cheese, 125

Chewy-textured cookies, 356

Chiffon cake, 384, 386

Chiffon pie fillings, 266

Chiffonade cut, 85

Child, Julia, 12

Chives, 160

Chocolate, 109-110
    arrival in Europe, 6
    carving, 420
    equipment list for, 403–404
    first manufacturing firm for, 7
    flavoring with, 372, 375
    history of, 394–395
    in ice cream, 326
    introduction to, 394–395
    manufacturing, 395–401
    paste, 419
    showpieces, 416–423
    stencils, 403
    storage, 402
    tempering machines, 403
    thickening of, 401
    warmer, 403

Chocolate box, making, 422–423

Chocolate decorations, 416–423
    piping, 391

Chocolate glazes, 376
    for cheesecake, 295

Chocolate molds, 403
    liquid fillings in, 411

Chocolate mousse, 302
    making, 303

Chocolate sauce, 350–351
    as border, 200

Chocolate truffles, making, 407

Chocolates, molded, 413–416

Choking, first aid for, 34

Cholesterol, 44

Chopping, coarse, 85

Choux paste, 274–277

Christmas melons, 141

Churn-frozen dessert, 328

Churros, 274

Chutney, 340

Ciabatta bread, 228

Cinnamon, 163

Citrus fruits, 136–139

Clarified butter, 127

Clarifying butter, 128

Clean-up stage in mixing, 221

Cleaning, 16

Cleaning kitchen equipment, 32

Cleaning products, 24

Clear flour, 96, 98

Coagulation of proteins in baking, 113

Coarse chopping, 85

Coarse sugar, 102

Cobblers, 340

Cocoa butter, 396

Cocoa liquor, 396

Cocoa painting, 420

Cocoa paste, flavoring with, 372, 375

Cocoa powder, 110

Coconuts, 148

Coffea arabica, 175

Coffea robusta, 175

Coffee, 175

Cognac, 183

Cold foaming method, 384

Cold paddles, 29

Cold storage, 27

Cold-water bath tempering method, 399, 400

Color-coding equipment, 25, 27

Color contrasts in plating, 198

Coloring chocolate, 421

Coloring pastillage, 429

Coloring sugar, 434–435

Columbus, Christopher, 6

Comfits, 5

Commercial glazes, 376

Common meringue, 311–312

Complementary proteins, 42

Complete proteins, 42

Complex carbohydrates, 39

Complex sugars, 39

Compote, 339

Compound chocolate, 109, 420

Compound extracts, 372

Compounds for flavoring, 107

Compressed yeast, 104, 215

Concentrated sweets, 38

Conching chocolate, 395, 396–397

Concord grapes, 147

Conduction, heat transfer by, 180

Confectioner's coating, 420

Confectioners' sugar, 102

Confections in Middle Ages, 5

Contamination, 16

Contemporary history of baking, 10

Convection cooking, 180

Convection ovens, 74
    for bread, 227

Conversion factor, 61

Cooked-fruit-and-juice pie filling method, 264

Cooked-juice pie filling method, 264, 265

Cookie crusts, 294

Cookie cutters, 76

Cookies, 356–363
    baking, 362
    cooling and storing, 363
    ingredients in, 356–357
    types of, 356

Cooking loss test, 62

Cooking techniques
    dry, 181–188
    moist, 188–191

Cookware, selecting, 80–82

Cooling food safely, 29

Cooling quick breads, 243

Cooling racks, 82

Corals, 446

Corn, 231

Corn syrup, 102

Cornstarch, 105

Cortés, 6

Cost controls, 58–60

Cost of sales, 58

Cost percentage, desired, 65

Costing form, 63

Costs, controlling, 65–69

Couche, 224

Coulis, 348

Coupe, 329

Couverture, 404

Couverture chocolate, 109

Cow's milk, 118

Cranberries, 142

Cranshaw melons, 140

Cream
    as liquid, 107
    receiving and storing, 120

Cream-based desserts, 299–307

Cream cheese, 125
    in cheesecake, 290
    icing, 373

Cream pie fillings, 266

Cream puffs, 274

Creaming, 110

Creaming method, 382
    for cheesecake, 290
    for cookies, 357–359
    of mixing, 240
    for pound cakes, 380

Creams, egg thickened, 304

Crème anglaise, 304–305, 349
    making, 350

Crème brûlée, 286, 288

Crème caramel, 286, 287

Crème Chiboust, 305–306

Crème fraîche, 121

Crème liqueurs, 172

Crème parisienne, 299

Crenshaw melons, 140

Crêpe pan, 80

Crêpes, 280
    making, 281

Crêpes Suzette, 344–345

Crisp-textured cookies, 356

Crisps, 341–342

Critical control points (CCPs), 26

Croissants, 248–249, 254
    shaping and finishing, 253

Croquembouche, 276, 447

Cross-contamination, 16

Crumb crusts, 262, 294

Crumbles, 341–342

Crunch in plated dessert, 195

Crust formation, 113–114

Crusts, cheesecake, 294–295

Crystal sugar, 447

Crystallization, avoiding, 432

Cultured butter, 100

Cultured dairy products, 121
    receiving and storing, 120

Cupcake tins, 81

Curds, 122

Currants, 142

Custard family, defining, 284

Custard pie fillings, 266

Custards, 286–290
    cold preparation method, 284–285
    hot preparation method, 285
    ingredients in, 285

Cuts, 31

Cutters, pastry, 76

Cutting pastillage, 427

D

Daily production reports, 69

Daily values, 53

Dairy products
    fermented, 121
    foodborne illness and, 121

Danish, 254
    shaping and finishing, 255

Danish dough, 249

Dark chocolate, 109, 396

Dates, 148

Dead dough, 233

Death phase of bacteria, 22

Deck oven, 73
    for bread, 226–227

Deco dough, 233

Décor, 374

Decorations, chocolate, 416–423
    finishing, 421
    storing, 421

Decorative breads, 233

Deep-fat fryer, 74

Deep-frying, 185–186

Deli-style cheesecakes, 291–292
    making, 293

Density of sugar syrup, 411

Deposit cookies, 359–360

Dessert sauces, 348–352
    decorating plate with, 196
    types of, 348–352

Dessert syrups, 352

Détrempe, 246, 247

Diastase, 97
Dicing, 86
Digestive enzymes, 39
Digital thermometers, 403
Dill, 161
Diplomat cream, 305
Dipping forks, 403
Dipping pralines, 412–413
Direct tempering method, 399
Disaccharides, 39
Dishwashing
    manual, 29–30
    mechanical, 30
Disposal of food, 29
Dividing bread dough, 223
Double-acting baking powder, 104
Dough cutter, 75
Dough dividers, 74
Dough hydration, 221
Dough temperature control, 216
Doughnuts, 238
Doughs, 110
    stretched, 277–279
Drawn butter, 127
Dried eggs, 131
Dried fruit, 156
Dropped cookies, 359
Drupes, 143
Dry active yeast, 104–105
Dry cooking techniques, 181–188
Dry measures, 78
Dry storage, 27
Dry yeast, 215
Drying equipment, 30
Dutch-process cocoa, 110

E

Earle, Timothy, 9
Early civilization, baking during, 4–5
Éclair paste, 274–277
Éclairs, 276
Economic aspect of buffet menu, 199
Economies of scale, 65

Edible portion, 59
Edible yield, 62
Edible yield percentage, 62, 63
Effector, 403
Egg substitutes, 131
Egg-thickened creams, 304
Egg white, 130, 131
    overwhipping, 310
Egg yolk, 130, 131
Egg yolk soufflé base, 319
    making, 319
Eggs, 129–133
    composition of, 129
    forms of, 130—131
    purposes in baking and pastry, 129
    receiving, 132
    safe preparing and cooking, 132
    separating, 133
    storing, 132
    testing for freshness, 132
    as thickener, 106
Eggshell, 129–130
Electronic scale, 78
Emergency procedures, 32
Emmental cheese, 123
Emperor grapes, 147
Emulsified shortening, 382
Endosperm, 96, 211
Enriched breads, 232–233
Enriched doughs, techniques for, 232
Enrichments in flour, 213
Enrobe, 375
Enrobing pralines, 412
Environmental Protection Agency (EPA), 30
Equipment, 72–75
    adjustments in formula conversion, 61
Equipment list for chocolate, 403–404
Escoffier, Auguste, 10
Essences, 107
Ethylene gas, 154
Excessive heat in chocolate, 401–402
Exotic fruits, 151–153
Expansion in baking, 113
Extensibility, 211
Extraction rate of flour, 212
Extracts, 107, 372, 375

F

Facultative, 22
Falling number of mill run, 213
Falls, 31
Fan sugar, 446
Fanning cut, 86
Fannings, 177
Farmer's cheese, 125
Fat and meringues, 310
Fat bloom, 398
Fat replacers, natural, 45
Fat-soluble vitamins, 46
Fats, 42, 99–101
    and gluten formation, 112
    melting in baking, 113
    selecting, 101
    storing, 101
Fatty acids, 42
Feijoa, 152
Fennel (seeds), 163
Fermentation, 214–215
    of bread dough, primary, 223
    by-products of, 214
    of cacao beans, 396
    effects of improper, 215
Fermented dairy products, 121
Fiber, 40
Figs, 152
Filberts, 166
Fillings for chocolates, 404–411
Final proof of bread dough, 224
Fine sugar, 102
Fire extinguishers, 32
Fire safety, 32
Firm cheeses, 122–123
First aid, 33
First-clear flour, 96
First In, First Out (FIFO), 27, 68
Flambéed desserts, 343
Flame sugar, 446
Flan, 289
Flat breads, 232
Flat icing, 373
Flavor contrasts in plating, 197

Flavoring cheesecakes, 292, 294

Flavorings, 107–108

Fleischmann, Charles, 8

Fleming, Claudia, 13

Fleximolds, 80

Floor time, 223

Flour, 210–213
    blends, 99
    and gluten formation, 112
    as thickener, 106

Flow icing, 430

Flow of food, 26

Fluting, 261

Foam cakes, 380, 383

Folding, 111

Folding bread dough, 223

Fondant, 374, 430–431
    cocoa painting on, 420

Fontal cheese, 123

Fontina cheese, 123

Food costs, 58
    managing, 58–59

Food energy, 38

Food Handlers Agreement, 35

Food processor, 75

Foodborne illness, 16, 17, 121

Foodservice contamination hazards, 16

Formula, 58

Formula conversion, 61–65

Formula cost, 64
    per unit, 64

Formula costing, 63

Formula modification, steps in, 52–53

Formula name, 62

Fourfold puff-pastry, 247

Fraisage, 268

Fraise, 173

Fraise des bois, 143

Framboise, 173

Frangelico, 172

Frangipane, 365

Franzipan, 365

Freezing laminated dough, 254

French buttercream, 370–371

French knife, 83
    using, 84–85

French meringue, 311–312
    in cakes, 386, 387

French method in puff pastry, 247

French short doughs, 268

French style cheesecake, 292

French-style ice cream, 324
    making, 325

Fresh egg, 130

Fresh soft cheese, 124

Fresh yeast, 215

Freshness testing for eggs, 132

Fritters, 238

Frozen desserts, plating, 329

Frozen eggs, 131

Frozen fruit, 155

Frozen soufflé, 320–321

Frozen storage, 27

Frozen yogurt, 324–325

Fructose, 102

Fruit
    brandies, 172
    canned, 155
    chutney, 340
    cuts, 85
    dried, 156
    for flavoring, 107
    frozen, 155
    gratins, 340
    mousse, 302–304
    pie fillings, 263–266
    preserved, 155–156
    roasting, 183
    salads, 339
    salsas, 339
    sauce, 349
    as sweetener, 50

Fruit, fresh, 136–154
    grading, 154
    purchasing and storing, 153–155
    receiving, 154
    ripening, 154–155

Fruit and fondant fillings, 411

Fruit desserts, 338–345
    traditional, 340–342

Fruit fillings for pralines, 410

Fruit-flavored liqueurs, 172

Fruit glaze for cheesecake, 295

Fruit purées in ice cream, 326

Fruit reduction soufflé base, 317

Fruit tarts, 267–268
    making, 269

Fruit toppings for cheesecake, 295

Fryer, deep-fat, 74

Fudge icing, 373

Functional foods, 54

Fungi, 21
    growth requirements of, 21

G

Ganache, 305, 404–405

Garnish in plated dessert, 195

Gas formation in baking, 113

Gastronomic aspect of buffet menu, 198

Gâteau Saint Honoré, 276, 306

Gelatin, 106
    creams stabilized with, 299–302

Gelatinization of starches in baking, 113

Gelato, 324

Gelling agents, 106

Genetically modified organisms (GMO), 54

Genoise sponge cake, 383–384
    making, 384

Germ, wheat, 96, 211

German buttercream, 368

Ghee, 127

Ghirardelli, Domingo, 8

Gianduja, 407–408, 420
    preparing filling for pralines, 409

Ginger, 163

Glazes, 375–377
    types of, 375–376

Glazing cakes, 377

Gliadin, 95, 112

Gloves, protective, 30–31

Glucose, 102, 103

Glucose units, 38

Gluing pastillage, 427

Gluing poured sugar pieces, 436

Gluten, 95, 211
    controlling, 112

Gluten-free flour mix, 51–52

Gluten intolerance, 51

Gluten window, 221

Glutenin, 95, 112, 211

Glycemic measure, 41

Goat cheese in cheesecake, 291

Godiva Chocolates introduced, 10

Good cholesterol, 44

Gooseberries, 142

Gorgonzola cheese, 123

Gouda cheese, 124

Gougères, 276

Gourmet coffees, 175

Granita, 334–335

Granulated sugar, 50, 102

Grapefruits, 137–138

Grapes, 147

GRAS (Generally Recognized As Safe) list of food additives, 54

Green tea, 176

Grigne, 227

Grilled fruit, 338–339

Grilling, 186–188

Grilling pineapple, 187

Gripping knives, 83–84

Groat, 231

Groundnuts, 166

Grunts, 341

Gruyère cheese, 123

Guavas, 152

H

Häagen Dazs introduced, 12

HACCP, 202

Hand tools, 75–77

Handle, knife, 83

Hard cheeses, 122

Hard rolls, 231–232

Hard wheat, 95, 211

Hard-wheat flour, 97–98
    in puff pastry, 246

Hardening of ice cream, 328–329

Hazard Analysis Critical Control Point (HACCP) system, 25
    seven steps of plan, 26

Hazardous foods, potentially, 17

Hazelnuts, 166

Hearth breads, 210

Heat transfer, 180–181

Heimlich maneuver, 34

Herbal liqueurs, 172

Herbs, 108, 160–162
    storing, 160
    types of, 160–162

Hershey, Milton, 9, 10

High-altitude baking, 388

High-density lipoproteins (HDL), 44

High-gluten flour, 97

High-intensity sweeteners, 50

High-ratio cakes, 382–383

High-ratio liquid shortening, 100

High-ratio shortening, 100

History of chocolate, 394–395

Holding food safely, 28–29

Homestyle pie filling method, 266

Homogenization, 119

Honey, 103

Honeydew melon, 140

Hood systems, 32

Hot air guns, 403

Hotel pans, 80

Hydration, 216

Hydrogenation, 43

Hydrometer, 79, 334, 352

Hygroscopic nougat fillings, 410

I

Ice carving, 203

Ice cream, 324–330
    ingredients in, 325–327
    production of, 328–329
    qualities of, 327
    sanitation standards for production, 329

Ice cream bombes, 330

Ice cream machine, electric, 74

Ice paddles, 29

Ice sugar, 446

Ice-water bath, 29

Icebox cookies, 360

Ices, 333
    avoiding problems with, 333–334
    types of, 334–335

Icings, 373–375
    for decorations, 429–430
    types of, 373

Immersion blenders, 403

Improved mix of bread dough, 223

Incomplete proteins, 42

Incorporating liquid praline fillings, 411

Individual ingredient cost, 63

Induction cooking, 180

Infection, 20

Infrared cooking, 181

Infusions in ice cream, 326

Ingredient cost, calculating, 63–64

Ingredients, 62
    adjustments in formula conversion, 61
    essential bakeshop, 95–110

Injuries, personal, 31

Insoluble fiber, 40, 41

Inspecting goods received, 67

Instant active yeast, 215

Instant dry yeast, 105

Intensive mix of bread dough, 222–223

Intoxication, 20

Inventory control, 68

Invert sugars, 102, 352

Invertase, 411–412

Invoice, 67
    checking purchase order to, 68

Invoice cost per unit, 63, 64

Ionizing radiation, 54

Irradiation, 54

Isomalt method of bubble sugar, 445–446

Italian buttercream, 369
    making, 371

Italian meringue, 314–315
    making, 314

J

Japonaise, 315

Johnson, Nancy, 7

Juan canary melons, 140–141

Julienne cut, 86

K

Kataifi, 277
Kentucky Bourbon, 175
Kirsch, 173
Kirschwasser, 173
Kissing crust, 225
Kiwis, 149
Kneading, 111
Knife construction, 82–83
Knife control, 84
Knife safety, 86–87
Knives, 82–89
    proper use of, 83–84
    storing, 87
Kugelhopf, 232
Kumquats, 138

L

Labor cost, 58
Lacerations, 33
Lactic acid, 214
Lactose, 102
Ladles, 79
Lag phase of bacteria, 22
Laminated dough, introduction to, 246–255
Laminating yeasted dough, 252
Lard, 101
    in pie dough, 258, 259
Lattice crust, 262–263
Layer cakes, 380
    assembling, 381
Leach, Richard, 13
Lean-dough breads, 228
Leavened flat breads, 232
Leavening agent, 103–105
    biological, 104–105
    chemical, 105
    in cookies, 357
    physical, 104
    storing, 105
Leaves from pulled sugar, 441
Lecithin, 129, 130, 396
Lemons, 136
    zesting, 137
Levain, 229

Levain de pâte, 230
Levain starter, 219
Limes, 137
Lindt, Rodolphe, 9, 395
Linoleic acid, 44
Linolenic acid, 44
Linzertorte, 268
    making, 270
Lipids, 42
Liqueurs, 172–173
    flavoring with, 373, 375
    in ice cream, 326–327
Liquid
    in baking, 107
    and gluten formation, 112
    measures, 78
    yeast 105
Liquid fillings for pralines, 411
    incorporating, 411–412
Live dough, 233
Loaf pans, 81
Log phase of bacterial growth, 22
Long-flake pie dough, 258
Loquats, 152
Low-density lipoproteins (LDL), 44
Lubin, Charles, 11
Lychee, 152

M

M&M candies introduced, 11
Macadamia nuts, 166
Mace, 164
Macédoine cut, 86
Machine tempering method, 399
Macronutrients, 38
Madeira wine, 174
Madeleine pans, 81
Magellan, Ferdinand, 6
Maillard Reaction, 226
Main item in plated dessert, 194
Maintaining equipment, 32
Major minerals, 47
Malt syrup, 103
Maltose, 102
Manchego cheese, 122

Mandarins, 138
Mandoline, 76
Mangoes, 149
    pitting and cutting, 149
Manufactured yeast, 214, 215
Maple syrup, 102–103
Marble slabs, 403
Margarine, 100, 128–129
    blends, 129
    receiving, 129
    storing, 129
Marsala wine, 174
Marzipan, 448–449
    cocoa painting on, 420
    fillings for pralines, 410
    storing and handling, 449
Marzipan alligator, making, 450
Marzipan roses, making, 451
Mascarpone cheese, 125
    in cheesecake, 291
Mass measurement, 92
Masters, Thomas, 8
Material Safety Data Sheets (MSDS), 24
Maturing flour, 213
Mealy pie dough, 258
Measuring equipment, 78–79
Measuring spoons, 79
Mechanical convection, 181
Mediterranean Diet, 43
Medium peaks, 311
Melba sauce, 349
Melons, 139–141
Melting chocolate, 402
Menu, buffet, 198
Meringue nests, piping, 316
Meringues, 310–316
    baking, 315
    categories of, 311–315
    stages of, 311
    storing, 315
Metal bars, 403
    for sugar work, 431
Metal cooling pins, 29
Micronutrients, 38
Microorganisms, growth requirements of, 21
Microwave, melting chocolate in, 403

Microwave cooking, 181

Microwave tempering method, 399

Middle Ages, baking during, 5–6
  confections in, 5

Milk, 118–121
  as liquid, 107
  receiving and storing, 120

Milk chocolate, 110, 396

Milkfat, removal of, 119–120

Milling process, 212
  for wheat, 96

Mincing, 85

Minerals, 47

Mint, 161

Mirror glazes, 376
  for cheesecake, 295

Mission figs, 152

Mitts, protective, 30–31

Mixers, 72

Mixing in bread making, 219

Mixing method
  and gluten formation, 112
  for quick breads, 238

Mixing techniques for bread dough, 222–223

Mixing yeasted laminated dough, 251

Modeling chocolate, 419

Modern period, baking during, 6

Modified food starch, 106

Moist cooking techniques, 188–191

Moisture in chocolate, 401

Molded chocolate, 417–418

Molded chocolates, 413–416

Molded cookies, 360

Molds, 21

Monosaccharides, 39

Monounsaturated fats, 42, 43

Mousse, 302

Mozzarella cheese, 125

Muffin tins, 81

Muskmelons, 139–140

N

Napoleons, 248

National Restaurant Association (NRA) *Accuracy in Menu* paper, 60

Natural convection, 180–181

Natural yeast, 229

Naturally leavened bread, 214, 229–230

Naturally leavened preferment, 217, 219

Nectarines, 144

Nestlé, Henri, 8

Neufchâtel cheese, 125
  in cheesecake, 290

New York-style/deli-style cheesecakes, 291–292
  making, 293

Nib, 396

No-bake-style cheesecake, 292

Nondairy toppings, 107

Nonedibles, 66

Nonperishable products, 66

Nonstick bake sheets, 80

Norwegian Omelet, 330

Nougat fillings, for pralines 410

NSF International, 75

Nut butters, 165

Nut fillings for pralines, 410

Nut liqueurs, 172

Nutmeg, 164

Nutrient-dense foods, 51

Nutrient density, increasing, 50–51

Nutrients, 38

Nutrition facts panel, reading, 53–54

Nutrition Labeling and Education Act of 1990, 53

Nutrition labels, 53

Nuts, 164–167
  for flavoring, 108
  storing, 164
  types of, 165–167

O

Oats, 231

Occupational Safety and Health Administration (OSHA), 30

Offset spatula, 77

Oils, 99–101
  storing, 101

Old dough, 218

Oleomargarine, 128

Omega-3 fatty acids, 44

Omega-6 fatty acids, 44

Omelet Surprise, 330

1-2-3 dough, 268
  for cheesecake, 294

Oolong tea, 176

Open market buying, 66

Operating expenses, 58

Orange, segmenting, 139

Oranges, 138

Oregano, 162

Organic acids in fermentation, 214

Organically grown food, 54

Osmotolerant yeast, 215

Oven, melting chocolate in, 402

Oven spring, 225

Ovens, 73–74
  bread, 226–227

Overcrystallized chocolate, 398

Overhead costs, 58

Overheated tempered chocolate, fixing, 398

Overproofed dough, 224

Overrun in ice cream, 328

P

Pain au chocolat, 249

Paintbrushes for sugar work, 431

Palmiers, 248

Pan breads, 232

Pan de mie, 232

Pancakes, 280

Pandowdy, 342

Pane francese, 228

Pane pugliese, 228

Panettone, 232

Panna cotta, 303

Panning
  cakes, 387
  pie dough, 260, 261
  quick breads, 242–243

Papayas, 150

Parasites, 20
  growth requirements of, 21

Parfait glacé, 331

Paring knife, 83
  using, 84

Paris-Brest, 276

Parmesan cheese, 122

Parmigiano-reggiano cheese, 122

Parstock, 68

Passion fruit, 150

Pasteur, Louis, 8, 104, 118, 214

Pasteurization, 118
    of eggs, 131

Pastillage, 426–429
    cocoa painting on, 420
    coloring, 429
    making a showpiece, 428
    storing a showpiece, 429

Pastry bags and tips, 76

Pastry brushes, 76

Pastry cream, 304–305
    making, 306–307
    as soufflé base, 317
    variations of, 305–307

Pastry cutters, 76

Pastry flour, 98, 258

Pastry platter, composition of, 206

Pastry wheels, 76

Pâte à Bombe, 304

Pâte à choux, 274–277
    making, 275

Pâte fermenté, 218

Patent flour, 96, 98

Pathogens, 20

Payard, François, 13

Paysanne cut, 86

Peach cobblers, making individual, 341

Peach melba, 329

Peaches, 144

Peanuts, 166

Pears, 146
    poaching, 189

Pecans, 166

Pecorino Romano cheese, 122

Pectin, 106

Peel, 76

Peel liqueurs, 172

Peeling techniques, 84

Pentosans, 231

Pepperidge Farm Bread Company, 11

Periodic ordering method, 68

Perishable products, 66

Perpetual inventory, 68

Persian melons, 141

Persimmons, 152

Personal hygiene, 23–24

Pesticides, 25

Peter, Daniel, 395

Petits fours, 363–365
    types of, 363, 365

Petits fours demi-secs, 363

Petits fours glacés, 365
    making, 364

Petits fours secs, 363

Petits fours varies, 365

pH level of bacteria, 22

Philadelphia brand cream cheese introduced, 9

Philadelphia-style ice cream, 324

Phyllo dough, 277, 278

Physical contamination hazards, 17, 25

Physical inventories, 68

Physical leaveners for quick breads, 237

Physical leavening, 215

Phytochemicals, 49

Pie dough
    ingredients, 258–259
    making, 259, 260
    types of, 258

Pie fillings, 263–267

Pièce montée, 203

Pies, cooling and storing, 271

Pignoli (pine nuts), 166

Pilferage, preventing, 69

Pillsbury & Co. established, 9

Pine nuts (pignoli), 166

Pineapple, 150–151
    grilling, 187
    peeling, coring, and trimming, 151

Piped sugar, 446

Piping chocolate, 391, 418–419

Pistachios, 167

Pithiviers, 248

Planetary mixers, 219

Plant proteins, 42

Plantains, 148

Plate cost, 65

Plate decoration, two sauces as, 199

Plate dusting, 202

Plated desserts, 194–198
    components of, 194–197
    techniques for saucing, 195

Plating
    contrasts in, 197–198
    types of, 196–197

Platter design for buffet, 203

Plums, 144–145

Poached fruit, 338

Poaching, 188

Poaching pears, 189

Poire Belle Hélène, 330

Poire Williams, 173

Polo, Marco, 6

Polymorphic cocoa butter, 397

Polysaccharide, 40

Polyunsaturated fats, 42, 43

Pomegranates, 152–153

Pomelo, 153

Pomes, 145–147

Poolish, 218

Popovers, 280–281

Poppy seeds, 167

Portion control, 59

Portion cost, 64

Potato starch as thickener, 106

Potentially hazardous foods, 17

Pots de crème, 289–290

Pound cakes, 380

Pourable batters, 280–281

Poured sugar
    gluing pieces, 436
    ingredients for, 432
    molds for, 435–436
    surfaces for, 435

Powdered sugar, 102

Pragmatic aspect of buffet menu, 199

Pralines, 404
    making, 413
    making molded, 415
    molded, 413–416
    storage and shelf life of, 416

Precrystallized chocolate, 397
Preferments, 214, 217–219, 251
Pregelatinized starches, 106
Prehistory of baking, 4–5
Preliminary selling price, 65
Presentation of buffet table, 201
Preserve-based sauce, 349
Preserved fruit, purchasing and storing, 155–156
Preserves, flavoring with, 372, 375
Preshaping bread dough, 224
Pressed cookies, 359
Pricing factor, 65
Prickly pears, 153
Primary fermentation
    of bread dough, 223
    of yeasted laminated dough, 251
Product specifications, 65
Profiteroles, 276
Proofing cabinet, 73
Proofing laminated dough, 254
Protease enzymes, 299
Protective clothing, personal, 30–31
Proteins, 41–42
Provolone cheese, 123
Puff-pastry
    ingredients, 246–247
    margarine, 129
    shortening, 100
    shortening as roll-in fat, 249
    storing, 248
Pulled sugar, 436, 438, 441
    cooking time and temperature for, 439
    function of ingredients in, 438–439
    items made from, 440–443
Pulled sugar leaves, making, 443
Pulled sugar ribbons and bows, making, 444
Pulled sugar roses, making, 442
Pulling sugar, 440, 441
Pullman pan, 232
Pumpkin seeds, 167
Puncture wounds, 34
Purchase order, 67
    checking to invoice, 68

Purchase quantities, 66
Purchasing, 65
Purchasing methods, 66
Purveyor relationships, 66
Putrefactive bacteria, 17

Q

Q factor, 64
Quark cheese in cheesecake, 291
Quiche, 267
Quick breads, 236–243
    mixing methods for, 238–241
    panning, 242–243
    types of, 238
Quick lime, 448
Quinces, 147

R

Rack oven, 73
Radiation cooking, 181
Ramekins, prepping for soufflés, 320
Rapid Kool, 29
Raspberries, 142
Receiving food, 27
Receiving system, 67
Red flame grapes, 147
Red wheat, 211
Reductions, 349
Reel oven, 74
Refined white flour, 211
Refining chocolate, 396
Refrigerator cookies, 360
Reheating chocolate, 402
Reheating food safely, 29
Repairing equipment, 32
Requisition, 69
Resting tempering method, 399
Retarder/proofer, 73
Retarding fermentation of bread dough, 223
Retarding laminated dough, 254
Retarding process, 223
Rhubarb, 153
Ribbon, yolks, 304

Ricotta cheese, 125
    in cheesecake, 291
Rings, 77
Ripened soft cheeses, 124
Ripening cheese, 126
Rivets, 83
Roasted fruit, 338–339
Roasting, 183–184
    fruit, 183
Rock candy, 447
Rock salt, 108
Rock sugar, 445
Rohwedder, Frederick, 11
Roll-in fats, types of, 249
Roll-in margarine, 128–129
Rolled cookies, 362
Rolled fondant, 430–431
    covering cake with, 430–431
Roller milling, 212
Rolling pastillage, 427
Rolling pie dough, 260, 261
Rolling pins, 77
Rombauer, Irma, 11
Roquefort cheese, 124
Rosemary, 161
Rotary rack oven, 73
Rounding bread dough, 224
Roux as soufflé base, 317–319
    making, 318
Royal icing, 373
    for showpieces, 429
    storing, 429–430
Rubber molds for sugar work, 431
Rubber spatula, 77
Rubbing, 111
Rubbing method, 239
    of mixing, 239
    in pie dough, 259
Rudkin, Margaret, 11
Run, 174
Run-out icing, 430
Russe, 80
Rye breads, 231
Rye flour, acidifying, 231

S

Sabayon, 304, 352
Saccharin, 50
Saccharometer, 79
Sacher, Franz, 7
Sacristans, 248
Sales controls, 69
Salt, 108–109
    in bread, 217
    functions of, 108
    and gluten formation, 112
Salted butter, 127
Sand casting, 447
Sanding pastillage, 427
Sanitizing, 16
    knives, 86
Santa Claus melons, 141
Sara Lee, 11
Saturated fats, 42
    reducing, 44
    replacing, 44
Sauce in plated dessert, 194
Saucepan, 80
Saucepot, 80
Sauces, 348–352
    making web pattern with, 201
Sauté pan, 80
Sautéing, 184–185
    apples, 184
Sautoir, 80
Savarin, 276
Scalds, 31, 285
Scales, 78, 92
Scaling, 92
    batter, 387
    in bread making, 219
Scoops, 79
Scoring bread dough, 225
Scotch Whiskey, 174
Scottish method in puff pastry, 247
Scrapers
    bench, 75
    bowl, 76
Seasonings, 108
Séchaud, Jules, 395
Seed-based liqueurs, 172

Seed crystals, 397
Seeds, 167–168
    storing, 164
    types of, 167–168
Segmenting an orange, 139
Semifreddo, 333
Semiperishable products, 66
Semisoft cheese, 123–124
Separating eggs, 133
Serum solids, 326
Service space for buffet, 200–201
Serving food safely, 29
Sesame seeds, 167
Shank, knife, 83
Shape contrasts in plating, 197
Shaping
    bread dough, 224
    laminated dough, 252
    pie dough, 259–263
Sharpening knives, 86, 88
Sharpening stone, 86
Sheet cookies, 362
Sheet pans, 80
Sheeter, 72
Sheeting laminated dough, 252
Shell eggs, 130
Shell method for liquid fillings, 411
Shell of egg, 129–130
Sherbet, 334
Sherry, 174
Shoes, 30
Short dough, 267–268
    for cheesecake, 294
    French, 268
Short-flake pie dough, 258
Short mix of bread dough, 222
Shortening, 99–100
    and gluten formation, 112
    melting in baking, 113
    as roll-in fat, 249
Showpieces, chocolate, 416–423
    finishing, 421
    storing, 421
Showpieces, introduction to, 426
Shrinkage, 62
Sickness in workplace or classroom, 34–35

Sieve, 77
Sifting, 111
Silica gel, 448
Silpats, 80
Simmering, 189–190
Simple carbohydrates, 38–39
Simple icing, 373
Simple sugars, 38
Simple syrup, 352
Single-crust pies, 261, 262
Single-source buying, 66
Slicer, 83
Slips, 31
Smallwares, 78–79
Sodium bicarbonate, 237
Soft cheese
    fresh, 124–125
    ripened, 124
Soft peaks, 311
Soft pie fillings, 266
Soft-textured cookies, 356
Soft wheat, 95, 211
Soft-wheat flour, 98–99
    in puff pastry, 246
Soluble fiber, 40–41
Sorbet, 334
Soufflé glace, 320–321, 331–333
    making, 332
Soufflés, 317–321
    bases for, 317–319
    serving, 320
Sour cream, 121
Sourdough bread, 214, 229
Sourdough starter, 105, 219, 229–230
Soy lecithin, 396
Spackle knives, 403
Spanakopita, 277
Sparkling wines, 174
Spatulas, 77
Specialty coffees, 175
Spelt, 231
Spice blends, 162
Spices, 108, 162–164
    storing, 162
    types of, 162–164

Spiral mixers, 219

Spirits, 174–175
    in ice cream, 326

Splash sugar, 446

Sponge cakes, 380, 383
    as crust, 295

Sponge preferment, 218

Spoom, 334

Spray gun for chocolate, 404, 421

Spreads, 100

Spring wheat, 211

Springform pans, 81

Sprinkler systems, 32

Spun sugar, 447

St. André cheese, 124

Stamps for sugar work, 432

Standard portions, 59, 62

Standard yield, 59, 62

Standardized formula, 59
    benefits of, 59–60

Star fruits, 153

Starch, 40

Starch method of incorporating liquid fillings, 411

Starch molds, 447

Starter, sourdough, 229–230

Stationary growth phase of bacteria, 22

Stationary mixer, 72

Steam as leavening agent, 103

Steaming, 191

Steaming oven, 225–226

Stiff peaks, 311

Still-frozen desserts, 331

Stilton cheese, 124

Stirred crème brûlée, 288

Stirring, 111

Stollen, 233

Stone fruit, 143–145

Stone grinding, 212

Storeroom controls, 68

Storing equipment, 30

Storing food, 27

Straight dough, 217

Straight flour, 96

Straight spatula, 77

Strains, 31–32

Strawberries 143

Stretched doughs, 277–279

Streusel topping, 261, 262

Strudel dough, 277–278
    making apple, 279

Sucralose, 50

Sucrose, 102

Sugar, 101
    coloring, 434–435
    fillings for pralines, 410–411
    and gluten formation, 112
    method of bubble sugar, 445
    mise en place for cooking, 432, 434
    pump, 431
    reducing, 50
    types of, 102

Sugar artistry, 431–448
    equipment for, 431–432
    storing, 448

Sugar bloom, 401

Sugar concentration, boiling point and, 434

Sugar showpiece, making, 437

Sugar syrup
    cooking procedure for, 433
    substituting for granulated, 50

Sundae, 329

Sunflower seeds, 168

Superfine sugar, 102

Swans, 276

Sweet cream butter, 100

Sweet dough, 232
    techniques for, 232

Sweeteners, 50, 101–103

Swiss apple flan, 268

Swiss buttercream, 369
    making, 370

Swiss cheese, 123

Swiss meringue, 313–314
    making, 313

Swivel peeler, 84

Syrups, 352–353

T

Table salt, 108

Table tempering method, 399

Tables, buffet, 200

Tableside desserts, 343–345

Tahini, 168

Tang, knife, 82

Tangelos, 138

Tangerines, 138

Tapioca as thickener, 106

Tart pans, 81

Tarte Tatin, 268

Tarts, cooling and storing, 271

Tea, 176–177

Temperature contrasts in plating, 197

Temperature danger zone (TDZ) of bacteria, 22

Tempered chocolate
    for showpieces and decorations, 416–417

Tempering chocolate, 397–398, 400

Tempering methods for chocolate, 399–401

Texture contrasts in plating, 197

Textured surfaces, 404

Texturing combs, 404

Texturing molds for sugar work, 432

Theme, buffet, 198

Thermometers, 23, 79

Thickeners, 105–106

Thompson seedless grapes, 147

3-2-1 dough, 258

Threefold puff-pastry, 247

Thyme, 161

Tips, pastry, 76

Titanium dioxide, 435

Toll House cookies, 11

Tomato concassée, 86

Torres, Jacques, 13

Torte, assembling, 385

Toxin-mediated infection, 20

Trace minerals, 47

Trans-fatty acids, 43

Transfer sheets, 404

Triglycerides, 99

Trim, 62

Trim loss, 62
Tropical fruits, 147–151
Trueing knives, 86, 89
Truth-in-menu, 60
Tube pans, 81
Tulip paste, working with, 205
Tunneling, 239
Turn in laminating, 252
Turnovers, 248
Turntable, 77
Two-crust pies, 262–263
Two-stage mixing method, 383
Two-step foaming mixing method, 386

U

Ultra high temperature (UHT) processing, 119
Ultrapasteurization, 118–119
Underproofed dough, 224
Uniforms, 30
Unleavened flat breads, 232
Unsalted butter, 127
Unsweetened chocolate, 109
Untempered chocolate, 398
Unyeasted laminated dough, 246–248
USDA grades of butter, 127

V

Vaccination tempering method, 399
Van Houten, Conrad J., 7
Vanilla sauce, 349
Varietal coffees, 175
Vegetable cuts, 85
Vegetable margarine, 128
Vegetable peeler, using, 84
Viruses, 20
    growth requirements of, 21
Vitamins, 46
Volume measurement, 92

W

Wafer cookies, 362
Waffles, 280
Wakefield, Ruth, 11
Walnuts, 167
Warm foaming method, 383
Warming cabinet, melting chocolate in, 403
Water, 50
    in bread making, 216
    as liquid, 107
Water activity in bacteria, 22
Water bath
    custard, 285–286
    melting chocolate in, 402
Water evaporation in baking, 113
Water icing, 373
Water-soluble vitamins, 46
Water Temperature Formula, 216–217
Watermelons, 141
Web pattern with sauces, making, 201
Wedding cake, assembling and decorating, 390–391
Weeping, 266
Weight measurement, 92
Wet slow baking, 284
Wheat berry, 211
Wheat flour, 95, 210–211
    treatments and additives, 97
Wheat kernel, 95–96
Whetstone, 86
Whey, 122
Whip, 77
Whipped butter, 127
Whipped cream, 298
Whipping, 111
    method for cookies, 359
    mixing method, 383
Whisk, 77
Whiskey, 174
White chocolate, 110, 396
White chocolate mousse, 302
White nougat, 410
White of egg, 130, 131
White wheat, 211
Whitening agents for cooked sugar, 435
Whole-grain breads, 230
Whole-grain flour, 211
Whole-wheat flour, 96, 98
Wild yeast, 214, 215, 217, 229
Williams-Sonoma, 12
Wine reduction, making, 190
Wines, 173–174
Winter wheat, 211
Withered tea leaves, 177
Wood grainers, 404
Work tables, 75
Wounds, first aid for, 33–35
Wright, William M., 9

X

X-acto knife, 404
Xanthan gum, 52

Y

Yeast, 21, 104–105, 213–215
Yeasted artisan hearth breads, 228
Yeasted laminated dough, 248–250
    making, 250–251
Yeasted preferment, 217, 218–219
Yield tests, 62
Yogurt, 121
    frozen, 324–325
Yolk of egg, 130, 131

Z

Zabaglione, 352
    making, 353
Zester, 77
Zesting a lemon, 137
Zones, buffet, 198, 199

Credits

All images are *Johnson & Wales*, except for the following:

Culinary Arts Museum at Johnson & Wales University: 2, 3 (top, bottom), 6, 7, 8, 9 (top, bottom), 10 (top, bottom), 11, 12, 236, 340 (bottom).

istock: Chapter 4 opener (printing calculator), bluestocking; Chapter 5 opener (copper pots), YinYang; Chapter 6 opener (wooden scoop/brown sugar), Copyright: Ewa Brozek; (eggs), Copyright: Claudia Cioncan; Chapter 7 opener (egg carton), Copyright Alex Kotlov; (milk pitcher), Copyright: Jozsef Szasz-Fabian; Chapter 8 opener (orange), Copyright: Daniel Gilbey; Chapter 9 opener (bundle of herbs), Copyright Sandra Caldwell; (almonds), Copyright: Branislav Senic; Chapter 10 opener (coffee cup), Copyright: Vasko Miokovic; Brandy snifter, Copyright: Fedor Patrakov; Chapter 13 opener (wooden scoop/flour), Copyright Adam Balatoni; Chapter 27 opener (chocolate & cinnamon), Copyright Andrey Kozachenko.

North American Miller's Association, 212.